Indirect Searches for New Physics

Indirect Searches for New Physics

Alexey A. Petrov

CRC Press is an imprint of the
Taylor & Francis Group, an **informa** business

Cover Image designed by CERN

First edition published [2021]
by CRC Press
6000 Broken Sound Parkway NW, Suite 300, Boca Raton, FL 33487-2742

and by CRC Press
2 Park Square, Milton Park, Abingdon, Oxon, OX14 4RN

Library of Congress Cataloging-in-Publication Data

Names: Petrov, Alexey A., 1971- author.
Title: Indirect searches for new physics / Alexey A. Petrov.
Description: First edition. | Boca Raton : CRC Press, 2021. | Includes bibliographical references and index.
Identifiers: LCCN 2020055207 | ISBN 9780815386049 (hardback) | ISBN 9780367765514 (paperback) | ISBN 9781351176019 (ebook
Subjects: LCSH: Particles (Nuclear physics)
Classification: LCC QC793.29 .P48 2021 | DDC 539.7/2--dc23
LC record available at https://lccn.loc.gov/2020055207

ISBN: 978-0-8153-8604-9 (hbk)
ISBN: 978-0-3677-6551-4 (pbk)
ISBN: 978-1-351-17601-9 (ebk)

Typeset in Computer Modern font
by KnowledgeWorks Global Ltd.

Dedication

To my family

Contents

Preface

How can one find New Physics without seeing it? Many of my colleagues in experimental particle and nuclear physics seek an answer to this question by looking for rare processes that might seem to be impossible to observe or measuring familiar quantiles with astonishing precision, trying to catch glimpses of something that is not explained by the reigning theory, the Standard Model. In such a case, theoretical calculations of observed processes must be done with controlled precision matching the experimental one. Or, if such clear predictions are not possible, methods should be devised to relate several experimental observables in such a way that the resulting relation eliminates pieces of the theory that are not precise enough. This is a daunting task.

In this field, there is a constant competition between theorists and experimentalists to reach a better precision. As experimental techniques, devices, and methods develop fast, experimental bounds on various observables quickly become obsolete, superseded by newer results. So do theoretical bounds on various fundamental parameters that are based on such experimental data. How can a book be relevant in the long term in such a situation? To mitigate this problem, I tried to put emphasis on theoretical and experimental methods employed to study various phenomena, instead of quoting numerical results. The current bounds, decay rates, and other relevant quantities would always be available on the website maintained by the Particle Data Group (PDG) [1].

This book is intended for advanced graduate students and postdocs in theoretical particle and nuclear physics, as well as for experimentalists who want to know what is going on in "theorist's kitchen". Some material included in this book was originally part of a course on Advanced Particle Physics given at Wayne State University over some years. While I tried to make the material self-contained, it would be highly recommended that the readers had some basic understanding of quantum field theory as given, for example, in the well-known graduate texts such as [147] or [163]. One of the basic tools of a particle phenomenologist is effective field theories. While no previous knowledge of this subject is required for understanding the content of this book, I can recommend several texts on that subject, including [3, 49, 149].

I wish to express my appreciation to my colleagues Marina Artuso, Robert Bernstein, Rob Harr, Gil Paz, and Nausheen Shah for carefully reading parts of the manuscript and for their insightful comments. I have also benefited greatly from various conversations with Thomas Mannel, Alex Khodjamirian, Matthias Neubert, Robert Bernstein, Marina Artuso, and Alex Lenz. Great thanks go to my current and former graduate students Mohammad AlFiky, Andriy Badin, Gagik Yeghiyan, Kristopher Healy, Y.G. Aditya, Derek Hazard, Cody Grant, and Renae Conlin, as well as my former postdocs Andrew Blechman, Dmitry Zhuridov, Bhujyo Bhattacharya, and James Osborne for collaborations on the projects related to this book. I would like to thank James Wells and other colleagues at the Leinweber Center for Theoretical Physics at the University of Michigan, where parts of this book were written, for their hospitality.

Last, but definitely not least, I would like to thank my wife, Tatiana, and my kids Anna and Danil for their help and understanding during the writing of this book. Parts of this book were written during the world-wide COVID-19 pandemic. I am thankful for my family's support in those challenging times. I am forever indebted to my mother Ludmila Petrova for everything she gave me.

Alexey A. Petrov
Detroit, Michigan, USA
October, 2020

1 Introduction

Understanding the most fundamental structure of the Universe remains the noble quest of modern particle physicists. Searches for New Physics constitute important components of experimental programs at many particle accelerators. What is New Physics and how does one search for it? Despite the name, the type of physics we are interested to find is not described exclusively in models proposed very recently. The term "New Physics" (NP) generally refers to a set of physical phenomena that is not described by the Lagrangian of the Standard Model (SM), the currently accepted theory of particle interactions.

The Standard Model describes a framework of how 17 fundamental particles: twelve fermions (six quarks and six leptons), four force-carrying gauge bosons, and the Higgs boson, interact to describe observable phenomena. The fermions are organized into three sequential groups or "generations" based on how they interact with the rest of the Standard Model. While all the visible matter in the Universe can be built from only two types of quarks (called "up" and "down"), one charged lepton (the electron), and three almost massless neutrinos, other Standard Model particles can be readily produced either at particle accelerators or in violent astrophysical processes. They were all also produced in the early Universe, less than a second after the Big Bang. It is not known why Nature chose to organize SM fermions into generations with similar properties. All heavier SM particles are unstable and quickly decay into the lighter particles, but it does not mean that they are less important. For instance, all three generations of quarks are needed to explain observed CP-violation in the quark sector.

The SM cannot be the final theory that describes the whole Universe. For starters, it does not include a description of gravity. Gravity is believed to be inessential until much higher energy scales, around the Planck mass

$$M_P = \sqrt{\frac{\hbar c}{8\pi G_N}} = 2.4 \times 10^{18}\ \mathrm{GeV/c^2} \tag{1.1}$$

are reached. In this book, we shall employ *natural* units, where the Planck constant $\hbar$ and speed of light c are both set to one, $\hbar = c = 1$, so M_P is defined by the dimensionful Newton's constant G_N. The SM also does not describe Dark Matter and Dark Energy, whose presence is required by current cosmological models. Finally, the construction of the SM itself is unsettling. For example, why are masses of fermions (quarks and leptons) so wildly different? Even within the quark sector of the SM the ratio of masses of the heaviest (top) to the lightest (up) quarks reaches 10^5! All in all, the Standard Model is very similar to the Periodic Table of elements in chemistry: it describes how particles interact, but is mute about why the model parameters happened to be what they are.

It is reasonable to believe that the Standard Model is an Effective Field Theory (EFT) [149], a low-energy approximation to some more fundamental theory that enters at a higher energy scale. However, the renormalizability of the SM makes it difficult to predict scales at which new degrees of freedom appear, which makes it quite distinct from other effective theories we know. This makes planning new high-energy experiments a rather complicated task, as no model-independent argument exists for which energy should be probed to discover those new particles.

Any particle, being lighter or heavier than any SM particle, but not included in the SM Lagrangian can be classified as New Physics. Indeed, the most straightforward way of discovering New Physics is to directly produce those new states at particle accelerators or in cosmic ray interactions. Numerous models for BSM physics, both introducing new fundamental symmetries and ad hoc, have been proposed to fix various shortcomings of the Standard Model. All of these models predict new particles that can be produced at high-energy particle accelerators with particular experimental signatures, which makes searching for those particles a worthwhile exercise.

The goal of this book is to describe physics concepts behind the searches for signals of New Physics without directly producing NP particles. This type of research goes under the name of *indirect searches for New Physics*, and it is complementary to direct "bump hunting" that we described above. At its core is the Heisenberg uncertainty relation that tells us that quantum fluctuations allow us to see glimpses of particles whose masses are larger than the scales associated with a process that is being investigated. Carefully studying processes that only involve SM particles can help to deduce the presence of new interactions. The catch is that we would not necessarily know what kind of New Physics is discovered, and we would have to know how to predict the processes we study in the SM with great accuracy. Depending on the masses and couplings of BSM particles to the SM ones we could see if the deviations from the SM could be seen. NP particle masses, as well as other phenomena, can define the *momentum scales* at which the NP operates.

1.1 Momentum scales and effective field theories

Description of physical phenomena around us, while complicated, can be simplified by noting one important fact: each phenomenon has one or several energy scales associated with it. Yet, for many problems, a thorough understanding of physics at higher energy scales is not important. For example, we do not need to know the mechanism for top quark mass generation to understand the generic structure of energy levels in a hydrogen atom. This is because the scales associated with those two phenomena are widely separated, as can be seen in Fig. 1.1, so any effect associated with the top quark in atomic physics would be (naively) suppressed by inverse powers of the top quark mass.

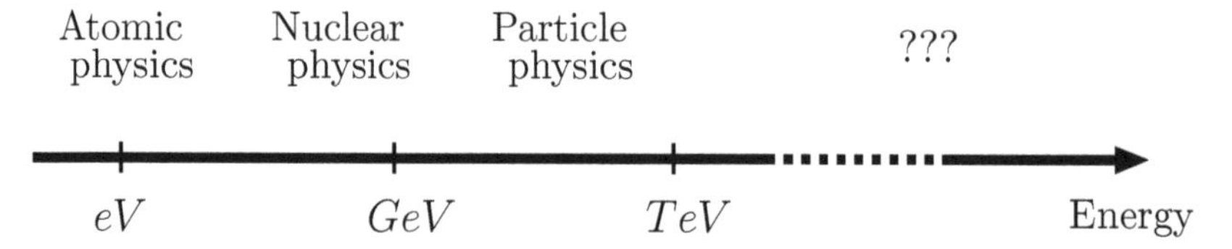

FIGURE 1.1 Disparate energy scales allow descriptions of physical phenomena in terms of only relevant degrees of freedom. Effects occurring at heavier energy scales are suppressed by inverse powers of those scales.

This observation can also be turned on its head. If we know enough about physics at a given scale that precision computations are possible, maybe one can find observables that are sensitive to effects of higher energy scales, even though they are suppressed by inverse powers of those scales! In fact, this statement, rigorously made, goes under the name of the Applequist-Carazzone theorem [8].

The language of *effective field theories* allows us to quantify this at the level of a Lagrangian by explicitly tracking the scales associated with the problem at hand. While we shall not discuss the powerful apparatus of EFTs in this book (see e.g. [149] for that), we will introduce the main concepts of EFT in this chapter.

1.1.1 Canonical dimensions of operators and New Physics

Building a model, be it the Standard Model or some other model with new interactions, requires a very useful concept that goes under the name of *operator dimension.* The gist of this concept is as follows. In natural units the action S is dimensionless. This allows us to determine the dimensional units of the Lagrangian $\mathcal{L}$ and, thus, all the fields that this Lagrangian is built from,

$$S = \int d^d x \; \mathcal{L}, \tag{1.2}$$

where $d = 4$ in four-dimensional space-time, but sometimes it is useful to leave it as a general parameter. In natural units, quantities of energy, momentum, and mass all have units of energy. This is so because velocity is dimensionless. For example, the famous relation $E = mc^2$ implies that mass and energy have the same units, say, electron-volts eV, $[E] = [m] = \text{eV}$. The uncertainty relation implies that distance has units of inverse energy, $[x] = [E^{-1}]$ and a derivative, $\partial_\mu = \partial/\partial x^\mu$ has units of energy, $[\partial_\mu] = [E]$. Equation (1.2) implies that the Lagrangian has units of energy to the fourth power, $[E^4]$, to make the action dimensionless. In d-dimensions the Lagrangian scales as $[E^d]$.

This allows us to determine *canonical dimension* of the fields that make up operators in a given Lagrangian. For example, let's look at the kinetic energy term for real scalar fields $\phi = \phi(x)$,

$$\mathcal{L}_{kin} = \frac{1}{2}\partial^\mu \phi \partial_\mu \phi. \tag{1.3}$$

Counting dimensions on both sides of Eq. (1.3) leads to a relation for the dimension of the scalar field, d_ϕ,

$$4 = 2 + 2d_\phi, \tag{1.4}$$

implying that $d_\phi = 1$ in energy units. This holds for both real and complex scalar fields. One can also check that a possible mass term for the scalar field, $m_\phi^2 \phi^2/2$, has the right dimension (four), provided that the mass parameter m_ϕ is of the (correct) dimension, one.

A similar exercise can be done for the fermion field, ψ. Employing the usual convention $\partial\!\!\!/ = \gamma^\mu \partial_\mu$, the relevant kinetic term is

$$\mathcal{L}_{kin} = \overline{\psi}(x) i \partial\!\!\!/ \psi(x), \tag{1.5}$$

leading to a relation for the dimension of a spinor field, d_ψ,

$$4 = 2d_\psi + 1, \tag{1.6}$$

implying that $d_\psi = 3/2$ in energy units. A possible mass term for the fermion field, $-m_\psi \overline{\psi}\psi$, implies that mass m has the expected dimension. We leave it as an exercise for the reader [149]

TABLE 1.1 Dimensions of fields and operators

Field or operator	Canonical dimension, [E]
Spin 0 field (ϕ, etc.)	1
Spin 1/2 field (ψ, etc.)	3/2
Spin 1 field (A_μ, etc.)	1
Derivative acting on a field (D_μ, etc.)	1
Mass term (m)	1
Lagrangian ($\mathcal{L}$)	4

to check that the dimension of a vector field A_μ, which we denote d_A, is one, just like for the scalar field. This holds whether the field is Abelian or non-Abelian.

As it will be useful for future discussions, we list all canonical dimensions of quantum fields and composite operators in Table 1.1. This table can be used to build operators whose dimensions are simply determined by adding dimensions of the fields and derivatives composing them.

We have already seen examples of operators of dimension four, like the ones we used to determine field dimensions in Eqs. (1.3) and (1.5) which represent kinetic energy operators. All those operators naturally have the same dimension as the Lagrangian. What would happen for higher dimensional operators, such as the ones entering the Fermi model of weak interactions? As an example, let us consider the operator that governs muon decay into an electron and two neutrinos,

$$\mathcal{L}_F = \frac{G_F}{\sqrt{2}} \overline{\psi}_\mu \gamma_\alpha (1-\gamma_5) \psi_\nu \overline{\psi}_\nu \gamma^\alpha (1-\gamma_5) \psi_e = \frac{4G_F}{\sqrt{2}} \overline{\psi}_{\mu,L} \gamma_\alpha \psi_{\nu,L} \overline{\psi}_{\nu,L} \gamma^\alpha \psi_{e,L}, \tag{1.7}$$

where ψ_f represent fermion fields for muon, electron, and neutrinos with $f = \mu$, e, and ν respectively, and G_F is the Fermi constant. Also, $\psi_{L,R} = (1/2)(1 \mp \gamma_5)\psi$ are the left (right) projections for the fermion fields above. This operator contains four fermion fields and no derivatives, which makes its dimension equal to six. Since the Lagrangian on the left-hand side of the equation has dimension four, we must conclude that the coupling constant has the dimension of $[E^{-2}]$. This is indeed so, as we know that the Fermi constant can be expressed in terms of a dimensionless weak coupling constant g_2 and the mass of the W boson, M_W,

$$\frac{G_F}{\sqrt{2}} = \frac{g_2}{8M_W^2}. \tag{1.8}$$

Physically, M_W can be interpreted as the scale above which we can no longer describe the weak interaction as a local product of two currents and need to take into account the effects of W propagation. This will be true in general: inverse powers of a scale parameter are needed to write an operator of arbitrary dimension d, such that the overall dimension of the operator and scale exponent n add up to four, $d - n = 4$.

It is interesting to note that the presence of dimension-6 operators leads to seemingly unfortunate high energy behavior of cross-sections. In particular, if we are to calculate a cross-section σ for the process $\mu^+ e^- \to \nu_e \nu_\mu$, which is governed by the Lagrangian Eq. (1.7), we find that it grows with the square of the collision energy E,

$$\sigma \sim G_F^2 E^2. \tag{1.9}$$

Since the cross-section grows with energy this way, it will eventually violate unitarity and the theory become sick. It can be shown [149] that the Fermi model of Eq. (1.7) is a good effective field theory up to energy scales Λ close to M_W. At higher scales a suitable full theory, or the so-called *ultraviolet completion* of the Fermi model has to be found. We now know that such UV completion is the Standard Model. In fact, the interaction described by the Lagrangian of Eq. (1.7) can be realized by an exchange of the W boson. Exactly the same thing can happen with BSM interactions: an exchange of some NP particle can be approximated at low energies by adding to the SM a set of higher-dimensional operators!

Since the ordinary space-time dimension (four) plays a special role in building the Lagrangian out of quantum fields, let us define some standard terminology for this case. All operators whose dimension is less than four are generically called *relevant* operators. Those whose dimension is four are called *marginal* operators. Finally, higher-dimensional operators with dimensions higher than four are referred to as *irrelevant* operators.

To illustrate the origin of this terminology, consider a toy model with real scalars described by the following Lagrangian,

$$\mathcal{L} = \frac{1}{2}\partial_\mu \phi \partial^\mu \phi - \frac{m_\phi^2}{2}\phi^2 - \frac{\lambda_4}{4!}\phi^4 - \frac{\lambda_6}{6!\Lambda^2}\phi^6 \tag{1.10}$$

This theory could model the behavior of neutral pions or Higgs bosons. Let us *rescale* the distances involved in the problem by a dimensionless parameter s and also redefine the field $\phi \to \phi'$ such that

$$x_\mu = s x'_\mu, \qquad \phi(x) = s^{(2-d)/2}\phi'(x'), \tag{1.11}$$

where, again, $d = 4$ in our four-dimensional spacetime. Taking $s \to \infty$ or $s \to 0$ allows us to study the long and short-distance behavior of the model. If we replace x and ϕ in the Lagrangian in Eq. (1.10) and insert it into the action of Eq. (1.2), we would get a "primed" action $S' = S'[\phi']$, i.e. the one written in terms of an x' and ϕ'

$$S' = \frac{1}{2}\int d^d x' \left[\partial'_\mu \phi' \partial'^\mu \phi' - s^2 m_\phi^2 \phi'^2 - s^{4-d}\frac{2\lambda_4}{4!}\phi'^4 - s^{6-2d}\frac{2\lambda_6}{6!\Lambda^2}\phi'^6\right]. \tag{1.12}$$

Examining Eq. (1.12), we see that the operators behave in three different ways upon rescaling. In the long-distance limit $s \to \infty$, the mass term grows with s and becomes more and more important when evaluating the Green functions. Thus the mass operator is a relevant operator. The ϕ^4 term does not depend on s in the rescaled action at all for $d = 4$, so has equal importance at any scale and is thus an example of a *marginal* operator. In general, all marginal operators have dimensions equal to that of the space-time dimension. Higher-dimensional operators, such as ϕ^6 term in Eq. (1.12), have coefficients that scale with s to negative powers, in the example $6 - 2d = -2$. They play a weaker role as $s \to \infty$ and thus are called irrelevant operators. They are the ones that could be generated by heavy BSM degrees of freedom at low energies.

We shall use this information later in the book when we discuss model building for BSM physics. A comment is in order: higher-dimensional operators, if induced by NP, can lead to new experimental signatures for SM particles. In the toy model of Eq. (1.12) such signatures include $2 \to 4$ decays induced by the ϕ^6 operator, provided that it was generated by some type of New Physics. The presence of irrelevant operators spoils the conventional renormalizability of a theory, in principle requiring an infinite number of counterterms to cancel UV divergences that appear in loop calculations with such operators. This problem is cured by using the powerful formalism of effective field theories [149].

1.1.2 Perturbative corrections and New Physics

Higher-dimensional operators are just one possible way that New Physics degrees of freedom can affect low energy observables. Another is quantum loop corrections. Recall from quantum field theory (QFT) that loop corrections induce a number of effects that can be experimentally measured. For example, comparing a measured cross-section for a certain process at several values of energy or momentum transfer allows us to see the running of coupling constants associated with a given model. That running can in principle be affected by the presence of other particles that are not included in the SM Lagrangian. This can be illustrated by comparing SM-predicted

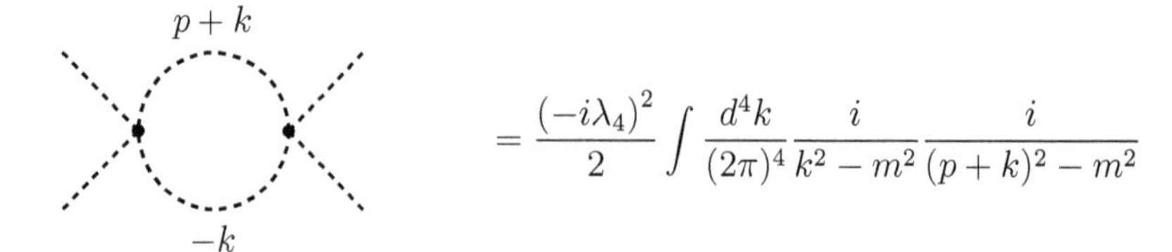

FIGURE 1.2 An example of a one loop correction in ϕ^4 theory. There are three diagrams corresponding to s, t, and u-channels. See [147] for a comprehensive discussion.

running of the QCD coupling constant $\alpha_s(\mu)$ to corresponding experimental measurements from energies around several GeVs to the energies around the Z-boson mass.

Let us illustrate it with an example of the already familiar toy model described by the Lagrangian of Eq. (1.10). Let us for a moment ignore (or set to zero) the term proportional to ϕ^6. This model is rather widely used as an illustration of many aspects of renormalizable perturbative QFT, yet goes under a rather mundane name a "ϕ^4 theory". The interaction term in the Lagrangian Eq (1.10) of this model describes a vertex with four fields ϕ interacting with a dimensionless coupling strength $\lambda_4 < 1$.

Imagine that we want to measure a coupling constant λ_4 at several values of p^2, which is the center of mass energy squared. To do so, we can measure a $2 \to 2$ particle cross-section, which in this model is completely defined by the interactions of particles ϕ. At tree-level the result is rather boring – we get the same constant λ_4. But if we calculate one-loop corrections, the results would show dependence on s and m. This dependence, after appropriate renormalization of the parameters, is completely specified by our ϕ^4 model and can be computed from three diagrams of the type shown in Fig. 1.2. We shall not present the calculations and resulting expressions here, they could be found in any QFT book, for example, in Chapter 10 of [147].

In order to model "New Physics" in our scenario, let us imagine a heavier field Φ with $M > m_\phi$ that is coupled to our field ϕ as

$$\mathcal{L}_\Phi = \frac{1}{2}\partial_\mu \Phi \partial^\mu \Phi - \frac{M^2}{2}\Phi^2 - \frac{\alpha}{2}\Phi^2\phi^2. \tag{1.13}$$

We might not produce the Φ-particles directly, but they would certainly affect the running of λ_4, as they would also run in the loops describing $2 \to 2$ scattering of ϕ particles in our model (see Fig. 1.3). The effect of the Φ particles on λ_4 could be deduced from the ϕ^4 result mentioned above by substituting $\lambda_4 \to \alpha$ and $m_\phi \to M$. In principle, by measuring the running of λ_4 we could infer useful information about properties Φ particles, in particular about α and M, even without directly producing Φ!

Similar methods could be utilized in the real world. We should just make sure that we could compute (and measure) the chosen observable to needed precision in the SM. Observables, such as electric dipole moments, $g-2$ of various particles, mass and width difference in meson-anti-meson oscillations could be affected by BSM particles running in the loops alongside the SM particles. The trick then is to find the observables where quantum effects due to new degrees of freedom are distinct and can be measured experimentally with suitable precision. We shall discuss those later in this book.

1.2 Low energy experiments and New Physics

What processes should we choose to maximize the sensitivity of searches to BSM degrees of freedom if those are not produced directly? In general, one can classify those searches as falling

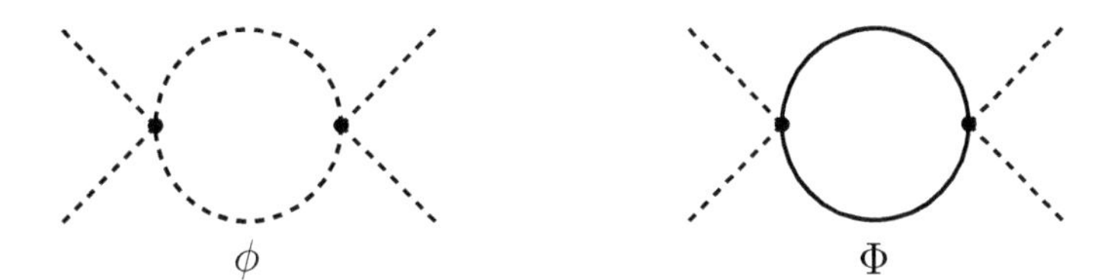

FIGURE 1.3 An example of a one loop correction in ϕ^4 theory and a "New Physics" contribution given by Φ fields. Dashed lines represent ϕ fields, while solid line denotes Φ field.

into three broad categories.

1. Searches in the processes that are *allowed* in the Standard Model.
   If a process is allowed in the SM at a tree level, it belongs to this category. In order to see the NP contributions, however, the process needs to be calculable in the SM with high precision, with all needed quantum corrections being under theoretical control. An example of such a process would be studies of leptonic decays of D or B-mesons. Another possibility for processes in this class includes NP searches that are testing relations among SM-allowed processes that are known to hold only in the SM, but not necessarily in models beyond the Standard Model. An example of such search includes testing Cabibbo-Kobayashi-Maskawa (CKM) triangle relations.
2. Searches in the processes that are *forbidden* in the Standard Model *at tree level.*
   Processes that involve flavor-changing neutral current (FCNC) interactions that change quark or lepton quantum number by one or two units do not occur in the Standard Model at tree level, as terms that mediate such interactions are absent from the SM Lagrangian. However, these transitions can happen in the Standard Model at one-loop level, which makes them rather rare. Processes like that can receive New Physics contributions from both tree-level interactions mediated by new interactions, and from loop corrections with NP particles. Processes of that type include meson-anti-meson mixing, or inclusive and exclusive radiative or semileptonic rare transitions. For instance, in the charm quark sector, they can be mediated by quark processes $c \to u\gamma$ or $c \to u\ell\bar{\ell}$. Lastly, measurements of CP-violating observables could be included here. A classic example, a dipole electric model of an electron that can only be generated at the four-loop level in the SM. On the contrary, it can be induced at one or two-loop level in many SM extensions.
3. Searches in the processes that are *forbidden* in the Standard Model.
   There are also processes that, while allowed by space-time symmetries, are forbidden in the Standard Model. Processes of that type are so rare that searches for their signatures require incredibly high statistics experiments. Their observation, however, would constitute a high-impact discovery, as it would unambiguously point towards physics beyond the Standard Model. Examples include searches for lepton- and baryon number-violating transitions, such as $n - \bar{n}$ oscillations.

There are many well-defined NP models that predict the existence of light, weakly-interacting particles. Such particles could be directly produced in meson or baryon decays. Since such light particles are not part of the SM particle spectrum, those processes do not occur in the Standard Model. Yet, they have similar experimental signatures as some SM transitions, such as mesonic or baryonic decays into neutrino final states. Neutrinos are invisible in general-purpose detectors that are used at particle accelerators. Similar to studies of processes with neutrinos in the final states, e^+e^- flavor factories (or kaon experiments) could use the hermeticity of their detectors to detect all other particles participating in transitions, thereby constraining the process with missing particles in the final state.

With the accumulation of large data sets in low energy experiments, we could test both very rare transitions and known transitions with the highest possible accuracy to infer the existence of New Physics degrees of freedom. Special experiments designed to look for processes that are forbidden in the Standard Model allow for direct access to effects of New Physics without ever producing an on-shell New Physics state.

1.3 Logic and organization of the book

It is not a secret that the field of precision measurements is the one that is driven by experimental data. New experimental results appear and become obsolete in a very short period of time, and so do theoretical bounds based on those experimental results. How can a book be relevant in the long term in such a situation?

An approach that is advocated here is to put emphasis on the theoretical methods and experimental techniques that are employed to study various phenomena instead of quoting the numerical results. The actual bounds, decay rates, and other relevant quantities are available on the website maintained by the Particle Data Group (PDG) [1] and can always be plugged into the expressions presented in the following chapters.

This book will describe many ways to hunt for heavy New Physics degrees of freedom without actually producing them in an experiment. We begin by describing the *tools of trade*, methods of effective field theories, symmetries, and basics of NP model building in Chapter 2. Then we delve into the theoretical and experimental bases behind indirect searches for New Physics. While we shall not describe the actual experimental apparata, we will provide a comprehensive description of ideas and methods that went into the construction of those devices.

Chapter 3 will deal with NP searches with charged leptons. As we shall see, many observables that are considered in those searches are actually quite clean and can be computed model-independently* in quantum electrodynamics (QED) with controlled accuracy. Ironically, the most important uncertainties in charged leptonic systems are related to strongly interacting quark contributions.

NP searches with quarks, hadrons, and nuclei are described in Chapter 4. While we follow the same template to describe the quantum systems as in the previous chapter, the presence of strong interactions plays a dominant role in determining the accuracy of predictions. In Chapter 5 we then turn to neutrinos in their terrestrial settings. In Chapter 6 we discuss indirect searches for New Physics at high energy accelerators done with Higgs and gauge bosons before concluding in Chapter 7.

Problems for Chapter 1

1. **Canonical dimensions.** Using methods developed in Section 1.1.1, determine canonical dimension for the spin-2 operators that describe gravitational fields. How would your answer change if we lived not in four, but eleven dimensions?
2. **Feynman rules.** Derive Feynman rules for the Lagrangians given by Eq. (1.10) and (1.13). Compute $2 \to 2$ scattering of ϕ particles at one loop to verify the statements made in Section 1.1.2. Consult Chapter 10 of [147] to verify your answer.

*We shall refer to a calculation as being *model-independent* when it can be done without describing strongly interacting systems (such as hadrons) in phenomenological models. Indeed, the Standard Model is also technically a model.

2

New Physics: light and heavy

2.1 Introduction

Why do we need New Physics? There are several observed phenomena in Nature that cannot be explained by currently known Standard Model physics. One such phenomenon is baryogenesis, i.e. the process that took place in the early universe that produced baryonic asymmetry that we observe today. In 1967, A. Sakharov proposed that in any theory, in order for the baryon asymmetry to appear, a set of three conditions must be satisfied: (a) the presence of baryon (or lepton) number violation, (b) the Universe must evolve out of thermal equilibrium, and (c) C and CP symmetries must be broken. While the SM appears to have all the needed ingredients for successful baryogenesis, its predictions appear to fall short of the experimental observations. Thus, new degrees of freedom are needed.

Another unexplained phenomenon is the existence of Dark Matter (DM). There is evidence that the amount of Dark Matter in the Universe by far dominates that of the luminous matter. It comes from a variety of cosmological sources such as the rotation curves of galaxies, gravitational lensing, Cosmic Microwave Background (CMB) features, and large-scale structures of the Universe. While the presence of DM is firmly established, its basic properties are still subject to debate. We see plenty of evidence for the presence of gravitating matter whose interactions with other force carriers of Nature seem to be strongly suppressed or even non-existent. There are many possible candidates for DM. The minimal assumption would be to posit that DM is a particle. Since it does not seem to radiate or absorb electromagnetic radiation, it appears to have a zero electric charge. The only SM particle that has such properties is a neutrino. It, however, appears that neutrinos cannot be DM particles due to their small mass and strength of their interactions with other visible matter. We must then conclude that the candidates for DM particles should lay outside the framework of the Standard Model.

Neutrinos themselves pose interesting questions. What is the origin of neutrino masses? Do we need to include (sterile) right-handed neutrino fields to keep the same mechanism of mass generation as the rest of the SM particles or build neutrino mass terms from the left-handed fields only at the expense of breaking the lepton number? Is there a mechanism that makes their masses so much smaller than those of the rest of known particles?

There are also plenty of more "aesthetical" reasons for the existence of beyond the Standard Model (BSM) physics. For instance, it is not known why there is such a large hierarchy of masses of quarks and leptons. If there is a dynamical reason for such disparate values of particle masses, it must lay outside the framework of the Standard Model. Similarly, the explanation for naturalness, the fact that Higgs' mass happened to be comparable to the electroweak scale, despite the fact that it is not stable against radiative corrections, must be found outside the SM framework. For many of those problems, SM plays the same role as the Periodic Table in chemistry: it provides the explanation for how different particles interact without providing the reasons for the numerical values of parameters of interaction.

Finally, (quantum) gravity is not part of the Standard Model! This, however, is a more complicated issue. While low-energy gravitational interactions can be readily incorporated in the effective field theory framework, effects around the Planck scale require a full quantum description of gravitational interactions. Fortunately, their effects are unobservably small at low energy, so we will not be discussing them in this book.

2.2 Continuous symmetries and new interactions

Modern quantum field theories are build using symmetry as a guiding principle. That is to say, a Lagrangian (density) for a theory is built in such a way that it respects all symmetries, global and internal, that are encoded in it. Both kinetic and interaction terms are encoded in such a way that the physics described by the theory is independent of the frame of reference. This codifies that the Lagrangian must be invariant under the global Poincare symmetry. Then a choice of the local gauge symmetry group and the representations for the matter fields is made, which locks in the way matter and gauge fields are described. Some of these symmetries, including space-time ones, can be broken in several different ways, which could be encoded in the Lagrangian as well. The beauty of such an approach is that only interaction terms of a certain structure could be added to the Lagrangian.

The Standard Model is not an exception. Its Lagrangian is invariant under translational symmetry, rotational symmetry, and the inertial reference frame invariance dictated by special relativity. The Standard Model describes the physics of electroweak interactions, which is posited to be described by $SU(2)_L \times U(1)_Y$ gauge group. Since quarks participate in both strong and electroweak interactions, QCD color group $SU(3)_c$ is also included, making the local $SU(3) \times SU(2) \times U(1)$ gauge symmetry the internal symmetry of the Standard Model [182].

We choose how matter fields transform under the symmetry group by assigning them to the multiplets. For the SM, the quarks are chosen in the following representations,

$$Q_L = \begin{pmatrix} u_L \\ d_L \end{pmatrix} = (3,2)_{+\frac{1}{6}}, \; u_R = (3,1)_{+\frac{2}{3}}, \; d_R = (3,1)_{-\frac{1}{3}}, \tag{2.1}$$

which states that the left-handed quark fields transform as triplets under $SU(3)_c$ color group of QCD, doublets under $SU(2)_L$ electroweak group and have the hypercharge of $Y = +1/6$ under $U(1)_Y$. The right-handed quarks are triplets under $SU(3)_c$, but singlets under electroweak $SU(2)_L$. Depending on their flavor, they have $Y = 2/3(-1/3)$ for up (down) type of quarks.

While leptons do not participate in strong interactions, which makes them singlets under $SU(3)_c$, their structure under the electroweak group is similar to that of the quarks,

$$L_L = \begin{pmatrix} \nu_L \\ e_L \end{pmatrix} = (1,2)_{-\frac{1}{2}}, \; e_R = (1,1)_{-1}. \tag{2.2}$$

Note that there are no right-handed neutrinos in what we define as the minimal Standard Model. We'll discuss massive neutrinos and their implications in Chapter 5.

It appears that there are three copies of quark and lepton doublet/singlets conventionally called *generations*. There is no theoretical reason for the number of generations to be three. Thus, the easiest *extension* of the SM could be to simply increase the number of generations. However, careful experimental studies of rare decays of K, D, and B mesons aided by the constraints from the Higgs production and decay patterns into $\gamma\gamma$ final state seem to push this possibility into a very unlikely corner.

The quark and lepton fields described above carry group indices. For example, $Q_{Lp}^{\alpha j}$ has $SU(3)$ color index α, $SU(2)$ index j, and generation index p. We shall suppress those indices for the brevity of the notations when displaying the equations where it is not necessary.

One popular idea driving extensions of the SM states that there might be particles that are still too heavy to be discovered at currently running particle accelerators. In that case, a New Physics model is usually characterized by a gauge group that includes the Standard Model as a subgroup. Such models are called grand unified theories (GUTs), as they hope to unify the electroweak and strong forces into a single force. This new interaction is characterized by a gauge symmetry group that includes both the SM and new matter particles and interactions, which are described by one unified coupling constant. To aid with such unifications, supersymmetric (SUSY) models could be considered. Since supersymmetry is broken, there is a hope to study SUSY partners at high-energy accelerators. As we see further in this book, their effects could also be studied at low energies.

If the GUT unification scale happens to be very high, say around $\Lambda_{\rm GUT} \sim 10^{16}$ GeV, only indirect effects of GUT theories can be seen. Those effects could be quite spectacular: since GUTs deal with multiplets that contain both quarks and leptons, transitions that involve baryon and lepton number violations, such as proton decay, become possible.

Another popular idea states that new particles that are not heavy – they could be light on the scale of already discovered SM particles or even massless. Since those particles are not discovered yet, they must have very small couplings to the SM particles. We shall discuss some examples later in this chapter.

In general, even terms that break Poincare symmetry can be added – which would lead to CPT-breaking – or other dimensions of space! But before we can talk about New Physics, let us finish the discussion of the Standard Model.

2.2.1 The Standard Model: symmetries and interactions

Particles of new physics, if heavier than the electroweak symmetry breaking scale, cannot be produced in low-energy experiments. Studies of quantum effects of new degrees of freedom require a complete understanding of experimental and theoretical backgrounds that are created by the Standard Model particles. It will then be useful to describe the Standard Model of particle interactions. Here we simply describe the construction of the SM, as it is usually extensively described in basic quantum field theory textbooks [65, 147, 186], and described in most University QFT courses. We shall only present general details that will be needed later in the book for seamless descriptions of BSM effects.

The SM Lagrangian $\mathcal{L}_{\rm SM} = \mathcal{L}_{\rm EW} + \mathcal{L}_{\rm QCD}$ contains quantum fields that describe SM particles and their interactions via electroweak and strong forces. Alternatively, it can be broken into parts that contain

$$\mathcal{L}_{\rm SM} = \mathcal{L}_{\rm gauge} + \mathcal{L}_{\rm kinetic} + \mathcal{L}_{\rm Higgs} + \mathcal{L}_{\rm Yukawa}, \tag{2.3}$$

with $\mathcal{L}_{\rm gauge}$ and $\mathcal{L}_{\rm kinetic}$ being kinetic terms of the gauge and fermion fields, terms that describe interactions of the Higgs field with itself and other bosonic particles in the Standard Model ($\mathcal{L}_{\rm Higgs}$), and those that describe Higgs interactions with fermions $\mathcal{L}_{\rm Yukawa}$.

Gauge symmetries

The choice of the local gauge group and representations for the matter fields dictates the way gauge fields and covariant derivatives are used in the Lagrangian. The covariant derivatives are defined as

$$D_\mu = \partial_\mu + ig_3 T^A G_\mu^A + ig_2 \frac{\tau^I}{2} W_\mu^I + ig_1 Y_f B_\mu, \tag{2.4}$$

where $T^A = \lambda^A/2$ with Gell-Mann matrices λ^A represent the generators of color $SU(3)$ and $\tau^I/2$ are Pauli matrices representing $SU(2)$ generators of weak interactions. Y_f represents the hypercharge of the fermion of type f. The covariant derivative acting on lepton fields can be obtained by setting $g_3 = 0$ in Eq. (2.4).

The fermion kinetic terms are given by

$$\mathcal{L}_{\text{kinetic}} = \overline{L}_L i\not{D} L_L + \overline{e}_R i\not{D} e_R + \overline{Q}_L i\not{D} Q_L + \overline{u}_R i\not{D} u_R + \overline{d}_R i\not{D} d_R. \tag{2.5}$$

Note that the matter fields in Eq. (2.5) implicitly bear weak isospin $j = 1, 2$, color $\alpha = 1, 2, 3$, and generation indices $p = 1, 2, 3$. We notice that the kinetic term, and thus the SM Lagrangian, contains $U(3)_L \times U(3)_e \times U(3)_Q \times U(3)_u \times U(3)_d$ flavor symmetry if quark and lepton mass terms are not included. In our notation, the gauge kinetic terms are

$$\mathcal{L}_{\text{gauge}} = -\frac{1}{4} G_{\mu\nu}^A G^{A\mu\nu} - \frac{1}{4} W_{\mu\nu}^I W^{I\mu\nu} - \frac{1}{4} B_{\mu\nu} B^{\mu\nu}, \tag{2.6}$$

where the gauge field strength tensors for SM fields are defined in the usual way. For the abelian, non-self-interacting weak hypercharge field,

$$B_{\mu\nu} = \partial_\mu B_\nu - \partial_\nu B_\mu, \tag{2.7}$$

while for the non-abelian weak isospin field

$$W_{\mu\nu}^I = \partial_\mu W_\nu^I - \partial_\nu W_\mu^I + g_2 f^{Ibc} W_\mu^b W_\nu^c. \tag{2.8}$$

As we can see, the electroweak gauge fields must be in the adjoint representation of $SU(2)_L$. They form a weak isospin triplet. The charged W-bosons are conventionally defined as

$$W_\mu^\pm = \frac{1}{\sqrt{2}} \left(W_\mu^1 + i W_\mu^2 \right), \tag{2.9}$$

while the neutral component of the triplet W_μ^3 mixes with the abelian hypercharge field B_μ to form the physical Z_μ and photon A_μ fields,

$$\begin{aligned} Z_\mu &= c_W W_\mu^3 + s_W B_\mu, \\ A_\mu &= -s_W W_\mu^3 + c_W B_\mu, \end{aligned} \tag{2.10}$$

where $c_W = \cos\theta_W$ and $s_W = \sin\theta_W$ are the cosine and sine of the (Weinberg) weak mixing angle. Finally, the QCD field is defined as

$$G_{\mu\nu}^A = \partial_\mu G_\nu^A - \partial_\nu G_\mu^A + g f^{ABC} G_\mu^B G_\nu^C. \tag{2.11}$$

The values of the third component of weak isospin and weak hypercharge are related to the electrical charge of the field through the Gell-Mann-Nishijima relation,

$$Q = I_W^3 + \frac{1}{2} Y_W. \tag{2.12}$$

The fact that the Standard Model Lagrangian is invariant under $SU(3) \times SU(2) \times U(1)$ commands that all gauge bosons must be massless. While this is a correct prediction for the QED and QCD parts of the Lagrangian, it fails miserably for weak interactions at low energies. There we know that the weak interactions are short-ranged, essentially point-like, which implies massive gauge boson exchanges. We have also not discussed the masses of fermions. While, in principle, mass generation mechanisms for the weak gauge bosons and fermions could be different, the SM offers an elegant solution to both problems via the Higgs mechanism. The gauge symmetries are broken by the Higgs-related part of the SM Lagrangian, which is defined as

$$\mathcal{L}_{\text{Higgs}} = (D_\mu H)^\dagger (D^\mu H) - V(H), \tag{2.13}$$

where we introduced the Higgs potential $V(H)$,

$$V(H) = -\mu^2 H^\dagger H - \lambda \left(H^\dagger H\right)^2. \tag{2.14}$$

The Higgs field in Eq. (2.13) is defined as an electroweak doublet,

$$H = \begin{pmatrix} H^+ \\ H^0 \end{pmatrix} = (1,2)_{\frac{1}{2}}, \tag{2.15}$$

which makes the terms in Eq. (2.13) singlets under the SM gauge symmetries. The covariant derivative acting on Higgs fields can be obtained by setting $g_3 = 0$ in Eq. (2.4). It would be needed to define the conjugated Higgs field, $\widetilde{H}_i \equiv \epsilon_{ik} \left(H^k\right)^*$. We define the two-dimensional totally antisymmetric tensor ϵ_{jk} with $\epsilon_{12} = +1$. The field $\widetilde{H}$ is then

$$\widetilde{H} = \begin{pmatrix} H^{0*} \\ -H^- \end{pmatrix}, \tag{2.16}$$

which also transforms as a doublet under $SU(2)$, but has an opposite hypercharge. Here we also used that $H^- = H^{+*}$. After spontaneous symmetry breaking (SSB), the Higgs doublet can be written in unitary gauge,

$$H = \frac{1}{\sqrt{2}} \begin{pmatrix} 0 \\ v + h(x) \end{pmatrix}, \tag{2.17}$$

where $v^2 = -\mu^2/\lambda$, and $h(x)$ represents the SM Higgs boson field. Mass terms for the gauge bosons can be obtained by setting $h(x) = 0$ in Eq. (2.17) and inserting it into Eq. (2.13). The result is

$$M_W = \frac{1}{2} g_2 v, \quad M_Z = \frac{v}{2}\sqrt{g_1^2 + g_2^2}, \text{ and } M_A = 0. \tag{2.18}$$

We invite the reader to consult [65, 147, 186] for a detailed discussion of how those mass terms appear.

Custodial symmetry

It turns out that there is a global $SU(2) \times SU(2)$ symmetry of the Higgs potential $V(H)$ in Eq. (2.13)! To see this, let us rewrite the Higgs part of the SM Lagrangian $\mathcal{L}_{\text{Higgs}}$ in a different, but equivalent to Eq. (2.13) form*. Let us define the Higgs *bi-doublet* field,

$$\Phi = \begin{pmatrix} H^{0*} & H^+ \\ -H^- & H^0 \end{pmatrix}. \tag{2.19}$$

Now, this definition of the Higgs field should change the way we define the Higgs Lagrangian. Since the physics should not change, we must redefine the covariant derivative,

$$D_\mu \Phi = \partial_\mu \Phi + i g_2 \frac{\tau^I}{2} W_\mu^I \Phi - i g_1 B_\mu \Phi \tau_3. \tag{2.20}$$

*In this section, we will mainly follow a discussion from [182].

It is important to note that the Pauli matrix τ_3 is added to the last term of Eq. (2.20) due to the fact that H and $\widetilde{H}$ have opposite hypercharges. The choices of Eqs. (2.19) and (2.20) lead to another form of $\mathcal{L}_{\text{Higgs}}$,

$$\mathcal{L}_{\text{Higgs}} = \text{Tr}\left(D_\mu \Phi\right)^\dagger \left(D^\mu \Phi\right) - V(\Phi), \tag{2.21}$$

where the Higgs potential $V(\Phi)$ is now written in terms of the matrix field Φ,

$$V(\Phi) = -\mu^2 \text{Tr}\ \Phi^\dagger \Phi - \lambda \left(\text{Tr}\ \Phi^\dagger \Phi\right)^2. \tag{2.22}$$

One can check that Eqs. (2.13) and (2.21) lead to the same Lagrangian when written in terms of component fields $H^\pm$ and H^0.

We can check that the new Lagrangian is invariant under the SM gauge group $SU(2)_L \times U(1)_Y$. If L is an $SU(2)_L$ transformation, the field Φ must change under it as $\Phi \to L\Phi$. Since covariant derivative of Eq. (2.20), by definition, transforms as $D_\mu \to L D_\mu L^\dagger$, the derivative acting on the field also transforms as $D_\mu \Phi \to L\left(D_\mu \Phi\right)$. With that, we could see that the Lagrangian of Eq. (2.21) is invariant under $SU(2)_L$. Likewise one can show that the Lagrangian is also invariant under $U(1)_Y$, as $\Phi \to \Phi e^{-(i/2)\theta\tau_3}$.

It is interesting to realize that in the limit $g_1 \to 0$ the Lagrangian Eq. (2.21) also possesses another (global) symmetry, which we can call $SU(2)_R$: $\Phi \to \Phi R^\dagger$,

$$\text{Tr}\left(D_\mu \Phi\right)^\dagger \left(D^\mu \Phi\right) \to \text{Tr} R\left(D_\mu \Phi\right)^\dagger \left(D^\mu \Phi\right) R^\dagger = \text{Tr}\left(D_\mu \Phi\right)^\dagger \left(D^\mu \Phi\right) \tag{2.23}$$

The cyclic property of traces and the fact that Higgs potential is written in terms of $\Phi^\dagger \Phi$ completes the proof. This means that the Higgs sector of the SM is invariant under the (accidental) global symmetry $SU(2)_L \times SU(2)_R$. The symmetry $SU(2)_R$ is often referred to as the *custodial* symmetry. It is then convenient to posit that the field Φ transforms as $\Phi \to L\Phi R^\dagger$.

It is interesting to see what happens when the Higgs field acquires vacuum expectation value. The matrix bi-doublet field changes to

$$\langle \Phi \rangle = \frac{1}{2}\begin{pmatrix} v & 0 \\ 0 & v \end{pmatrix}. \tag{2.24}$$

This breaks the $SU(2)_L \times SU(2)_R$ down to $SU(2)_{L+R}$, which sometimes is also referred to as custodial symmetry. In the limit $g_1 \to 0$, $W_\mu^\pm$ and Z_μ transform as a triplet under $SU(2)_{L+R}$. This implies that all of those particles have the same masses. Examining Eq. (2.18) shows that it is indeed the case!

More importantly, custodial symmetry helps to understand how our theory behaves beyond the leading order, both perturbatively and non-perturbatively! We shall discuss it in Chapter 6 where we introduce S, T, and U parameters.

Flavor symmetries

Finally, the Yukawa part of the SM Lagrangian is

$$-\mathcal{L}_{\text{Yukawa}} = Y^e \overline{L}_L H e_R + Y^u \overline{Q}_L \widetilde{H} u_R + Y^d \overline{Q}_L H d_R + \text{h.c.}, \tag{2.25}$$

where Y^f are the matrices of the Yukawa coupling constants for flavor f.

Quark and charged lepton masses also appear after spontaneous symmetry breaking when the Higgs field acquires a vacuum expectation value. Setting $h(x) \to 0$ in Eq. (2.17), and plugging it into Eq. (2.25), we realize that all quark and charged lepton masses are proportional to the product $Y^f v$. In particular, for the quark mass matrices,

$$-\mathcal{L}_{\text{Yukawa}} = \left(M_u\right)_{ij} \overline{u}_{Li}^f u_{Rj}^f + \left(M_d\right)_{ij} \overline{d}_{Li}^f d_{Rj}^f + \text{h.c.}, \tag{2.26}$$

where the two mass matrices are proportional to the matrices of Yukawa couplings,

$$(M_q)_{ij} = \frac{v}{\sqrt{2}} \left(Y^f\right)_{ij}. \tag{2.27}$$

Since the Yukawa matrices are not diagonal, the mass matrices are not diagonal either. However, we have some freedom to perform a bi-unitary transformation to diagonalize the mass matrix,

$$V_{qL} M_q V_{qR}^\dagger = M_q^{\text{diag}}, \quad \text{with } q_{Li} = (V_{qL})_{ij}\, q_{Lj}^f, \text{ and } q_{Ri} = (V_{qR})_{ij}\, q_{Rj}^f, \tag{2.28}$$

Note that the V_{qL} and V_{qR} are not related to each other, so $V_{qL} V_{qR}^\dagger \equiv V_{\text{CKM}} \neq 1$, the famous Cabibbo-Kobayashi-Maskawa (CKM) matrix. We also notice that the presence of the Yukawa (and thus mass) terms in the SM Lagrangian breaks the flavor symmetry in the quark sector to one global baryon symmetry $U(1)_B$,

$$U(3)_Q \times U(3)_u \times U(3)_d \to U(1)_B. \tag{2.29}$$

Similarly, in the lepton sector of the minimal Standard Model, the flavor symmetry is broken down to the lepton number $U(1)_L$

$$U(3)_L \times U(3)_e \to U(1)_L. \tag{2.30}$$

While the right-handed neutrinos are not part of the minimal Standard Model, they can be included in the Lagrangian. If their masses are generated in the same way as the quark masses, they are called *Dirac* neutrinos. Alternatively, one can only use left-handed fields to generate neutrino masses, in which case we explicitly require lepton-number violating terms to be added to the electroweak Lagrangian. This constitutes another possibility for New Physics to modify the SM Lagrangian. We shall discuss neutrinos in Chapter 5.

It is rather unsettling that a large hierarchy of quark and lepton masses is simply parameterized by an unnaturally large hierarchy of the values of Yukawa couplings. While technically natural, this situation is aesthetically not pleasing and goes under the name of the Standard Model *flavor problem**. Its solution is often addressed in various BSM scenarios.

2.3 Discrete symmetries: C, P, and T

Studies of discrete symmetries play an important role in searches for New Physics in low-energy experiments. Since we know how those symmetries affect the Standard Model fields, learning patterns of the breaking of those symmetries allow for an extra handle in studies of New Physics.

2.3.1 Classical limit

Before considering the implications of C, P, and T symmetries in quantum measurements, let us introduce them classically. P-transformation is a transformation $\vec{r} \to -\vec{r}$. In Cartesian coordinates it is achieved by reflection of all coordinates: $\{x, y, z\} \to \{-x, -y, -z\}$. Physically, in three dimensions, this transformation is equivalent to a reflection through a mirror followed by a rotation of π radians around an axis defined by the mirror plane, as shown in Fig. 2.1. This is why this transformation is often referred to as "reflection". T-transformation flips the arrow of

*It might well be that the solution of the SM flavor problem is of the "just so" type. After all, there are other hierarchies in physics that do not require additional symmetries for the explanation. For example, a hierarchy of masses of planets in our Solar System is simply accepted.

time, $t \to -t$, while C-transformation changes particles to antiparticles, i.e. flips the signs of a particle's charges, including an electric charge. Although technically, C-transformation does not belong in classical physics, as particle-antiparticle pair creation is a purely quantum process, it is still convenient to introduce such transformation classically. If a law of physics – or a particular interaction – does not change under the above discrete transformations (or a combination of thereof), then it is said that the interaction is *invariant* under such transformation.

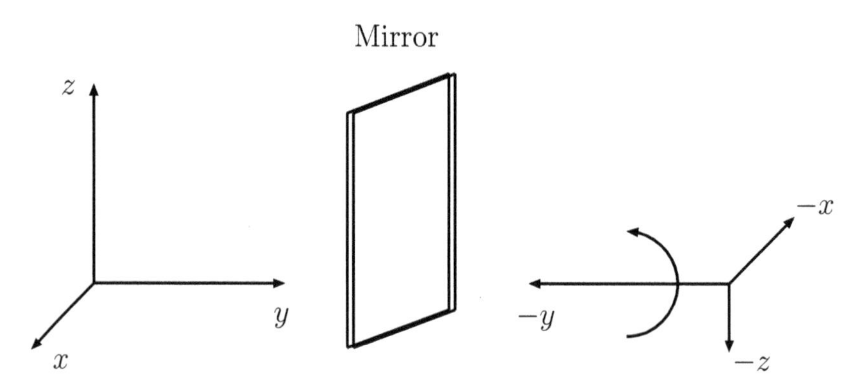

FIGURE 2.1 Parity transformation $\vec{r} \to -\vec{r}$ is equivalent to a reflection through a mirror followed by a rotation of π radians around an axis defined by the mirror plane.

To see how classical quantities change under C, P, and T transformations, we exploit definitions of various dynamical variables and equations of motion. For example, velocity

$$\vec{v} = \frac{d\vec{r}}{dt}, \tag{2.31}$$

is odd under both P and T transformations, as P flips the sign of the numerator, leaving the denominator intact. T-transformation does the opposite, flipping the sign of the denominator while leaving the numerator intact. Similarly, we can see that the momentum, $\vec{p} = m\vec{v}$ is also odd under both P and T transformations.

It is interesting to see that the force,

$$\vec{F} = \frac{d\vec{p}}{dt}, \tag{2.32}$$

behaves differently under P and T: it is odd under P and even under T (as both numerator and denominator change sign under T). Angular momentum, $\vec{L} = \vec{r} \times \vec{p}$, is even under P and odd under T. While spin represents a purely quantum effect, we should expect that it behaves the same way as angular momentum under those transformations.

Since we established how any force would behave under P and T transformation from Eq. (2.32), we can study how particular interactions behave under discrete transformations. For electrodynamics this can be done by examining a definition of the Lorentz force $\vec{F}_L$ that describes the motion of an electrically charged particle in electric $\vec{E}$ and magnetic $\vec{B}$ fields,

$$\vec{F}_L = q\left(\vec{E} + \vec{v} \times \vec{B}\right), \tag{2.33}$$

where q represents the particle's electric charge. Examining Eq. (2.33) and using what we learned from Eq. (2.32), we conclude that $\vec{E} \to -\vec{E}$ and $\vec{B} \to \vec{B}$ under P, as both $\vec{F}$ and $\vec{v}$ are odd under reflection. At the same time, $\vec{E} \to \vec{E}$ and $\vec{B} \to -\vec{B}$ under T.

TABLE 2.1 Maxwell equations and discrete symmetries

Equation	P	T	C	CPT
$\nabla \cdot E = 4\pi\rho$	+	+	−	−
$\nabla \cdot B = 0$	−	−	−	−
$\nabla \times B - \frac{\partial E}{\partial t} = 4\pi j$	−	−	−	−
$\nabla \times E + \frac{\partial B}{\partial t} = 0$	+	+	−	−

Now that electric charge has been introduced into consideration, we can describe the effects of a C-transformation on particles and fields. In particular, since C-transformation by definition changes $q \to -q$, we would expect that a scalar density of charges $\rho(\vec{r}, t)$ would also change a sign under C: $\rho(\vec{r}, t) \to -\rho(\vec{r}, t)$, and so would the current $\vec{j}$: $\vec{j}(\vec{r}, t) \to -\vec{j}(\vec{r}, t)$. We should add that the continuity equation for the conserved current implies that $\vec{j}$ is also odd under P and T transformations. Finally, one could argue that since C changes the signs of charges, i.e. the sources of the fields, the fields should change their signs too under C: $\vec{E} \to -\vec{E}$ and $\vec{B} \to -\vec{B}$.

Armed with what we learned, we can now see how the laws of electrodynamics, i.e. the Maxwell equations, behave under C, P, T, and their combinations. The results could be found in Table 2.1. As we could see, in each case both the left-hand side and right-hand side of any Maxwell equation had the same transformation properties under all discrete symmetries.

Before we discuss how to generalize the classical concepts discussed here to quantum mechanics, let us see how helicity h, a projection of a particle's spin along the direction of its momentum,

$$h = \frac{\vec{s} \cdot \vec{p}}{|\vec{s}||\vec{p}|}, \tag{2.34}$$

transforms under C, P, and T. As we know, helicity is a frame-dependent concept, as one can always choose a frame in which the observer moves faster than the particle. The only case when this is not possible is when the particle is massless and moving with the speed of light. In this case the helicity is a good (conserved) quantum number and is equivalent to chirality. While helicity is related to spin, i.e. is not a classical quantity from the start, we can still infer its transformation properties from what we now know about spin and momentum. It can be seen that $h \to h$ under C, $h \to -h$ under P, and $h \to h$ under T.

2.3.2 Quantum mechanics of C, P, and T

We clearly need to modify our concepts of C, P, and T transformations as we switch over to quantum mechanics. Indeed, symmetry transformations in quantum mechanics are achieved by unitary operators. This is true for C and P, which are promoted to unitary operators $\mathcal{C}$ and $\mathcal{P}$,

$$\mathcal{C}^\dagger = \mathcal{C}^{-1}, \ \mathcal{P}^\dagger = \mathcal{P}^{-1}. \tag{2.35}$$

If those transformations preserve the laws of physics and are good symmetries of Nature, the Sommerfield theorem requires that the operators that represent such transformations commute with the Hamiltonian $\mathcal{H}$, as we would expect from our discussion of electromagnetism in Section 2.3.1.

$$[\mathcal{C}, \mathcal{H}] = 0, \ [\mathcal{P}, \mathcal{H}] = 0. \tag{2.36}$$

This could be a good point to proceed by building the explicit form of those operators. However, we must take a pause to realize that our statement above is not exactly true. It is known experimentally that both C and P are not respected by weak interactions! Thus,

$$[\mathcal{C}, \mathcal{H}_{\rm W}] \neq 0, \ [\mathcal{P}, \mathcal{H}_{\rm W}] \neq 0, \tag{2.37}$$

so parity and charge conjugation cannot be good quantum numbers!

To solve this problem we must rely on the fact that weak interactions are short-ranged (and weak), so we can still prepare the initial and final states in the states of definite P and C. This fact allows us to build (interpolating) operators with definite P and C quantum numbers.

So far we described the properties of the unitary operators corresponding to $\mathcal{P}$ and $\mathcal{C}$ transformations. We also need to be careful in how we define time-reversal operator $\mathcal{T}$, as can be clearly seen by examining the Schrödinger equation,

$$i\frac{\partial \psi}{\partial t} = -\frac{\vec{\nabla}^2}{2m}\psi, \tag{2.38}$$

where, if we apply the concepts introduced in Section 2.3.1, we realize that the left-hand side of Eq. (2.38) is odd, while the right-hand side is even! Since we don't expect T-violation to be present for all quantum-mechanical processes, we should require that time reversal also involves changing $i \to -i$ and $\psi \to \psi^*$. This leads us to the fact that, contrary to P and C, T-transformation is achieved by an *anti-unitary* transformation*. This leads to a different action of a time reversal onto a scattering matrix S,

$$\mathcal{C}S\mathcal{C}^{-1} = S, \quad \mathcal{P}S\mathcal{P}^{-1} = S, \quad \mathcal{T}S\mathcal{T}^{-1} = S^\dagger. \tag{2.39}$$

In other words, time reversal transformation also interchanges the in- and out- states of the S-matrix.

In quantum field theory the fields are also represented by operators. Thus, one can study transformations of field operators under C, P, T, and their combinations. Derivations of such transformations are integral parts of standard courses in quantum field theory [29, 147] so we would not repeat them here. For completeness, we simply list them here, leaving the reader to check their derivations. The actions of all C, P, and T transformations are derived in [29]. For the scalar fields,

$$\begin{aligned} \mathcal{P}\phi(t,\vec{r})\mathcal{P}^\dagger &= e^{-i\alpha_P}\phi(t,-\vec{r}), \\ \mathcal{C}\phi(t,\vec{r})\mathcal{C}^\dagger &= e^{i\alpha_C}\phi^\dagger(t,\vec{r}), \\ \mathcal{T}\phi(t,\vec{r})\mathcal{T}^{-1} &= e^{i\alpha_T}\phi(-t,\vec{r}), \end{aligned} \tag{2.40}$$

where α_i are arbitrary phases. For the vector fields, in particular the QED photon A_μ,

$$\begin{aligned} \mathcal{P}A_\mu(t,\vec{r})\mathcal{P}^\dagger &= A^\mu(t,-\vec{r}), \\ \mathcal{C}A_\mu(t,\vec{r})\mathcal{C}^\dagger &= -A_\mu(t,\vec{r}), \\ \mathcal{T}A_\mu(t,\vec{r})\mathcal{T}^{-1} &= A^\mu(-t,\vec{r}), \end{aligned} \tag{2.41}$$

as A_μ is a Hermitian operator. Finally, for the fermion fields,

$$\begin{aligned} \mathcal{P}\psi(t,\vec{r})\mathcal{P}^\dagger &= e^{i\beta_P}\gamma^0\psi(t,-\vec{r}), \\ \mathcal{C}\psi(t,\vec{r})\mathcal{C}^\dagger &= e^{i\beta_C}\psi^c(t,\vec{r}), \\ \mathcal{T}\psi(t,\vec{r})\mathcal{T}^{-1} &= e^{i\beta_T}\gamma_0^*\gamma_5^* C^* A\psi(-t,\vec{r}), \end{aligned} \tag{2.42}$$

where, again, β_i are the arbitrary phases, and $\psi^c = CA^T\psi^{\dagger T}$.

While the field transformations described by Eqs. (2.40), (2.41), and (2.42) are useful, of greater interest are the Lorentz-covariant objects that are built out of scalar, fermion, and vector fields, and the ways they respond to C, P, and T transformations. These objects are the main building blocks of SM and BSM Lagrangians and therefore are of practical interest. We collect the results in the Table 2.2.

---

*Recall that antiunitary transformation $\hat{A}$ can be thought of as a combination of a unitary transformation $\hat{U}$ and complex conjugation $\hat{K}$, i.e. $\hat{A} = \hat{U}\hat{K}$.

TABLE 2.2 Discrete symmetries and fermionic currents.

Current	P	T	C	CP	CPT
$\overline{\psi}\chi$	$\overline{\psi}\chi$	$\overline{\psi}\chi$	$\overline{\chi}\psi$	$\overline{\chi}\psi$	$\overline{\chi}\psi$
$\overline{\psi}\gamma_5\chi$	$-\overline{\psi}\gamma_5\chi$	$\overline{\psi}\gamma_5\chi$	$\overline{\chi}\gamma_5\psi$	$-\overline{\chi}\gamma_5\psi$	$-\overline{\chi}\gamma_5\psi$
$\overline{\psi}\gamma_\mu\chi$	$\overline{\psi}\gamma_\mu\chi$	$\overline{\psi}\gamma_\mu\chi$	$-\overline{\chi}\gamma_\mu\psi$	$-\overline{\chi}\gamma_\mu\psi$	$-\overline{\chi}\gamma_\mu\psi$
$\overline{\psi}\gamma_\mu\gamma_5\chi$	$-\overline{\psi}\gamma_\mu\gamma_5\chi$	$\overline{\psi}\gamma_\mu\gamma_5\chi$	$\overline{\chi}\gamma_\mu\gamma_5\psi$	$-\overline{\chi}\gamma_\mu\gamma_5\psi$	$-\overline{\chi}\gamma_\mu\gamma_5\psi$
$\overline{\psi}\sigma_{\mu\nu}\chi$	$\overline{\psi}\sigma_{\mu\nu}\chi$	$-\overline{\psi}\sigma_{\mu\nu}\chi$	$-\overline{\chi}\sigma_{\mu\nu}\psi$	$-\overline{\chi}\sigma_{\mu\nu}\psi$	$\overline{\chi}\sigma_{\mu\nu}\psi$

The information presented in Table 2.2 can be used to devise how left- and right-handed currents respond to discrete transformations. For example, for a vector current of left-handed fields,

$$j_{L,\mu} = \overline{\psi}_L\gamma_\mu\chi_L = \frac{1}{2}\left(\overline{\psi}\gamma_\mu\chi - \overline{\psi}\gamma_\mu\gamma_5\chi\right), \tag{2.43}$$

from which it follows that it transforms into itself under $\mathcal{T}$ and into a right-handed current $j_{R,\mu} = \overline{\psi}_R\gamma_\mu\chi_R$ under $\mathcal{P}$. It also transforms into a negative of a Hermitian-conjugated version of $j_{R,\mu}$ under $\mathcal{C}$. Using Eqs. (2.40), (2.41), and (2.42), as well as Table 2.2, transformation laws of all of the terms in a Lagrangian can be obtained. One simply needs to remember that under space reflections we should have $\partial^\mu \to \partial_\mu$, and under time reversal we expect that $\partial^\mu \to -\partial_\mu$.

2.4 Flavor and CP-violation in the Standard Model

It is known experimentally that CP is a broken symmetry in the Standard Model. However, it is the *pattern* of CP symmetry breaking in the SM that makes its study interesting for New Physics searches. A possibility for CP-violation exists when the processes and their CP-counterparts have different experimental rates. Examining the SM Lagrangian of Eq. (2.3), we could notice several possible places where CP-violation can potentially arise.

First, we notice that if the Lagrangian is built by simply writing out all possible terms of dimension four or less than terms like

$$\mathcal{L} = \frac{g^2}{32\pi^2}\theta G_{\mu\nu}\widetilde{G}_{\mu\nu}, \tag{2.44}$$

where $\widetilde{G}^{\mu\nu} = (1/2)\epsilon^{\mu\nu\alpha\beta}G^{\alpha\beta}$ is the dual field tensor, are simply absent! One can argue that similar terms like $B\widetilde{B}$ for the abelian gauge fields can be written in terms of a full derivative and thus removed from the action, and $SU(2)$ terms $W\widetilde{W}$ do not have any observable consequences. Yet, the QCD term in Eq. (2.44) can not be easily removed. In fact, the presence of this term leads to an observable effect: a non-zero value of the electric dipole moment (EDM) of the neutron d_n. Experimentally, neutron's EDM is bounded to be very small, $|d_n| < 10^{-24}$ e·cm. This implies that $\theta \ll 1$ and is usually taken as an indication that CP is a good symmetry of strong interactions, so such terms are not needed. This "just so" explanation is somewhat unsatisfactory but provides an interesting way to modify the minimal SM Lagrangian by the inclusion of additional dimension-four operators. We shall discuss the strong CP problem in Chapter 4.

Second, we could potentially break CP-symmetry the same way we broke chiral symmetry, i.e. spontaneously. This could be achieved by acquiring complex phases in the broken phase. The vacuum expectation value of the Higgs field would then take the form

$$H = \frac{1}{\sqrt{2}}\begin{pmatrix} 0 \\ ve^{i\theta} \end{pmatrix}. \tag{2.45}$$

This mechanism could, in principle, lead to observable CP-violating effects, but not in the minimal Standard Model. As we can see from Eq. (2.40), the Higgs field transforms under a CP-symmetry transformation as

$$[CP]\, H(\vec{r},t)\, [CP]^\dagger = e^{i\alpha} H^\dagger(-\vec{r},t). \tag{2.46}$$

Thus, by choosing $\alpha = 2\theta$ we can always make the SM vacuum invariant under CP-transformation. Thus, CP cannot be broken spontaneously in the Standard Model. However, this can be achieved in BSM models with several Higgs doublets, which is yet another possible extension of the SM.

Finally, let us take a closer look at SM Yukawa terms of Eq. (2.25) to show this is the only place where CP can be broken in the Standard Model. A rough argument goes as follows. Consider the Lagrangian

$$\mathcal{L}_Y = Y_{ij}\overline{\psi}_{Ri}\chi_{Lj}\phi + Y_{ij}^\dagger \overline{\chi}_{Lj}\psi_{Ri}\phi^\dagger, \tag{2.47}$$

where we explicitly wrote out the Hermitian-conjugated part. The CP-conjugated Lagrangian is

$$[CP]\, \mathcal{L}_Y\, [CP]^\dagger = Y_{ij}\overline{\chi}_{Lj}\psi_{Ri}\phi^\dagger + Y_{ij}^\dagger \overline{\psi}_{Ri}\chi_{Lj}\phi. \tag{2.48}$$

Comparing it to Eq. (2.47), one can see that, unless the matrix of the Yukawa couplings is real, CP is broken. This is exactly what happens in the SM. The physical effect of the rotation from the gauge to mass basis in Eq. (2.28) can be seen in charged weak currents described by the term in the SM Lagrangian

$$\mathcal{L}_W = \frac{g_2}{\sqrt{2}}\overline{u}_{Li}\gamma^\mu \left[V_{uL}V_{dR}\right]_{ij} d_{Lj}W_\mu^+ + h.c. = \frac{g_2}{\sqrt{2}} V_{ij}\ \overline{u}_{Li}\gamma^\mu d_{Lj}W_\mu^+ + h.c \tag{2.49}$$

which is a part of the kinetic term of Eq. (2.5). Here V_{ij} is the CKM matrix introduced in Eq. (2.28). It has a generic form

$$V = \begin{pmatrix} V_{ud} & V_{us} & V_{ub} \\ V_{cd} & V_{cs} & V_{cb} \\ V_{td} & V_{ts} & V_{td} \end{pmatrix}. \tag{2.50}$$

We cannot predict the numerical values of the matrix elements of the CKM matrix within the SM. However, we can figure out the number of parameters that are needed for its complete parameterization. It might be instructive to consider a generic case of N generations. First, a generic complex $N \times N$ matrix contains $2N^2$ real parameters. Since the CKM matrix is unitary, the unitarity relation $VV^\dagger = 1$ gives N^2 relations among the parameters of the matrix, i.e. sums of the products of matrix elements that are either 0 or 1. This results in $2N^2 - N^2 = N^2$ angles or phases parameterizing the matrix. Further, since quark wave functions are only defined up to a phase, we can rotate the phases of up and down quarks to remove additional $2N-1$ parameters*. This means that there remains only $N^2 - (2N-1) = (N-1)^2$ parameters of a CKM matrix for N generations of quarks.

We can figure out how many of those parameters are angles. In case of an N-dimensional space, one can perform ${}_N C_2 = N(N-1)/2$ rotations in each plane, which means that the rest of the parameters are phases numbering $(N-1)^2 - N(N-1)/2 = (N-1)(N-2)/2$. To summarize, for N generations of quarks, there are

$$\begin{array}{ll} \dfrac{N(N-1)}{2} & \text{angles} \\ (N-1)(N-2)/2 & \text{phases.} \end{array} \tag{2.51}$$

*There is one less than $2N$ rephasings needed because V is always multiplied by a quark bilinear, which means that for at least one matrix element it is sufficient to rephase only one quark field. This can be seen in the case of one up and one down quark: the CKM phase is removed by rotating the phase of *either* up or down quarks.

This implies that for the case of $N = 2$ generations the CKM matrix has no phases, i.e. real. Now, for the $N = 3$ generations of quarks, which seems to be favored by Nature, there are three angles and one phase that completely parameterize the CKM matrix, in which case the CKM matrix alone can encode CP-violation in the Standard Model. The "standard" parameterization of the CKM matrix can be written in terms of those angles and a phase,

$$\begin{aligned} V &= \begin{pmatrix} 1 & 0 & 0 \\ 0 & c_{23} & s_{23} \\ 0 & -s_{23} & c_{23} \end{pmatrix} \begin{pmatrix} c_{13} & 0 & s_{13}e^{-i\delta} \\ 0 & 1 & 0 \\ -s_{13}e^{i\delta} & 0 & c_{213} \end{pmatrix} \begin{pmatrix} c_{12} & s_{12} & 0 \\ -s_{12} & c_{12} & 0 \\ 0 & 0 & 1 \end{pmatrix} \\ &= \begin{pmatrix} c_{12}c_{13} & s_{12}c_{13} & s_{13}e^{-i\delta} \\ -s_{12}c_{23} - c_{12}s_{13}s_{23}e^{i\delta} & c_{12}c_{23} - s_{12}s_{13}s_{23}e^{i\delta} & c_{13}s_{23} \\ s_{12}s_{23} - c_{12}s_{13}c_{23}e^{i\delta} & -c_{12}s_{23} - s_{12}s_{13}c_{23}e^{i\delta} & c_{13}c_{23} \end{pmatrix}, \end{aligned} \tag{2.52}$$

where $c_{ij} = \cos\theta_{ij}$, $s_{ij} = \sin\theta_{ij}$, and δ is the CKM phase.

This is a rather remarkable result! It implies that all CP-violating effects in the Standard Model are related to a single phase of the CKM matrix! It also reinforces the fact that studies of CP-violation are important tool for searches for New Physics, as any inconsistency of the observed CP-violating signals with the CKM picture would automatically lead to discovery of physics beyond the Standard Model.

As we do not have a way to predict the numerical values of the CKM matrix elements, we can still measure them experimentally. We shall discuss the peculiarities of these measurements in Chapter 4. But it would be appropriate to point out that it appears that the CKM matrix has a very special form, with the diagonal matrix elements being almost of order one and a gradual diminishing of the sizes away from the main diagonal. In fact, L. Wolfenstein proposed a very convenient way to parameterize the CKM matrix [184], which resembles a perturbative expansion in parameter $\lambda \simeq |V_{us}| \sim 0.23$,

$$V = \begin{pmatrix} 1 - \frac{\lambda^2}{2} & \lambda & A\lambda^3(\rho - i\eta) \\ -\lambda & 1 - \frac{\lambda^2}{2} & A\lambda^2 \\ A\lambda^3(1 - \rho - i\eta) & -A\lambda^2 & 1 \end{pmatrix} + \mathcal{O}(\lambda^4), \tag{2.53}$$

where the other parameters are $A \simeq 0.81$, $\rho \simeq 0.14$, and $\eta \simeq 0.35$. They can be related to the standard parameterization of Eq. (2.52) as $\lambda = s_{12}$, $A\lambda^2 = s_{23}$, and $A\lambda^3(\rho - i\eta) = s_{13}e^{-i\delta}$.

The current values of λ, A, ρ, and η can be found in [1]. The parameterization of Eq. (2.53) is called the *Wolfenstein parameterization**. It provides a very convenient way to estimate sizes of different semileptonic and nonleptonic transitions based on their scaling with powers of λ.

2.5 Heavy New Physics and model building

As we already pointed out at the beginning of this chapter (see Sec. 2.1), it appears that the Standard Model needs additional help in order to describe the Universe we happen to live in.

New matter fields: quarks and leptons

The simplest extension of the SM would be to add more generations of *chiral* quarks and leptons. This extension, however, does not appear to get support from experimental results. Alternatively,

*A variant of the parameterization in Eq. (2.53) is often used and is sometimes referred to as *Buras-Wolfenstein* parameterization. It can be obtained from Eq. (2.53) by substituting $\rho \to \bar{\rho} = \rho(1 - \lambda^2/2)$ and $\eta \to \bar{\eta} = \eta(1 - \lambda^2/2)$.

vector-like quarks and leptons could be added. The beauty of extensions like that is in the fact that one does not have to worry about anomaly cancellations.

Integrating out heavier quarks and leptons can also lead to a completely new operator structures, especially of other new light degrees of freedom are present. For example, an extra abelian vector V_μ field could mix with the photon A_μ at low energies if both couple to some electrically-charged heavy new fermion states. Integrating out those states at low energy would lead to structures like

$$\mathcal{L} \supset \frac{\kappa}{2} V_{\mu\nu} F^{\mu\nu}, \tag{2.54}$$

where the coefficient κ depends on the logarithms of the masses of those new heavy fermion states. This structure is gauge invariant and can lead to very interesting consequences for low-energy experiments. Because of the similarity with the gauge boson kinetic term such structures go under the name *kinetic mixing*. We will discuss kinetic mixing in Sec. 2.7.2. Other possibilities include supersymmetric extensions of the SM, where new fermions appear as SUSY partners of the bosonic states of the SM.

New scalar particles

Another possible extension of the Standard Model involves enlarging the number of the Higgs doublets, with the simplest one – the two-Higgs-doublet model (2HDM). This extension is sufficiently well-motivated because 2HDM can offer a way to generate a baryon asymmetry of the Universe that is consistent with today's observations. This is so because 2HDM provides additional scalar degrees of freedom, as well as new possible sources of CP-violation [28]. Many models of New Physics also use its features, such as axion models or supersymmetric models, where two Higgs doublets is a minimal scheme in which electroweak symmetry breaking can be accommodated. In this book, we will be interested in multi-Higgs models (and 2HDM in particular) from the point of view of a low-energy effective field theory.

The Higgs sector of the model that contains several Higgs doublets can be obtained by careful modifications of Eq. (2.13). While the kinetic part receives an obvious extension, the potential part of Eq. (2.14) requires some attention. Even in the simplest extension in the number of Higgs doublets, the 2HDM, one must be careful in choosing the physical parameters, as some can be rotated away or redefined by selecting field bases. The most general Higgs potential contains 14 parameters and can have CP-conserving, CP-violating, and charge-violating minima [28]. Assuming that CP is conserved in the Higgs sector and CP is not broken spontaneously, the most general Higgs potential for the 2HDM with doublets Φ_1 and Φ_2 with hypercharge +1 can be written as [28]

$$\begin{aligned} V(\Phi_1,\Phi_2) &= m_{11}^2\Phi_1^\dagger\Phi_1 + m_{22}^2\Phi_2^\dagger\Phi_2 - m_{12}^2\left(\Phi_1^\dagger\Phi_2 + \Phi_2^\dagger\Phi_1\right) \\ &+ \frac{\lambda_1}{2}\left(\Phi_1^\dagger\Phi_1\right)^2 + \frac{\lambda_2}{2}\left(\Phi_2^\dagger\Phi_2\right)^2 + \lambda_3\Phi_1^\dagger\Phi_1\Phi_2^\dagger\Phi_2 \\ &+ \lambda_4\Phi_1^\dagger\Phi_2\Phi_2^\dagger\Phi_1 + \frac{\lambda_5}{2}\left[\left(\Phi_1^\dagger\Phi_2\right)^2 + \left(\Phi_2^\dagger\Phi_1\right)^2\right], \end{aligned} \tag{2.55}$$

Some terms in the potential $V(\Phi_1, \Phi_2)$ can be reduced by imposing ad-hoc symmetries. Keeping in mind the simplifying assumptions we made earlier, each Higgs doublet will take the following form after spontaneous symmetry breaking

$$\Phi_i = \begin{pmatrix} \phi_i^+ \\ \frac{1}{\sqrt{2}}(v_i + \rho_i + i\eta_i) \end{pmatrix}, \tag{2.56}$$

for $i = 1, 2$. The v_i are two vacuum expectation values of the Higgs fields. Indeed not all of the fields in Eq. (2.56) are physical. Out of the total eight Goldstone fields, three become longitudinal degrees of freedom of Z^0 and $W^{\pm}$ fields. The remaining five fields are physical and constitute real Higgs fields. Among those are two scalars, one pseudoscalar, and a charged scalar field. Integrating them out can give us new operators structures at low energy, not previously identified in the Fermi model. Looking at the patterns of the Wilson coefficients of those structures in low-energy experiments can, in principle, tell us how the UV completion of the model looks like.

We should make a couple of comments before discussing other possible types of SM extensions. As one might note, substituting Eq. (2.56) into Eq. (2.55) would lead to non-diagonal mass terms for ϕ_i^+, ρ_i, and η_i. For example, for the charged scalars [28],

$$\mathcal{L}_{\phi^\pm} = \left(m_{12}^2 - (\lambda_4 + \lambda_5)v_1 v_2\right)\left(\phi_1^-, \phi_2^-\right)\begin{pmatrix} v_2/v_1 & -1 \\ -1 & v_1/v_2 \end{pmatrix}\begin{pmatrix} \phi_1^+ \\ \phi_2^+ \end{pmatrix}. \tag{2.57}$$

This matrix needs to be diagonalized, which can be achieved by the rotation

$$\begin{aligned} H_1 &= \Phi_1 \cos\beta + \Phi_2 \sin\beta, \\ H_2 &= -\Phi_1 \sin\beta + \Phi_2 \cos\beta, \end{aligned} \tag{2.58}$$

where the rotation parameter β is defined as

$$\tan\beta = \frac{v_2}{v_1}. \tag{2.59}$$

The rotation of Eq. (2.58) diagonalizes the charge scalar mass matrix, resulting in a zero-mass state $G^{\pm}$ that becomes the longitudinal degree of freedom of the $W^{\pm}$ bosons, which makes it consistent with the Goldstone theorem. Additionally, there is a massive physical state $H^{\pm}$ with the mass $m_{\pm}^2 = \left(m_{12}^2/v_1 v_2 - (\lambda_4 + \lambda_5)\right)\left(v_1^2 + v_2^2\right)$. The same rotation diagonalizes the pseudoscalar mass matrix, while similar rotation with the angle α defines new states h and H,

$$\begin{aligned} h &= \rho_1 \sin\alpha - \rho_2 \cos\alpha, \\ H &= -\rho_1 \cos\alpha - \rho_2 \sin\alpha. \end{aligned} \tag{2.60}$$

The h is usually referred to as the "light" Higgs state, while $H-$ as the "heavy" one. To add to the confusion, the "Standard Model" Higgs state, i.e. the one with the mass of 125 GeV, is given by $H_{\rm SM} = h\sin(\alpha - \beta) - H\cos(\alpha - \beta)$.

The low-energy structures that appear due to 2HDM depend on how the scalar degrees of freedom couple to quarks and leptons. This leads to possible phenomenological issues with 2HDM related to the existence of the FCNC terms in the Lagrangian that appear after SSB. Since now two Higgs doublets are present, quarks and leptons mass matrices can not, in general, be diagonalized, leading to FCNC. Those could be removed by special arrangements of how the Higgs doublets couple to fermions. Possibilities include coupling of one Higgs doublets to leptons and the other one to quarks only (Type I 2HDM), or coupling of one doublet to up-type and the other – to the down type fermions (Type II 2HDM). Such models are known as models with *natural flavor conservation.* Those choices could be arranged by introducing various discrete symmetries on the Higgs sector [28]. Alternatively, one might learn to live with FCNCs and do not require any special coupling arrangement. Such models are now known as Type III 2HDMs.

Other complications of this scenario are also possible. For example, $SU(2)$ gauge singlet scalars can also be added, which would mix with the scalars remaining after the SSB. They might or might not lead to new operator structures at low energies.

New symmetries

One more way, which follows the logic of unification of distinct interactions into different manifestations of one is to find a larger symmetry group that includes the Standard Model's

$SU(3) \times SU(2) \times U(1)$ as a subgroup. There are several such groups: $SU(5)$, $SO(10)$, to name a couple. There are also many ways that those (and other) symmetry groups can be combined with other symmetries, including supersymmetry, to ensure unification at some high scale. Since we are interested in the bottom-up approach of discovering New Physics in this book, we shall not describe those models in detail here. The interested reader should consult wonderful reviews [162, 187].

It is nonetheless useful to provide some useful examples here. We will touch upon symmetry structures of the simplest GUT models, such as Georgi-Glashow's $SU(5)$. We shall see that choices of symmetry groups and matter representations lead to interesting consequences for low energy phenomenology.

$SU(5)$ is the smallest group that contains the Standard Model gauge group as a subgroup. It contains 24 generators, which would correspond to 24 gauge bosons. Thus, we should expect new gauge fields, which, upon breaking of $SU(5)$ down to $SU(3) \times SU(2) \times U(1)$, would be significantly heavier than the characteristic electroweak symmetry breaking scale. If we are to unify strong and electroweak interactions, it would make sense to put all matter fields such that they all would transform under $SU(5)$ symmetry transformations within the same representation. There are two issues with that. First, the fundamental representation, **5**, only contains five fields. For the case of a single generation, one needs to include 3 (color) × 2 (chirality) × 2 (flavor) = 12 quark fields and 2 (chirality) × 2 (flavor) − 1 (RH neutrino) = 3 lepton fields. Suggestively (or maybe miraculously), adding the antisymmetric **10** to the **5** representation accounts for all needed fields. Second, since gauge transformations commute with the Lorentz group, we should work with fields of the same chirality within each multiplet. A clever and convenient way of doing so is suggested by the fact that the charge conjugation of a chiral field flips its chirality (see Sec. 2.3 for more detail). Thus, for example, a right-handed down quark in Eq. (2.1) would change into a left-handed antiquark,

$$d_R = (3, 1)_{-\frac{1}{3}} \to \bar{d}_L = (3^*, 1)_{\frac{1}{3}} \tag{2.61}$$

which makes it possible to put both down quarks and leptons in the same $\mathbf{5}^*$ multiplet,

$$\psi^\alpha = \begin{pmatrix} \bar{d}_i \\ \bar{d}_j \\ \bar{d}_k \\ \nu \\ e \end{pmatrix}, \tag{2.62}$$

Note that charge conjugation also flips the sign of the hypercharge. The Latin indices on the down-quark field indicate color. The rest (ten) of the fields are conveniently placed in the **10** [187],

$$\psi^{\alpha\beta} = \begin{pmatrix} 0 & \bar{u} & -\bar{u} & u & d \\ -\bar{u} & 0 & \bar{u} & u & d \\ \bar{u} & -\bar{u} & 0 & u & d \\ -u & -u & -u & 0 & \bar{e} \\ -d & -d & -d & -\bar{e} & 0 \end{pmatrix}, \tag{2.63}$$

where we suppressed color indices on the quark fields. One important observation that follows from Eqs. (2.62) and (2.63) is that quarks and leptons sit in the same reps of the symmetry groups, which implies that there are generators of $SU(5)$ that transform one into another. Physically, this means that there are interactions that change quarks into leptons. These interactions would break baryon and lepton numbers and, in principle, lead to low energy processes such as proton decay or neutron-antineutron oscillations. Since there are no such interactions in the

SM*, they would belong to the set of generators broken at the GUT scale. This implies that the corresponding gauge bosons would be very heavy, making baryon-number violating interactions very rare. However, the mere presence of such interactions means that we can add low energy effective operators that break baryon and lepton number to compute experimental observables.

An alternative approach to unification is provided by Pati-Salam models [162, 187]. These models extend the QCD symmetry group to $SU(4)$, treating the lepton number as a "fourth color". The model also has left-right symmetry and predicts the existence of right-handed interactions with heavy W' and Z' bosons at high energy.

There is a great variety of models that employ various symmetry groups that include the Standard Model as the low-energy *effective field theory*. If we are only interested in finding the physics beyond the Standard Model, we do not really need to know what exactly the ultraviolet completion of the SM really is. While in principle, the traces of broken symmetries might imply relations between the coefficients of the Wilson coefficients of effective operators written in terms of the SM degrees of freedom, low-energy studies should excel in computing observables or relations between the observables that are only expected to exist in the Standard Model.

2.6 Heavy New Physics without model building

If we take the notion of the Standard Model being a low-energy effective theory for some yet unknown fundamental theory, then we can employ the whole machinery of EFT [149] to parameterize all possible NP effects on low-energy observables. The SM is an unusual EFT: it is perturbatively renormalizable, which means that it does not have an internal scale that signals where it breaks down. We have to introduce it *externally*, i.e. postulate that additional degrees of freedom will appear at some scale Λ, which also becomes a parameter of our EFT description. These additional degrees of freedom might include new matter and gauge fields that build up the true UV completion of the SM. Or they might include something that we cannot anticipate at the moment. It is not important for the description of low energy observables.

What is important is that introduction of the scale Λ will allow us to write out higher-dimensional operators. While this makes our theory non-renormalizable, it offers a way to parametrize all possible effects of New Physics with only SM degrees of freedom.

2.6.1 Integrating out new particles: effective Lagrangians

Before constructing the most general SM Lagrangian with higher dimensional operators, let's look at an example of a known effective field theory in particle physics: the Fermi model [149]. In Eq. (1.7) we have already mentioned that EFT, where it was used to describe muon decay. We can generalize it to include quarks in low-energy weak interaction processes. For example, taking the Lagrangian describing decays of strange particles, such as $K \to \pi\pi$, we can write that in the Fermi model

$$\mathcal{L}_{\text{eff}}^{(6)} = -\frac{4G_F}{\sqrt{2}} V_{us} V_{ud}^* (\bar{u}\gamma^\mu P_L s)(\bar{d}\gamma_\mu P_L u), \tag{2.64}$$

where $P_L = (1 - \gamma_5)/2$ is the left-handed projection operator and V_{ij} are the CKM matrix elements. Note that $\mathcal{L}_{\text{eff}}^{(6)}$ only contains left-handed fields as a rule for describing the charged current interactions. The Lagrangian in Eq. (2.64) describes a non-renormalizable theory that has problems with its ultraviolet behavior. Nevertheless, the Fermi model has tremendous success in describing low energy weak interactions and can be generalized to include all quarks and

*Baryon number can only be broken in the Standard Model via nonperturbative sphaleron process.

leptons. For arbitrary quarks q_i

$$\mathcal{L}_{\text{eff}}^{(6)} = -\frac{4G_F}{\sqrt{2}} \sum_{q_i} V_{q_1 q_2} V^*_{q_4 q_3} (\bar{q}_1 \gamma^\mu P_L q_2)(\bar{q}_3 \gamma^\mu P_L q_4), \tag{2.65}$$

where $q_i \to q_k$ quark transition changes quark electric charge by one unit. Furthermore, even neutral currents can be successfully described,

$$\mathcal{L}_{\text{eff}}^{(\text{neutral})} = -\frac{4G_F}{\cos^2 \theta_W} \sum_{A,B=L,R} (a_q^A a_Q^B)(\bar{q}\gamma^\mu P_A q)(\bar{Q}\gamma^\mu P_B Q), \tag{2.66}$$

where $a_q^A = \frac{1}{2}\delta_{AL} - \frac{2}{3}\sin^2\theta_W$. All (initially) unknown coefficients, such as G_F or $V_{q_i}V^*_{q_k}$ can be determined by comparing computed expressions of various decay rates and cross-sections to experimental data. In line with the EFT framework, predictions can be made once all unknown coefficients are determined from experiments. Then, the presence of physical phenomena that is not described by $\mathcal{L}_{\text{eff}}$ can be attributed to "physics beyond the Fermi model".

Following the EFT paradigm, we can even include higher-dimensional operators, which in this case should come at dimension eight. The most general effective Lagrangian for the charged current interactions that would generate such terms is

$$\begin{aligned} \mathcal{L}_{\text{eff}}^{(8)} &= -\frac{4G_F}{\sqrt{2}} V_{us} V^*_{ud} \Big[C_1 (\bar{u}\gamma^\mu P_L D^2 s)(\bar{d}\gamma_\mu P_L u) \\ &+ C_2 (\bar{u}\gamma^\mu P_L s)(\bar{d}\gamma_\mu P_L D^2 u) \\ &+ C_3 (\bar{u}\gamma^\mu P_L D^\nu s)(\bar{d}\gamma_\mu P_L D_\nu u) \Big], \end{aligned} \tag{2.67}$$

where D_μ is the usual gauge covariant derivative that includes QCD and QED gauge fields, and we only included left-handed fields per our rule.

Alternatively, one can *assume* that there exists a proper UV completion of Eq. (2.64). In that case, a *matching* procedure can be performed: one calculates the same Green function with full and effective theories and then match the results. Since the physics that is not properly described by the Fermi model Lagrangian appears at short distances, it does not matter what we choose as initial and final states for the transition amplitude (Green function). It is convenient to choose free quark states. This way one can determine the Wilson coefficients (e.g. G_F) of a low-energy effective field theory, provided that we choose a particular UV completion to which we can match it.

It was proposed in the '70s of the 20th century that a renormalizable theory of electroweak interactions should be used in place of the Lagrangian of Eq. (2.64), which is now known as the Standard Model. In the SM a dimension-4 Lagrangian that describes interactions of quark currents with a W-boson should be used,

$$\mathcal{L}_W = \frac{g_2}{\sqrt{2}} V_{ij} \bar{q}_i \gamma^\mu P_L q_j W^\pm_\mu + \cdots . \tag{2.68}$$

We should emphasize that the Standard Model, which provides us with the Lagrangian of Eq. (2.68), is only one possible ultraviolet completion of the Fermi model. As we know now, it is indeed a correct one.

To match the Fermi model in the SM in the $\Delta S = 1$ sector, let's compute a matrix element for $s \to \bar{u}ud$ transition in the SM,

$$\mathcal{M}^{\Delta S=1} = \left(\frac{ig_2}{\sqrt{2}}\right)^2 V_{us} V^*_{ud} (\bar{u}\gamma^\mu P_L s) \, \frac{-ig_{\mu\nu}}{p^2 - M_W^2} \, (\bar{d}\gamma^\nu P_L u) \tag{2.69}$$

where we work in the Feynman gauge, neglecting the contributions from gauge artifact fields, as they will not affect our arguments here and are small for the light quarks anyway. In the limit that $p^2 \ll M_W^2$ we can expand the denominator of our matrix element of Eq. (2.69) in the Taylor series,

$$\frac{1}{p^2 - M_W^2} = -\frac{1}{M_W^2}\left(1 + \frac{p^2}{M_W^2} + \cdots\right). \tag{2.70}$$

The matrix element then becomes

$$\mathcal{M}^{\Delta S=1} = \frac{i}{M_W^2}\left(\frac{ig_2}{\sqrt{2}}\right)^2 V_{us}V_{ud}^*(\bar{u}\gamma^\mu P_L s)(\bar{d}\gamma_\mu P_L u) + \mathcal{O}\left(\frac{1}{M_W^4}\right). \tag{2.71}$$

Computing the same matrix element computer with the aid of the Fermi model Lagrangian of Eq. (2.68), we find

$$\mathcal{M}_{FM}^{\Delta S=1} = \frac{i}{M_W^2}\left(\frac{ig_2}{\sqrt{2}}\right)^2 V_{us}V_{ud}^*(\bar{u}\gamma^\mu P_L s)(\bar{d}\gamma_\mu P_L u). \tag{2.72}$$

Matching Eqs. (2.71) and (2.72) and ignoring terms of the order higher than $1/M_W^2$, we arrive at the matching condition

$$\frac{4G_F}{\sqrt{2}} = \frac{g_2^2}{2M_W^2}, \tag{2.73}$$

which is equivalent to Eq. (1.8). The Wilson coefficient G_F is called the Fermi constant. Both the tree level use of the Lagrangian in Eq. (2.68) and the effective theory in Eq. (2.64) give the same answer up to $\mathcal{O}(1/M_W^4)$ corrections.

We can keep going, computing contributions from the terms suppressed by higher powers of $1/M_W$. For example, if we include the next term in the Taylor expansion of the propagator, the matrix element becomes

$$\Delta\mathcal{M}^{\Delta S=1} = \frac{ip^2}{M_W^4}\left(\frac{ig_2}{\sqrt{2}}\right)^2 V_{us}V_{ud}^*\left(\bar{u}\gamma^\mu P_L s\right)\left(\bar{d}\gamma_\mu P_L u\right). \tag{2.74}$$

This expression can be matched to the expression for the matrix element computed from the Lagrangian of Eq. (2.67), as the presence of a p^2 in the amplitude tells us that there must be a derivative in our effective Lagrangian. By comparing the matrix elements of the three operators from Eq. (2.67) with the full theory result of Eq. (2.74), we can obtain the C_i. Notice that the Fermi constant we matched earlier has been explicitly factored from these coefficients – this is standard practice for the calculations within the Fermi model.

We can now apply the same logic to the Standard Model! If the SM Lagrangian of Eq. (2.3) is to be considered as the leading term in $1/\Lambda$ expansion, we can write out the most general Lagrangian, written in terms of the SM degrees of freedom, that contains higher-dimensional operators. For this procedure to work we need to make an additional assumption about the properties of possible New Physics.

2.6.2 Standard Model as an effective theory or how to build an EFT without knowing the nature of New Physics

A proper expansion for a physical observable requires that the power series is convergent or at least asymptotic [16]. If New Physics appears at some high scale Λ, then the expansion parameter that one expects to have is $\epsilon = v/\Lambda$, where $v \simeq 246$ GeV is taken as a representative electroweak scale. In principle, any masses of electroweak gauge bosons or the Higgs mass can be used in place of v. We, therefore, should expect that there is a decent separation of scales between the Standard Model and New Physics energy scales, so that meaningful description of the SM as an effective field theory is possible [149]. If this condition is satisfied, then an effective Lagrangian can be written such that $\mathcal{L}_{\rm SM}$ represents the lowest order in $1/\Lambda$ expansion,

$$\begin{aligned}\mathcal{L} &= \mathcal{L}_{\rm SM} + \mathcal{L}^{(5)} + \mathcal{L}^{(6)} + \cdots \\ &= \mathcal{L}_{\rm SM} + \frac{1}{\Lambda}\sum_k C_k^{(5)} Q_k^{(5)} + \frac{1}{\Lambda^2}\sum_k C_k^{(6)} Q_k^{(6)} + \mathcal{O}\left(\frac{1}{\Lambda^3}\right). \end{aligned} \tag{2.75}$$

This EFT goes under the name of Standard Model Effective Field Theory or SMEFT. In building it one has to follow the usual rules for EFT building. The effective field theory that we are about to compose should (i) be invariant under the same set of symmetries as the SM, that is, $SU(3)_{\rm C} \times SU(2)_{\rm L} \times U(1)_Y$, and (ii) should contain only the SM degrees of freedom. This assumes that whatever the full theory is, the SM degrees of freedom are incorporated into it as fundamental or composite fields. We are going to assume in what follows that the electroweak symmetry breaking is realized linearly, as it is done in the Standard Model. Other than that, all possible operators of relevant dimension should be included.

The building of SM EFT operators proceeds from the SM scalar, fermion, and gauge fields, as well as their derivatives, $\{H, \psi, X, D\}$, whose dimensions can be read off Table 1.1. The operators can then be classified not only by dimension but also by the operator content. Conventionally, the operator classes are named by their field content, $\psi^k H^l D^m X^n$. For instance, the class $\psi^2 H^2 D^2$ is a class of dimension-seven operators containing two fermion fields, two Higgs fields, and two derivatives.

We should point out that the operators containing Higgs fields appearing in the SM EFT Lagrangian are naturally defined in the unbroken phase of the theory, where the vacuum expectation value of the Higgs field vanishes. At low energy one introduces Higgs vacuum expectation value and field excitation. At the scales below the electroweak symmetry breaking scale, one additionally integrates out Higgs and massive gauge fields, so most studies discussed in this book would center around ψ^4 and $\psi^2 X$ classes of operators.

Once all possible operators are written out, one should ask a question if the set of operators is minimal. That is to say if a relation can be found among the operators, some of them can be considered redundant and should be excluded from the list. There are several relations that can be employed to eliminate redundant operators, such as Fierz identities or identities among Pauli matrices. Finally, equations of motions (EOM) can be employed to make field redefinitions and place chosen operators into different classes. Conventionally, Hermitian-conjugated counter-parts of the operators are not counted. It is believed that SM EFT discussed below contains a minimal set of operators. Organized by the canonical operator dimension, we expect that there is one operator of dimension-two $(H^\dagger H)$, 13 operators of dimension-four, one operator of dimension-five, 63 operators of dimension-six, and 20 operators of dimension-seven. One can clearly see the proliferation of operators with increasing dimension, especially viewing separately the operators of odd and even dimensions.

Operators of dimension five

Operators of dimension five could be built out of two fermion and two Higgs fields, both of which transform as doublets under the electroweak $SU(2)$ group. Thus, this class of operators can be named $\psi^2 H^2$. As it turns out, there is only one operator that can be built out of those degrees of freedom if all SM gauge symmetry constraints are imposed on $\mathcal{L}^{(5)}$ of Eq. (2.75),

$$Q^{(5)} = \epsilon_{jk}\epsilon_{mn} H^j H^m \left(L_p^k\right)^T \mathcal{C} L_r^n \equiv \left(\widetilde{H}^\dagger L_p\right)^T \mathcal{C} \left(\widetilde{H}^\dagger L_r\right). \tag{2.76}$$

This operator actually has a name: it is commonly referred to as the Weinberg operator [178]. Note that we introduced a charge-conjugation operation $\mathcal{C}$ to Eq. (2.76), where $\mathcal{C} = i\gamma^2\gamma^0$ in the Dirac representation of gamma-matrices. As it turns out, the Weinberg operator violates the lepton number. One might recall that baryon and lepton numbers are accidental symmetries in the Standard Model Lagrangian*. Therefore, experimental discovery of lepton- (or baryon-) number violating processes would indeed provide evidence for physics beyond the Standard Model. It is reassuring that SMEFT captures and quantifies this phenomenon, as well as the fact that it is suppressed by powers of ϵ compared to the SM electroweak processes.

It is interesting to notice that the first operator that violates the lepton (or baryon) number appears at the odd dimension. One can actually prove that in SMEFT all operators of any odd dimension must violate lepton or baryon number [159], while operators of even dimension can have both lepton- and baryon-number conserving and violating operators.

The presence of the Weinberg operator has several curious consequences. One is the possibility to generate Majorana masses for neutrinos. The leading term in SMEFT, $\mathcal{L}_{\text{SM}}$ does not contain any right-handed neutrino fields. This leads to the fact that neutrinos remain massless in the SM, as the only dimensionally-appropriate term would be of the type $m_\nu^D \overline{\nu}_L \nu_R +$ h.c. Since there are no right-handed neutrino fields, such a term is not allowed. Of course, if we simply postulate the existence of such fields, we could extend the usual Yukawa mechanism to neutrinos to generate "Dirac masses." Alternatively, if neutrinos are their own antiparticles, i.e. they are the Majorana fermions, it is possible to write a mass term with purely left-handed fields, which would violate the lepton number

$$\mathcal{L}_\nu = \frac{1}{2} m^{(M)} \overline{\nu}_L^c \nu_L + \text{h.c.} \tag{2.77}$$

where $\nu_L^c = C\overline{\psi_L}^T$ stands for the charge-conjugated neutrino field. It is interesting that, after spontaneous symmetry breaking, the Weinberg operator of Eq. (2.76) leads to Majorana neutrino masses,

$$(m)_{pq}^{(M)} \sim C_{pq}^{(5)} \frac{v^2}{\Lambda}. \tag{2.78}$$

This relation immediately implies that $\Lambda > 10^{14}$ GeV to generate neutrino masses $M_\nu < 1$ eV, if we set $C^{(5)} \sim 1$. This is a rather high scale that is unlikely to be probed in experiments at particle accelerators. While this scale can be somewhat lowered by making the Wilson coefficients smaller, a question can be asked if the scale Λ is indeed a unique scale for all operators in the effective theory. We shall return to this question at the end of this chapter.

Alternatively, a requirement of lepton number conservation can be placed on the effective operators, assuming that neutrino masses are generated by some other mechanism, such as the Dirac mechanism discussed above. In this case, the first New Physics correction would be parameterized by the operators of dimension six.

Operators of dimension six

The set of operators of dimension six contains both baryon (lepton)-number conserving and baryon (lepton)-number violating operators. The dimension-six terms of the SM EFT Lagrangian can be built the same way as any effective Lagrangian discussed previously [149]: by combining relevant fields (degrees of freedom) consistent with gauge and space-time symmetries of the theory and then using equations of motion, field redefinitions, and other operator identities to eliminate redundant operators to find the minimal basis.

Following [36] and [106], we present a complete set of non-redundant dimension-6 operators in the so-called "Warsaw basis". This basis contains 59 baryon-number conserving

*Baryon number violation can be achieved in the Standard Model via non-perturbative sphaleron mechanism, which is not represented by a single local term in $\mathcal{L}_{\text{SM}}$.

TABLE 2.3 Operators with H^n, sets X^3, H^6, H^4D^2, and $\psi^2 H^3$.

	X^3		H^6 and H^4D^2		$\psi^2 H^3$+ h.c.
Q_G	$f^{ABC}G_\mu^{A\nu}G_\nu^{B\rho}G_\rho^{C\mu}$	Q_H	$\left(H^\dagger H\right)^3$	Q_{eH}	$\left(H^\dagger H\right)\left(\overline{L}_p e_r H\right)$
$Q_{\widetilde{G}}$	$f^{ABC}\widetilde{G}_\mu^{A\nu}G_\nu^{B\rho}G_\rho^{C\mu}$	$Q_{H\Box}$	$\left(H^\dagger H\right)\Box\left(H^\dagger H\right)$	Q_{uH}	$\left(H^\dagger H\right)\left(\overline{Q}_p u_r \widetilde{H}\right)$
Q_W	$\epsilon^{IJK}W_\mu^{I\nu}W_\nu^{J\rho}W_\rho^{K\mu}$	Q_{HD}	$\left(H^\dagger D^\mu H\right)^*\left(H^\dagger D_\mu H\right)$	Q_{dH}	$\left(H^\dagger H\right)\left(\overline{Q}_p d_r H\right)$
$Q_{\widetilde{W}}$	$\epsilon^{IJK}\widetilde{W}_\mu^{I\nu}W_\nu^{J\rho}W_\rho^{K\mu}$				

and 5 baryon-number violating operators, which are organized into twelve classes (see Tables 2.1-2.6). Among those classes are five that contain various four-fermion operators (Tables 2.5 and 2.6), which give the most important contributions in low energy experiments. Spelling out different flavors of quarks and leptons gives a total of 2499 baryon-number conserving operators.

Since any terms suppressed as $1/\Lambda^3$ do not contribute to the dimension-six terms, only EOMs obtained from the SM operators (i.e. from $\mathcal{L}_{\rm SM}$ of Eq. (2.3)) would be needed. They are

$$\begin{aligned}
(D^\mu D_\mu H)^j &= m^2 H^j - \lambda\left(H^\dagger H\right)H^j - \bar{e}Y_e^\dagger l^J + \epsilon_{jk}\bar{q}^k Y_u u - \bar{d}Y_d^\dagger q^j, \\
(D^\alpha G_{\alpha\mu})^A &= g_3\left(\overline{Q}\gamma_\mu T^A Q + \overline{u}\gamma_\mu T^A u + \overline{d}\gamma_\mu T^A d\right), \\
(D^\alpha W_{\alpha\mu})^I &= \frac{g_2}{2}\left(H^\dagger i\overleftrightarrow{D}_\mu^I H + \bar{l}\gamma_\mu\tau^I l + \overline{Q}\gamma_\mu\tau^I Q\right), \\
\partial^\alpha B_{\alpha\mu} &= g_1 Y_H H^\dagger i\overleftrightarrow{D}_\mu H + g_1\sum_{f=l,e,q,u,d} Y_f\bar{f}\gamma_\mu f.
\end{aligned} \tag{2.79}$$

These EOMs can be used to move operators between different classes. Following [106], let us briefly discuss the classes of operators under consideration.

At dimension-six one can build the effective operators composed entirely of bosonic fields, i.e. the Higgs field, gauge fields $X_{\mu\nu} \in \{B_{\mu\nu}, W_{\mu\nu}^I, G_{\mu\nu}^A\}$, and covariant derivatives acting on them. Due to invariance under $SU(2)_L$, those operators must contain an even number of H fields, as each such field transforms as a doublet under $SU(2)_L$. Due to Lorentz invariance, they must also contain an even number of covariant derivatives. Some possible classes of operators also do not exist due to Lorentz symmetry, such as XH^4*, due to asymmetry of $X_{\mu\nu}$ with respect to $\mu \leftrightarrow \nu$ interchanges.

It is also possible to move the entire classes of operators into other classes. For example, operators belonging to possible classes H^2D^2, H^2XD^2, and X^2D^2 can be recast into the operators from the classes X^3, H^2X^2, H^6, H^4D^2, and operators containing fermion currents using equations of motion of Eq. (2.79), the Bianchi identity $D_{[\alpha}X_{\mu\nu]} = 0$, and the fact that $[D_\mu, D_\nu] \sim X_{\mu\nu}$. The remaining independent operators are presented in Table 2.3 and Table 2.4. Notice that all bosonic operators – that is X^3, X^2H^2, H^4, and H^2D^2 sets – are automatically Hermitian.

There are several possibilities for the operators containing fermion currents. Operators containing one current, i.e. ψ^2, must also contain Higgs and gauge fields. They are also presented in Table 2.3 and Table 2.4. The minimal set presented in those tables can be obtained using the equation of motion,

$$\begin{aligned}
i\not{D}L &= Y_e eH, \quad i\not{D}e = Y_e^\dagger H^\dagger L, \\
i\not{D}Q &= Y_u u\widetilde{H} + Y_d dH, \quad i\not{D}u = Y_u^\dagger \widetilde{H}^\dagger Q, \quad i\not{D}d = Y_d^\dagger H^\dagger Q,
\end{aligned} \tag{2.80}$$

and identities among Dirac matrices such as,

$$\epsilon_{\alpha\beta\mu\nu}\sigma^{\mu\nu} = 2i\sigma_{\alpha\beta}\gamma_5. \tag{2.81}$$

*We shall omit Lorentz indices of the gauge fields $X_{\mu\nu}$ when discussing classes of operators, i.e. $X \equiv X_{\mu\nu}$.

TABLE 2.4 Operators with H^n, sets X^2H^2, ψ^2XH, and ψ^2H^2D.

	X^2H^2		ψ^2XH+ h.c.		ψ^2H^2D
Q_{HG}	$H^\dagger H G^A_{\mu\nu}G^{A\mu\nu}$	Q_{eW}	$\left(\overline{L}_p\sigma^{\mu\nu}e_r\right)\tau^I H W^I_{\mu\nu}$	$Q^{(1)}_{Hl}$	$\left(H^\dagger i\overleftrightarrow{D}_\mu H\right)\left(\overline{L}_p\gamma^\mu L_r\right)$
$Q_{\widetilde{HG}}$	$H^\dagger H \widetilde{G}^A_{\mu\nu}G^{A\mu\nu}$	Q_{eB}	$\left(\overline{L}_p\sigma^{\mu\nu}e_r\right)HB_{\mu\nu}$	$Q^{(3)}_{Hl}$	$\left(H^\dagger i\overleftrightarrow{D}^I_\mu H\right)\left(\overline{L}_p\tau^I\gamma^\mu L_r\right)$
Q_{HW}	$H^\dagger H W^I_{\mu\nu}W^{I\mu\nu}$	Q_{uG}	$\left(\overline{Q}_p\sigma^{\mu\nu}T^Au_r\right)\widetilde{H}G^A_{\mu\nu}$	Q_{He}	$\left(H^\dagger i\overleftrightarrow{D}_\mu H\right)\left(\overline{e}_p\gamma^\mu e_r\right)$
$Q_{\widetilde{HW}}$	$H^\dagger H \widetilde{W}^I_{\mu\nu}W^{I\mu\nu}$	Q_{uW}	$\left(\overline{Q}_p\sigma^{\mu\nu}u_r\right)\tau^I\widetilde{H}W^I_{\mu\nu}$	$Q^{(1)}_{Hq}$	$\left(H^\dagger i\overleftrightarrow{D}_\mu H\right)\left(\overline{Q}_p\gamma^\mu Q_r\right)$
Q_{HB}	$H^\dagger H B_{\mu\nu}B^{\mu\nu}$	Q_{uB}	$\left(\overline{Q}_p\sigma^{\mu\nu}u_r\right)\widetilde{H}B_{\mu\nu}$	$Q^{(3)}_{Hq}$	$\left(H^\dagger i\overleftrightarrow{D}^I_\mu H\right)\left(\overline{Q}_p\tau^I\gamma^\mu Q_r\right)$
$Q_{\widetilde{HB}}$	$H^\dagger H \widetilde{B}_{\mu\nu}B^{\mu\nu}$	Q_{dG}	$\left(\overline{Q}_p\sigma^{\mu\nu}T^Ad_r\right)HG^A_{\mu\nu}$	Q_{Hu}	$\left(H^\dagger i\overleftrightarrow{D}_\mu H\right)\left(\overline{u}_p\gamma^\mu u_r\right)$
Q_{HWB}	$H^\dagger\tau^I H W^I_{\mu\nu}B^{\mu\nu}$	Q_{dW}	$\left(\overline{Q}_p\sigma^{\mu\nu}d_r\right)\tau^I HW^I_{\mu\nu}$	Q_{Hd}	$\left(H^\dagger i\overleftrightarrow{D}_\mu H\right)\left(\overline{d}_p\gamma^\mu d_r\right)$
$Q_{\widetilde{HWB}}$	$H^\dagger\tau^I H \widetilde{W}^I_{\mu\nu}B^{\mu\nu}$	Q_{dB}	$\left(\overline{Q}_p\sigma^{\mu\nu}d_r\right)HB_{\mu\nu}$	Q_{Hud}	$i\left(\widetilde{H}^\dagger D_\mu H\right)\left(\overline{u}_p\gamma^\mu d_r\right)$

TABLE 2.5 Four-fermion operators, classes $(\overline{L}L)(\overline{L}L)$, $(\overline{R}R)(\overline{R}R)$, and $(\overline{L}L)(\overline{R}R)$.

	$(\overline{L}L)(\overline{L}L)$		$(\overline{R}R)(\overline{R}R)$		$(\overline{L}L)(\overline{R}R)$
Q_{ll}	$\left(\overline{L}_p\gamma^\mu L_r\right)\left(\overline{L}_s\gamma^\mu L_t\right)$	Q_{ee}	$\left(\overline{e}_p\gamma^\mu e_r\right)\left(\overline{e}_s\gamma^\mu e_t\right)$	Q_{le}	$\left(\overline{L}_p\gamma^\mu L_r\right)\left(\overline{e}_s\gamma^\mu e_t\right)$
$Q^{(1)}_{qq}$	$\left(\overline{Q}_p\gamma^\mu Q_r\right)\left(\overline{Q}_s\gamma^\mu Q_t\right)$	Q_{uu}	$\left(\overline{u}_p\gamma^\mu u_r\right)\left(\overline{u}_s\gamma^\mu u_t\right)$	Q_{lu}	$\left(\overline{L}_p\gamma^\mu L_r\right)\left(\overline{u}_s\gamma^\mu u_t\right)$
$Q^{(3)}_{qq}$	$\left(\overline{Q}_p\gamma^\mu\tau^I Q_r\right)\left(\overline{Q}_s\gamma^\mu\tau^I Q_t\right)$	Q_{dd}	$\left(\overline{d}_p\gamma^\mu d_r\right)\left(\overline{d}_s\gamma^\mu d_t\right)$	Q_{ld}	$\left(\overline{L}_p\gamma^\mu L_r\right)\left(\overline{d}_s\gamma^\mu d_t\right)$
$Q^{(1)}_{lq}$	$\left(\overline{L}_p\gamma^\mu L_r\right)\left(\overline{Q}_s\gamma^\mu Q_t\right)$	Q_{eu}	$\left(\overline{e}_p\gamma^\mu e_r\right)\left(\overline{u}_s\gamma^\mu u_t\right)$	Q_{qe}	$\left(\overline{Q}_p\gamma^\mu Q_r\right)\left(\overline{e}_s\gamma^\mu e_t\right)$
$Q^{(3)}_{lq}$	$\left(\overline{L}_p\gamma^\mu\tau^I L_r\right)\left(\overline{Q}_s\gamma^\mu\tau^I Q_t\right)$	Q_{ed}	$\left(\overline{e}_p\gamma^\mu e_r\right)\left(\overline{d}_s\gamma^\mu d_t\right)$	$Q^{(1)}_{qu}$	$\left(\overline{Q}_p\gamma^\mu Q_r\right)\left(\overline{u}_s\gamma^\mu u_t\right)$
		$Q^{(1)}_{ud}$	$\left(\overline{u}_p\gamma^\mu u_r\right)\left(\overline{d}_s\gamma^\mu d_t\right)$	$Q^{(8)}_{qu}$	$\left(\overline{q}_p\gamma^\mu T^A q_r\right)\left(\overline{u}_s\gamma^\mu T^A u_t\right)$
		$Q^{(8)}_{ud}$	$\left(\overline{u}_p\gamma^\mu T^A u_r\right)\left(\overline{d}_s\gamma^\mu T^A d_t\right)$	$Q^{(1)}_{qd}$	$\left(\overline{q}_p\gamma^\mu q_r\right)\left(\overline{d}_s\gamma^\mu d_t\right)$
				$Q^{(8)}_{qd}$	$\left(\overline{Q}_p\gamma^\mu T^A Q_r\right)\left(\overline{d}_s\gamma^\mu T^A d_t\right)$

Notice also that $\gamma_\mu\gamma_\nu$ can always be written as a sum of a symmetric and antisymmetric parts, $\gamma_\mu\gamma_\nu = g_{\mu\nu} - i\sigma_{\mu\nu}$, and we do not need to insert γ_5 into the fermion currents, as we work in chiral representation, $\gamma_5\psi_{L,R} = \mp\psi_{L,R}$. Finally, the four-fermion operators can be reduced to the basis displayed in Table 2.5 and Table 2.6 by using Fierz identities and color algebra relations such as

$$T^A_{ij}T^A_{k\ell} = \frac{1}{2}\delta_{i\ell}\delta_{kj} - \frac{1}{2N_c}\delta_{ij}\delta_{k\ell}. \tag{2.82}$$

As a result, both baryon-number conserving and baryon-number violating operators are obtained. It is useful to remember that Lorentz-invariant combinations of spinors do not contain terms with an odd number of ψ or $\bar{\psi}$, like $\psi\psi\psi\bar{\psi}$. Finally, we note that the matrix of anomalous dimensions for the dimension-six SMEFT operators has been computed in papers [5, 118, 119] for the baryon-number conserving operators.

Operators of dimension seven

There are twenty operators of dimension seven [126]. To derive the minimal set of those operators one should use the same methods as in the derivation of the basis for dimension-six operators

TABLE 2.6 Four-fermion operators, classes $(\overline{L}R)(\overline{R}L)$, and B (baryon-number) violating.

	$(\overline{L}R)(\overline{R}L)$		B-violating
Q_{ledq}	$\left(\overline{L}^j_p e_r\right)\left(\overline{d}_s Q^j_t\right)$	Q_{duq}	$\epsilon^{\alpha\beta\gamma}\epsilon_{jk}\left[\left(d^\alpha_p\right)^T Cu^\beta_r\right]\left[\left(Q^{\gamma j}_s\right)^T CL^k_t\right]$
$Q^{(1)}_{quqd}$	$\left(\overline{Q}^j_p u_r\right)\epsilon_{jk}\left(\overline{Q}^k_s d_t\right)$	Q_{qqu}	$\epsilon^{\alpha\beta\gamma}\epsilon_{jk}\left[\left(Q^{\alpha j}_p\right)^T CQ^{\beta k}_r\right]\left[(u^\gamma_s)^T Ce_t\right]$
$Q^{(8)}_{quqd}$	$\left(\overline{Q}^j_p T^A u_r\right)\epsilon_{jk}\left(\overline{Q}^k_s T^A d_t\right)$	$Q^{(1)}_{qqq}$	$\epsilon^{\alpha\beta\gamma}\epsilon_{jk}\epsilon_{mn}\left[\left(Q^{\alpha j}_p\right)^T CQ^{\beta k}_r\right]\left[(Q^{\gamma m}_s)^T CL^n_t\right]$
$Q^{(1)}_{lequ}$	$\left(\overline{L}^j_p e_r\right)\epsilon_{jk}\left(\overline{Q}^k_s u_t\right)$	$Q^{(3)}_{qqq}$	$\epsilon^{\alpha\beta\gamma}\left(\tau^I\epsilon\right)_{jk}\left(\tau^I\epsilon\right)_{mn}\left[\left(Q^{\alpha j}_p\right)^T CQ^{\beta k}_r\right]\left[(Q^{\gamma m}_s)^T CL^n_t\right]$
$Q^{(3)}_{lequ}$	$\left(\overline{L}^j_p\sigma_{\mu\nu}e_r\right)\epsilon_{jk}\left(\overline{Q}^k_s\sigma^{\mu\nu}u_t\right)$	Q_{duu}	$\epsilon^{\alpha\beta\gamma}\left[\left(d^\alpha_p\right)^T Cu^\beta_r\right]\left[(u^\gamma_s)^T Ce_t\right]$

TABLE 2.7 The dimension-seven operators. See [126] for derivation.

	$\psi^2 H^4$		$\psi^2 H^2 D^2$
Q_{LH}	$\epsilon_{ij}\epsilon_{mn}\left(L^i \mathcal{C} L^m\right) H^j H^n \left(H^\dagger H\right)$	$Q_{LHD}^{(1)}$	$\epsilon_{ij}\epsilon_{mn} L^i \mathcal{C}\left(D^\mu L^j\right) H^m \left(D_\mu H^n\right)$
		$Q_{LHD}^{(2)}$	$\epsilon_{im}\epsilon_{jn} L^i \mathcal{C}\left(D^\mu L^j\right) H^m \left(D_\mu H^n\right)$
	$\psi^2 H^3 D$		$\psi^2 H^2 X$
Q_{LHDe}	$\epsilon_{ij}\epsilon_{mn}\left(L^i \mathcal{C}\gamma_\mu e\right) H^j H^m D^\mu H^n$	Q_{LHB}	$\epsilon_{ij}\epsilon_{mn}\left(L^i \mathcal{C}\sigma_{\mu\nu} L^m\right) H^j H^n B^{\mu\nu}$
		Q_{LHW}	$\epsilon_{ij}\left(\tau^I \epsilon\right)_{mn}\left(L^i \mathcal{C}\sigma_{\mu\nu} L^m\right) H^j H^n W^{I\mu\nu}$
	$\psi^4 D$		$\psi^4 H$
$Q_{LL\bar{d}uD}^{(1)}$	$\epsilon_{ij}\left(\bar{d}\gamma_\mu u\right)\left(L^i \mathcal{C} D^\mu L^j\right)$	$Q_{LLL\bar{e}H}$	$\epsilon_{ij}\epsilon_{mn}\left(\bar{e}L^i\right)\left(L^j \mathcal{C} L^m\right) H^n$
$Q_{LL\bar{d}uD}^{(2)}$	$\epsilon_{ij}\left(\bar{d}\gamma_\mu u\right)\left(L^i \mathcal{C}\sigma^{\mu\nu} D_\nu L^j\right)$	$Q_{LLQ\bar{d}H}^{(1)}$	$\epsilon_{ij}\epsilon_{mn}\left(\bar{d}L^i\right)\left(Q^j \mathcal{C} L^m\right) H^n$
$Q_{\bar{L}QddD}^{(1)}$	$\left(Q\mathcal{C}\gamma_\mu d\right)\left(\bar{L}D^\mu d\right)$	$Q_{LLQ\bar{d}H}^{(2)}$	$\epsilon_{im}\epsilon_{jn}\left(\bar{d}L^i\right)\left(Q^j \mathcal{C} L^m\right) H^n$
$Q_{\bar{L}QddD}^{(2)}$	$\left(\bar{L}\gamma_\mu Q\right)\left(\bar{d}\mathcal{C} D^\mu d\right)$	$Q_{LL\bar{Q}uH}$	$\epsilon_{ij}\left(\bar{Q}_m u\right)\left(L^m \mathcal{C} L^i\right) H^j$
$Q_{ddd\bar{e}D}$	$\left(\bar{e}\gamma_\mu d\right)\left(\bar{d}\mathcal{C} D^\mu d\right)$	$Q_{\bar{L}QQdH}$	$\epsilon_{ij}\left(\bar{L}_m d\right)\left(Q^m \mathcal{C} Q^i\right)\widetilde{H}$
		$Q_{\bar{L}dddH}$	$\left(d\mathcal{C}d\right)\left(\bar{L}d\right) H$
		$Q_{\bar{L}uddH}$	$\left(\bar{L}d\right)\left(u\mathcal{C}d\right)\widetilde{H}$
		$Q_{Leu\bar{d}H}$	$\epsilon_{ij}\left(L^i \mathcal{C}\gamma_\mu e\right)\left(\bar{d}\gamma^\mu u\right) H^j$
		$Q_{\bar{e}QddH}$	$\epsilon_{ij}\left(\bar{e}Q^i\right)\left(d\mathcal{C}d\right)\widetilde{H}^j$

discussed previously. Additionally, the Schouten identity,

$$\epsilon_{in}\epsilon_{jm} = \epsilon_{im}\epsilon_{jn} - \epsilon_{ij}\epsilon_{mn}, \tag{2.83}$$

and the $SU(2)$ group identity,

$$\tau^I_{jk}\tau^I_{mn} = 2\delta_{jn}\delta_{mk} - \delta_{jk}\delta_{mn}, \tag{2.84}$$

are useful in obtaining the set of dimension-seven operators reported in Table 2.7.

Contrary to the set of dimension-six operators, one can prove that it is not possible to construct dimension-seven (or any odd-dimensional) operators without using fermion fields. This immediately follows from the fact that since X is a dimension two object, an odd number of either Higgs fields or covariant derivatives D_μ must be present to construct an odd-dimensional operator. However, in the absence of fermionic currents, Lorentz invariance requires an even number of D's. Similarly, the fact that H has hypercharge 1/2 ensures that an even number of Higgs fields must be present in an operator. Thus, no odd-dimensional operator can be constructed without employing fermion currents, each of which has mass dimension three.

With fermionic operators present, a multitude of operator classes is possible. In fact, the maximum number of fermionic fields possible for the construction of dimension-seven operators is four. Adding a Higgs field or derivative results in the classes $\psi^4 H$ and $\psi^4 D$. They are all listed in Table 2.7.

Operator bases

We should probably finish this section with a word of caution. Consistent use of effective field theory requires a definition of a complete basis of non-redundant operators. Yet, just like in the world of vectors in three-dimensional space, there is no unique choice of the basis. As an equivalent description of particle dynamics in 3D space is possible in Cartesian, spherical, cylindrical, and countless numbers of other coordinate bases, the same equivalent description of the physics is possible with other choices of base operators in effective field theory compared to what we introduced above. What we shall call SM EFT in this book will implicitly refer to the "Warsaw basis" described above in Sec. 2.6.2.

There are, indeed, other bases available. To name a few, the SILH basis introduced in [54,68, 87], or the EGGM basis described in [69,70]. All those bases represent equivalent descriptions of the theory and can be related to each other by appropriate field redefinitions. Thus, predictions for the observables made in any of those bases will still be model-independent.

A problem, however, might arise if the basis is truncated, i.e. an assumption* is made that an operator or a set of operators gives the dominant contribution to a certain process. In such a case, field redefinitions could take one outside of the selected basis. We will encounter such cases in Chapter 6.

2.7 Light New Physics

New Physics particles do not have to be heavy. In fact, such particles can be arbitrary light, even massless, provided that their interactions with SM particles are suppressed to the point of being unobservable in the current round of experiments. We shall refer to such particles as the "dark sector" (DS). It is interesting that those properties fit the bill for such BSM particles to play the role of Dark Matter.

Taking DM as a primary motivation for the existence of physics beyond the Standard Model, we realize that if it is comprised of some fundamental particle or particles, their experimentally-measured properties, such as its relic abundance or production cross-sections can be predicted. For example, measurements of the abundance $\Omega_{DM}h^2 \sim 0.12$ by WMAP collaboration can be used to place constraints on the masses and interaction strengths of those DM particles. Indeed, the relation

$$\Omega_{DM}h^2 \sim \langle \sigma_{ann} v_{rel} \rangle^{-1} \propto \frac{M^2}{g^4}, \tag{2.85}$$

with M and g being the mass and the interaction strength associated with DM annihilation, implies that for a DM particle with the mass M around electroweak scale – typically less than 1 TeV – the interaction strength g should also be comparable to that of electroweak interactions! Such DM particles are commonly referred to as "weakly-interacting massive particles" or WIMPs. This coincidence is often referred to as the "WIMP miracle". Yet, difficulties in understanding small-scale gravitational clustering in numerical simulations with WIMPs may lead to preference being given to much lighter DM particles. Particularly there has been interest in studying models of light dark matter particles with masses in the keV range. According to Eq. (2.85), the light mass of dark matter particles then implies an extremely weak interaction between the Dark Matter and the SM sector.

One of the main features of the light NP models is that DM particles do not need to be stable against decays to even lighter SM particles. This implies that one does not need to impose an ad-hoc Z_2 symmetry when constructing an effective Lagrangian for DM interactions with the Standard Model fields, so DM particles can be emitted and absorbed by SM particles. Due to their extremely small couplings to the SM particles, experimental searches for such low-mass DM particles must be performed at experiments where large statistics is available. In addition, the experiments must be able to resolve signals with missing energy. Such signals can be studied at any low energy e^+e^- colliders or flavor factories.

*We assume that such assumption is external to the defined power counting of EFT.

2.7.1 How new light particles can escape detection: axions, dark photons, etc.

There are several possibilities to arrange for dark sector particles to interact with the SM. Those possibilities are conventionally called "portals" and refer to particular terms in the low energy effective Lagrangian, which can be once again realized in terms of a canonical dimension expansion,

$$\mathcal{L}_{\text{eff}} = \mathcal{L}_{\text{SM}} + \mathcal{L}_{\text{DS}} + \mathcal{L}^{(4)}_{\text{portal}} + \mathcal{L}^{(5)}_{\text{portal}} + \mathcal{L}^{(6)}_{\text{portal}} + \dots, \tag{2.86}$$

where $\mathcal{L}_{\text{SM}}$ is the Standard Model Lagrangian and $\mathcal{L}_{\text{DS}}$ is a Lagrangian that describes particles living in the dark sector and various portal terms connect DS to the SM. The construction of such terms usually follows the same rules as the construction of SMEFT described above, with DS particles transforming as singlets under the SM gauge group.

Operators of dimension four

At dimension four, there are only three possibilities to include dark sector particles,

$$\mathcal{L}^{(4)}_{\text{portal}} = \mathcal{L}_{\text{Higgs}} + \mathcal{L}_{\text{neutrino}} + \mathcal{L}_{\text{vector}} \tag{2.87}$$

where it should be understood that either one or several terms of that type could be present. These "portals" of Eq. (2.87) represent the only way dark sector particles can be connected to the SM via renormalizable interactions. In particular,

$$\mathcal{L}_{Higgs} = \epsilon_H^S \left(H^\dagger H\right) |S|^2 + \epsilon_H^V \left(H^\dagger H\right) V_\mu V^\nu, \tag{2.88}$$

where the dark sector scalar S could be a singlet of some dark sector gauge group or belong to its higher reps if such exists. The dark sector vector field V_μ could be gauged or not. This portal uses a fact that $H^\dagger H$ is a scalar, SM hypercharge zero combination. After spontaneous symmetry breaking the only way to connect S to the SM fermions is via the Higgs exchange.

Another possibility includes the so-called "neutrino portal",

$$\mathcal{L}_{neutrino} = \epsilon_\nu \left(\bar{L} H\right) \psi, \tag{2.89}$$

where the dark sector fermion ψ is coupled to the SM particles the same way one would introduce a right-handed Dirac neutrino into the SM. After SSB the new DS particle would mix into the SM neutrinos.

Finally, the vector portal,

$$\mathcal{L}_{vector} = -\frac{\epsilon_V}{2} V_{\mu\nu} F^{\mu\nu}, \tag{2.90}$$

uses the fact that the $U(1)$ field of the SM is gauge-invariant by itself, so a combination in Eq. (2.90) is also gauge invariant and couples the DS vector field $V_{\mu\nu}$ to the SM. Note that the strength tensor $F^{\mu\nu}$ in general represents the hypercharge field $B^{\mu\nu}$ of Eq. (2.7). It is however often convenient for practical applications to directly couple $V_{\mu\nu}$ to the QED field $F^{\mu\nu}$.

The smallness of the DS-SM particle interactions in Eq. (2.87) is insured by fine-tuning the couplings ϵ_H, ϵ_ν, and ϵ_V, which can be made as small as needed. Alternatively, various dynamical mechanisms can be constructed to keep those constants sufficiently small.

Operators of dimension five

Higher-dimensional operators that constitute $\mathcal{L}^{(5)}_{\text{portal}}$, $\mathcal{L}^{(6)}_{\text{portal}}$, and possibly even other terms in Eq. (2.86) have a natural way to suppress the mixing between the SM particles and the dark sector. A common name for such a mechanism is "hidden valleys", which stems from the point that higher-dimensional operators always come suppressed by a scale Λ. While there is no theoretical reason for this scale to be large, it provides natural suppression of such operators.

FIGURE 2.2 Hidden valley analogy (see text) of the smallness of interaction strength between the light NP particle and the SM particles. People living in the left valley can only learn about the existence of people in the right valley by expending a lot of energy by climbing the mountain to cross over. Illustration by Anna A. Petrov.

It might be somewhat surprising to realize that there could be undiscovered particles with masses comparable to those of the SM particles. Yet, it is entirely possible and could be achieved if their couplings to the SM sector are very tiny. Such construction might appear quite unnatural, but it does not have to be. The following analogy might be helpful. Imagine people living in two valleys separated by a high mountain row. The only way the people in one valley could meet the people in the other one is by climbing the mountain to cross over, expending a lot of energy in the process (see Fig. 2.2). In the same matter, if there is a very heavy particle that mediates the interactions of the SM and "hidden sector" particles, the only way to easily produce them would be to try to produce them in a high-energy collision with collision energy comparable to the mass of the mediator. At low energy (and in complete agreement with the principles of EFT), such interactions will be suppressed by a large scale corresponding to the mediator mass.

Again, there are several possible constructions that allow to couple BSM fields to the SM ones,

$$\mathcal{L}^{(5)}_{\text{portal}} = \mathcal{L}^{(5)}_{\text{Higgs}} + \mathcal{L}^{(5)}_{\text{fermion}} + \mathcal{L}^{(5)}_{\text{axion}}, \tag{2.91}$$

where the Lagrangians are written below. For the Higgs portal.

$$\mathcal{L}^{(5)}_{\text{Higgs}} = \sum_{\Gamma} \frac{C^{(5)}_{1,\Gamma}}{\Lambda} \left(H^\dagger H\right) \overline{\chi}\Gamma\chi + \frac{C^{(5)}_{2}}{\Lambda} \left(H^\dagger H\right)^2 S, \tag{2.92}$$

where $\chi(S)$ represents a dark sector fermion (singlet boson), respectively. Throughout, $C_i^{(n)}$ are the effective Wilson coefficients that characterize the strength of SM-dark sector interactions of dimension n in the effective theory and Λ characterizes the scale at which the EFT description breaks down. The fermion portal,

$$\mathcal{L}^{(5)}_{\text{fermion}} = \sum_{\Gamma'} \frac{C_3^{(5)}}{\Lambda} \bar{L}_L \gamma_\mu L_L X^\mu S + \frac{C_4^{(5)}}{\Lambda} \bar{L}_L H e_R S + \cdots , \tag{2.93}$$

where ellipses represent similar structures with quark operators, and X^μ stands for the SM gauge field. Finally, there is an axion portal,

$$\mathcal{L}^{(5)}_{\text{axion}} = \frac{C_6^{(5)}}{\Lambda} a F_{\mu\nu} \widetilde{F}^{\mu\nu}, \tag{2.94}$$

where a is the axion-like field, and it is conventional to introduce the axion coupling constant $f_a = \Lambda$. Yet higher-dimensional operators can be written as needed.

2.7.2 Effective Lagrangians with light particles

Since new light degrees of freedom cannot be integrated out, they are considered on the same footing as the Standard Model fields. This means that we have to consider explicit models of how such light new particles are coupled to the SM. Let us discuss some of the popular models of light New Physics.

Axion-like particles

Axions are light particles that were proposed as a way to solve a "strong CP problem", i.e. answer the question of why QCD seems to preserve CP-symmetry. An elegant solution, proposed by R. Peccei and H. Quinn [146], called for the promotion of the coupling constant θ of Eq. (2.44) to a scalar field and introduction of a new global $U(1)_{\text{PQ}}$ symmetry under which the scalar field is charged. If the scalar field develops a vacuum expectation value, it breaks $U(1)_{\text{PQ}}$ spontaneously. The axion field is then the Goldstone boson of this broken symmetry. Due to instanton effects in QCD such Goldstone field can acquire a small mass. For sufficiently small values of its mass, this state itself can play the role of the light DM particle.

Generalizing the notion of an axion particle, one can introduce its couplings to leptons as well as quarks. Such particles are usually called the *axion-like* particles. Their attractive features as the light dark matter particles made such models quite popular. We can study the tree-level interactions with the Standard Model fermions. The most general Lagrangian consists of a combination of dimension-five operators,

$$\mathcal{L}_a = -\frac{\partial_\mu a}{f_a} \bar{\psi} \gamma^\mu \gamma_5 \psi + \frac{C_\gamma}{f_a} a F_{\mu\nu} \widetilde{F}^{\mu\nu}, \tag{2.95}$$

where a is the axion-like particle. The coupling constant f_a has units of mass. Taking into account the chiral anomaly we can substitute the second term with a combination of vector and axial-vector fermionic currents,

$$\mathcal{L}_a = -\left(\frac{1}{f_a} + \frac{4\pi C_\gamma}{f_a \alpha}\right) \partial_\mu a \; \bar{\psi} \gamma^\mu \gamma_5 \psi - i m_\psi \left(\frac{8\pi C_\gamma}{f_a \alpha}\right) a \bar{\psi} \gamma_5 \psi. \tag{2.96}$$

A generic axion-like DM considered in the previous section was an example of a simple augmentation of the Standard Model by an axion-like dark matter particle. A somewhat different picture can emerge if those particles are embedded in a more elaborate BSM scenario. For example, in models of heavy dark matter of the "axion portal"-type, spontaneous breaking of the Peccei-Quinn (PQ) symmetry leads to an axion-like particle that can mix with the CP-odd Higgs A^0 of a two Higgs Doublet model (2HDM).

An interesting feature of this model is the dependence of the light DM coupling upon the quark mass. This means that the decay rate would be dominated by the contributions enhanced by the heavy quark mass. This would also mean that the astrophysical constraints on the axion-like DM parameters might not probe all of the parameter space in this model.

In a concrete model [143], the PQ symmetry $U(1)_{PQ}$ is broken by a large vacuum expectation value $\langle S \rangle \equiv f_a \gg v_{EW}$ of a complex scalar singlet Φ. In the so-called *interaction basis* the axion state appears in Φ as

$$\Phi = f_a \exp\left[\frac{ia}{\sqrt{2}f_a}\right] \tag{2.97}$$

and A^0 appears in the Higgs doublets in the form

$$\Phi_u = \begin{pmatrix} v_u \exp\left[\frac{i\cot\beta}{\sqrt{2}v_{EW}}A^0\right] \\ 0 \end{pmatrix}, \qquad \Phi_d = \begin{pmatrix} 0 \\ v_d \exp\left[\frac{i\tan\beta}{\sqrt{2}v_{EW}}A^0\right] \end{pmatrix}, \tag{2.98}$$

where we suppress the charged and CP-even Higgses for simplicity and define $\tan\beta = v_u/v_d$ and $v_{EW} = \sqrt{v_u^2 + v_d^2} \equiv m_W/g_2$. We choose the operator that communicates PQ charge to the Standard Model to be of the form*

$$\mathcal{L} = \lambda \Phi^2 \Phi_u \Phi_d + h.c. \tag{2.99}$$

This term contains the mass terms and, upon diagonalizing, the physical states in this basis are given by

$$a_p = a\cos\theta - A^0 \sin\theta \tag{2.100}$$

$$A_p^0 = a\sin\theta + A^0 \cos\theta \tag{2.101}$$

where $\tan\theta = (v_{EW}/f_a)\sin 2\beta$. Here a_p denotes the "physical" axion-like state.

In a type II 2HDM, the relevant Yukawa interactions of the CP-odd Higgs with fermions are given by

$$\mathcal{L}_{A^0 f\bar{f}} = \frac{ig\tan\beta}{2m_W} m_d \bar{d}\gamma_5 d A^0 + \frac{ig\cot\beta}{2m_W} m_u \bar{u}\gamma_5 u A^0 \tag{2.102}$$

where $d = \{d, s, b\}$ refers to the down type quarks and $u = \{u, c, t\}$ refers to the up-type quarks. The interaction with leptons is the same as above with $d \to \ell$ and $u \to \nu$.

In the axion portal scenario, the axion mass is predicted to lie within a specific range of $360 < m_a \leq 800$ MeV to explain the galactic positron excess.

Dark photons

Another possibility for a light weakly-interacting particle is a light (keV-range) vector dark matter boson (LVDM) coupled to the SM solely through kinetic mixing with the hypercharge field strength. This can be done consistently by postulating an additional $U(1)_V$ symmetry. The relevant terms in the Lagrangian are

$$\mathcal{L} = -\frac{1}{4}F_{\mu\nu}F^{\mu\nu} - \frac{1}{4}V_{\mu\nu}V^{\mu\nu} - \frac{\kappa}{2}V_{\mu\nu}F^{\mu\nu} + \frac{m_V^2}{2}V_\mu V^\mu + \mathcal{L}_{h'}, \tag{2.103}$$

where $\mathcal{L}_{h'}$ contains terms with, say, the Higgs field which breaks the $U(1)_V$ symmetry, κ parameterizes the strength of kinetic mixing, and, for simplicity, we directly work with the photon field

*This is the case of the so-called Dine-Fischler-Srednicki-Zhitnitsky (DFSZ) axion, although other forms of the interaction term with other powers of the scalar field Φ are possible.

A_μ. In this Lagrangian only the photon A_μ fields (conventionally) couples to the SM fermion currents.

It is convenient to rotate out the kinetic mixing term in Eq. (2.103) with field redefinitions

$$A \to A' - \frac{\kappa}{\sqrt{1-\kappa^2}} V', \qquad V \to \frac{1}{\sqrt{1-\kappa^2}} V'. \tag{2.104}$$

The mass m_V will now be redefined as $m_V \to m_V/\sqrt{1-\kappa^2}$. Also, both A'_μ and V'_μ now couple to the SM fermion currents via

$$\mathcal{L}_f = -eQ_f A'_\mu \bar{\psi}_f \gamma^\mu \psi_f - \frac{\kappa e Q_f}{\sqrt{1-\kappa^2}} V'_\mu \bar{\psi}_f \gamma^\mu \psi_f, \tag{2.105}$$

where Q_f is the charge of the interacting fermion thus introducing our new vector boson's coupling to the SM fermions. Calculations can be now carried out with the approximate modified charge coupling for $\kappa \ll 1$,

$$\frac{\kappa e}{\sqrt{1-\kappa^2}} \approx \kappa e. \tag{2.106}$$

As we can see, in this case, the coupling of the physical photon did not change much compared to the original field A_μ, while the DM field V'_μ acquired small gauge coupling κe.

2.8 Non-trivial extensions of the Standard Model. Lorentz violation

So far we discussed constructions with heavy or light New Physics that can help to alleviate some problems that the Standard Model has, describe the observed phenomena that have no explanation in the SM, or simply could be present in the low energy description of physics that we have based on the way we build the SM Lagrangian. Are there phenomena that are forbidden by existing symmetries or other mechanisms, but could still be observed?

Admittedly, this is a rather odd – and for all that matters, not a well-defined question. We are trying to build a framework to describe the physical Universe, so the foundation of the employed methods must be correct. Yet, it does not mean that it should not be asked. And while it would not be possible to even begin answering it completely, we can provide an example of a step in that direction.

One of the pillars of any local quantum field theory that provides the framework for building phenomenologically viable constructions that include the Standard Model is the CPT theorem. It states that, with some mild technical conditions, any unitary, local, Lorentz-invariant point-particle quantum field theory in flat Minkowski space must be CPT invariant. That was implicit in our discussion above, while we did discuss violations of charge, parity, and time-reversal (and some of their combinations), the constructed theories were explicitly CPT-invariant. What if CPT invariance is broken?

A way to attack the CPT-breaking is to find a consistent example of invalidating one of its initial assumptions. A convenient way to assault CPT-theorem is to imagine situations when Lorentz invariance of a theory is broken. It appears a plausible way to do so, as there exist many successful constructions of non-relativistic effective field theories. So it might be possible to build an EFT that includes Lorentz violation.

One of the takeaways from basic quantum field theory classes is that there are two essential ways to break an existent symmetry in QFT. It could be done explicitly or spontaneously. As it turns out, explicit Lorentz violation leads to incompatibility of the Bianchi identities with the covariant conservation laws for the energy-momentum and spin-density tensors [124]. There are, however, successful constructions of EFTs that break Lorentz symmetry spontaneously [51, 52].

To illustrate the construction of such EFT, let us restrict our attention to the QED sector of the Standard Model with electrons e and photons A_μ only. The CPT-conserving QED Lagrangian is given by

$$\mathcal{L}_{\text{QED}} = \bar{e} i \not{D} e - m_e \bar{e} e - \frac{1}{4} F_{\mu\nu} F^{\mu\nu}, \tag{2.107}$$

where $F_{\mu\nu} = \partial_\mu A_\nu - \partial_\nu A_\mu$. There are several possible terms that can be added to Eq. (2.107) that break Lorentz symmetry [51]. There are possible CPT-even terms,

$$\mathcal{L}_{\text{even}} = -\frac{1}{2} H_{\mu\nu} \bar{e} \sigma^{\mu\nu} e + \frac{i}{2} c_{\mu\nu} \bar{e} \gamma^\mu D^\nu e + \frac{i}{2} d_{\mu\nu} \bar{e} \gamma_5 \gamma^\mu D^\nu e - \frac{1}{4} (k_F)_{\alpha\beta\mu\nu} F^{\alpha\beta} F^{\mu\nu}. \tag{2.108}$$

In addition, there are possible CPT-odd terms,

$$\mathcal{L}_{\text{odd}} = -a_\mu \bar{e} \gamma^\mu e + b_\mu \bar{e} \gamma^\mu \gamma_5 e + \frac{1}{2} (k_{AF})^\alpha \epsilon_{\alpha\beta\mu\nu} A^\beta F^{\mu\nu}, \tag{2.109}$$

where the real coupling coefficients a, b, c, d, and H are the constant coefficients that inherit the Lorentz properties corresponding to the number of indices they are carrying. The dimensions of the operators are such that the overall dimension of the effective Lagrangian is four.

While these operators are all possible in the CPT-violating version of QED, not all of them are unique. Just like in any effective field theory, field redefinitions eliminate some redundant operators. In particular, redefining

$$e(x) \to \exp(-ia \cdot x) e(x) \tag{2.110}$$

eliminates the operator $-a_\mu \bar{e} \gamma^\mu e$. This means that there are no observable effects of non-zero a_μ. Another possible class of field redefinitions can be generically written as

$$e \to (1 + C \cdot \Gamma)\, e, \tag{2.111}$$

where $\Gamma = \{\gamma^\mu, \gamma^\mu \gamma_5, \sigma_{\mu\nu}\}$, and C represents couplings or combination of couplings $H_{\mu\nu}$, etc. with appropriate Lorentz indices. The result of such field redefinition is elimination of the combination of couplings $\epsilon^{\mu\nu\alpha\beta} H_{\alpha\beta} + m(d^{\mu\nu} - d^{\nu\mu})$. At the end of the day, the only observable effects one can expect would be proportional to b_μ, $H_{\mu\nu}$, and the symmetric components of $c_{\mu\nu}$ and $d_{\mu\nu}$ [51].

There are, in principle, other terms that could be added to Eqs. (2.108) and (2.109), but, besides breaking Lorentz symmetry, they are also incompatible with the electroweak structure of the SM. Thus, they should be additionally suppressed.

2.9 Notes for further reading

This chapter sets up a stage for what will be discussed later on. We want to emphasize that searches for New Physics at low energies do not actually require searches for a particular model of New Physics. In places where checks of the SM predictions are not viable, fits to the Wilson coefficients of SMEFT operators may provide a way forward. There are many sources that can be recommended for further studies of the issues touched upon in this chapter. For example, explicit BSM model building is described, among other references, in [162, 187] with the discussion of various gauge groups available in [172, 187]. The Standard Model of particle physics is described in most textbooks of quantum field theories, we can recommend [65, 182]. A good introduction to effective field theories can be found in [49, 134, 149]. A complete discussion of the minimal SMEFT basis in "Warsaw" representation was first given in [106]. The renormalization group properties of dimension-six operators, in particular, the matrix of anomalous dimensions has been computed in [5, 118, 119] for the baryon-number conserving operators. A thorough discussion of discrete symmetries and their experimental studies can be found in [29]. A rather pedestrian discussion of the CPT theorem and consequences of its breaking with references to the original works can be found in [127].

Problems for Chapter 2

1. **Problem 1.** Check that Eqs. (2.13) and (2.21), written in terms of component fields, lead to the same Lagrangian.
2. **Problem 2.** Perform the matching for the Lagrangian of Eq. (2.67) to obtain the Wilson coefficients C_i by matching the matrix elements at with the full theory result of Eq. (2.74) at next-to-leading order in $1/M_W$. What other operators need to be included in Eq. (2.67) if flavor-changing neutral current weak decays with the photon emissions are to be described?

3

New Physics searches with charged leptons

3.1 Introduction

Leptons provide incredible opportunities for precision experiments aimed at uncovering New Physics effects because they are not charged under the gauge group of strong interactions. This does not mean that leptonic observables do not have uncertainties associated with non-perturbative strong interaction effects, as both quarks and leptons are charged under electroweak interactions. It *does*, however, mean that those effects appear at higher orders in perturbative QED, suppressed by powers of the fine structure constant α. This significantly improves constraints on the parameters of possible New Physics interactions, provided that New Physics particles do couple to leptons and their couplings are not significantly suppressed.

As was mentioned in Chapter 1, both flavor-conserving and flavor-changing observables can be studied. While flavor-changing observables are thought to be convenient for studies of NP because SM does not contain elementary FCNC vertices, NP does not have to contain FCNC interactions. Thus, both flavor-violating and flavor-conserving observables are important.

It is also best to study observables that are related to broken symmetries, especially if such breaking is, for some reason, parametrically small, but in BSMs is not. Finally, SM EFT-motivated parameterizations of experimental observables allow for model-independent studies of New Physics [76].

3.2 Flavor-conserving observables: $(g-2)$ and lepton EDMs

The main strategy for searching for New Physics in flavor conserving interactions involves probing particle properties in response to some external forces. As electric and magnetic fields can be created in laboratory conditions, and their strength could be measured with excellent precision, studies of electromagnetic properties of leptons play the leading role in searches for New Physics in flavor-conserving interactions.

Alternatively, the pair creation of leptons of the same flavor by either photons or Higgs and Z-bosons can probe similar quantities. We shall defer our discussion of Higgs and Z decays to leptons to Chapter 6 of this book.

3.2.1 Anomalous magnetic moment

A charged, spin-1/2 particle has a magnetic dipole moment which can be written as

$$\vec{\mu} = g\frac{e}{2m}\vec{s}, \tag{3.1}$$

where μ is the spin magnetic moment of the particle, m is its mass, and $e > 0$ is the elementary charge. For a structureless, point particle the g-factor is $g = 2$. For an electron, $\mu_B = e/2m_e$ is called the Bohr magneton. We know that in nonrelativistic quantum mechanics the interactions between Dirac particle's spin and external magnetic field can be described by Pauli's Hamiltonian,

$$\mathcal{H}_M = -\vec{\mu}\cdot\vec{B} = -\mu\,\vec{\sigma}\cdot\vec{B}. \tag{3.2}$$

Here σ_i are the Pauli matrices and $\vec{\mu}$ is the magnetic dipole moment defined in Eq. (3.1). It must be oriented along the direction of the particle's spin, as it is the only direction available for a spin-1/2 particle.

It is instructive to see how the Hamiltonian of Eq. (3.2) transforms under discrete C, P, and T symmetries. Since both $\vec{\sigma}$ and $\vec{B}$ are axial vectors, they do not change under space inversion parity P (by definition). Both of them change sign under time reversal transformation, T. Thus, the Hamiltonian $\mathcal{H}_M$ is *invariant* under a combined PT transformation, as it is a product of those two axial vectors. It is also invariant under a charge conjugation transformation C since all local quantum field theories that respect Lorentz invariance are invariant under a combined CPT-transformation.

Quantum corrections change the value of g. The probe, a photon, can interact with one of the virtual particles that produce quantum corrections to the photon-lepton vertex instead of the lepton itself, and thereby explore the "quantum structure" of the vertex. It would then be convenient to define the magnetic moment in terms of general form-factors defined in lepton-photon interactions. A matrix element of the electromagnetic current j^μ can be written as

$$\langle p_i|j^\mu|p_f\rangle = \overline{u}(p_i)\Gamma^\mu(P,q)u(p_f) = (-ie)\,\overline{u}(p_i)\left[F_1(q^2)\gamma^\mu + \frac{iF_2(q^2)}{2m}\sigma^{\mu\nu}q_\nu\right]u(p_f), \tag{3.3}$$

where $q = p_f - p_i$, $P = p_f + p_i$, and $u(p_i)$ and $u(p_f)$ are the on-shell 4-spinors normalized such that $\overline{u}(p)u(p) = 2m$. The magnetic moment can then be defined as $\mu = (F_1(0) + F_2(0))/(2m)$. Now, it follows from Eq. (3.1) that $g = F_1(0) + F_2(0)$.

The form of Eq. (3.3) follows from the fact that a generic matrix element of a vector current can only depend on two form-factors*. Since we are interested in a "static" quantity, the magnetic moment, it would be useful to use the projection technique introduced in [13] to rewrite the matrix element of Eq. (3.3) as

$$\Gamma_\mu(P,q) \approx \Gamma_\mu(P,0) + q^\nu\frac{\partial}{\partial q_\nu}\Gamma_\mu(P,q)\bigg|_{q=0} = V_\mu(p) + q^\nu T_{\nu\mu}. \tag{3.4}$$

While quantum corrections renormalize the electromagnetic vertex of Eq. (3.3), gauge invariance protects the value $F_1(0) = 1$. As can be then seen from Eq. (3.1), this results in deviation of the g-factor from its non-renormalized value of two. It is then reasonable to introduce a quantity that quantifies this deviation,

$$a_f = \frac{g-2}{2} = F_2(0), \tag{3.5}$$

*We will describe how to parameterize various form factors in Chapter 4.

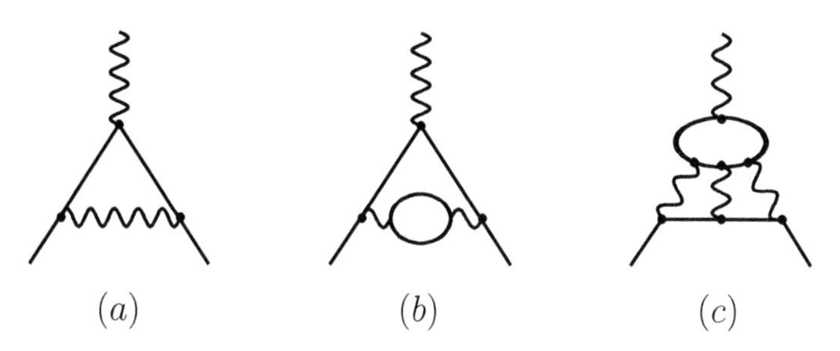

FIGURE 3.1 Examples of SM contributions to anomalous magnetic moment: (a) lowest order Schwinger term, (b) vacuum polarization, and (c) light-by-light scattering.

where the index f indicates the lepton flavor. a_f is usually referred to as *anomalous magnetic moment.* We can average over the direction of P with $P \cdot q = 0$. It is then possible to show [13,117] that a_f can be written as

$$a_f = \operatorname{Tr}\left[G_\mu V^\mu(p) + H_{\mu\nu} T^{\nu\mu}(p)\right]_{p^2=m_f^2}, \tag{3.6}$$

where the tensors G_μ and $H_{\mu\nu}$ are defined as

$$\begin{aligned} G_\mu &= \frac{1}{4(d-1)m_f^2}\left[m_f^2\gamma_\mu - \left((d-1)m_f + d\not{p}\right)p_\mu\right], \\ H_{\mu\nu} &= \frac{1}{8(d-2)(d-1)m_f}\left(\not{p}+m_f\right)\left[\gamma_\mu,\gamma_\nu\right]\left(\not{p}+m_f\right). \end{aligned} \tag{3.7}$$

The formula (3.7) is written in d dimensions, implying that dimensional regularization will be used for the calculations of higher order contributions. Note that $V^\mu(p)$ and $T^{\nu\mu}(p)$ only depend on $p = P/2$.

Since the anomalous magnetic moment of leptons comes from radiative corrections, it might be useful to write an effective Lagrangian whose Wilson coefficients are proportional to a_f. Following our discussion above, we see that the form factors $V^\mu(p)$ and $T^{\nu\mu}(p)$ can, in principle, develop imaginary parts. The effective Lagrangian can be written as [117]

$$\mathcal{L}_{\text{eff}} = -\frac{1}{2}\overline{\psi}_{\ell_f}\sigma^{\mu\nu}\left[D_{\ell_f}P_R + D^*_{\ell_f}P_L\right]\psi_{\ell_f}F_{\mu\nu}, \tag{3.8}$$

where ψ_{ℓ_f} represents a lepton field of flavor f. One can see that

$$\text{Re } D_{\ell_f} = \frac{a_f e}{2m_f}, \qquad \text{Im } D_{\ell_f} = d_f = \frac{\eta_f}{2}\frac{e}{2m_f}. \tag{3.9}$$

We see that the imaginary part of the D_{ℓ_f} (and thus $F_2(0)$) corresponds to an *electric* dipole moment (or EDM), d_f, of a lepton. We will discuss EDMs in Sec. 3.2.2.

Muon's anomalous magnetic moment is usually denoted as a_μ. We shall concentrate on a_μ from now on and comment on the anomalous moments of electron and tau at the end of the section.

The Standard Model prediction for a_μ

The Standard Model prediction for the muon anomalous magnetic moment [108] can be written as a sum of three terms,

$$a_\mu^{SM} = a_\mu^{QED} + a_\mu^{EW} + a_\mu^{had} \tag{3.10}$$

The leading order QED contribution, known as the Schwinger term, was one of the first radiative corrections ever computed (back in 1948),

$$a_\mu^{QED,LO} = \frac{\alpha}{2\pi}, \tag{3.11}$$

is depicted in Fig. 3.1 (a) and happens to be finite. Higher-order terms are usually calculated in a particular renormalization scheme.

The QED contribution, which includes all the photonic and leptonic loops, and electroweak contribution, which includes the loops involving Z, W or Higgs bosons, are known incredibly accurately [7],

$$\begin{aligned} a_\mu^{QED} &= 116584718.951(0.009)(0.019)(0.007)(.077) \times 10^{-11}, \\ a_\mu^{EW} &= 153.6(1.0) \times 10^{-11}. \end{aligned} \tag{3.12}$$

The numerical results presented in Eq. (3.12) include QED corrections calculated up to a five loop order, and electroweak corrections calculated to two-loop order with leading three-loop contributions also included.

The main uncertainty of theoretical prediction of a_μ lies in the final term of Eq. (3.10), which includes all hadronic loop contributions. This term, in turn, may also be subdivided into two major parts.

$$a_\mu^{had} = a_\mu^{hvp} + a_\mu^{hlbl}, \tag{3.13}$$

where a_μ^{hvp} denotes hadronic vacuum polarization contribution, an example of which is given in Fig. 3.1 (b), while a_μ^{hlbl} denotes hadronic light-by-light scattering contribution, whose example is depicted on Fig. 3.1 (c).

Hadronic vacuum polarization contribution

As can be seen from Fig. 3.1 (b), a particular contribution to a_μ includes a vacuum polarization loop $\Pi(q^2)$. As follows from its definition

$$\Pi_{\mu\nu}(q) = i \int d^4x e^{ix\cdot q} \langle 0| T \left[J_\mu(x) J_\nu(0) \right] |0\rangle = \left(q_\mu q_\nu - q^2 g_{\mu\nu} \right) \Pi(Q^2), \tag{3.14}$$

where it is convenient to introduce $Q^2 = -q^2$ and $J_\mu(x)$ denotes electromagnetic current of the light quarks,

$$J_\mu(x) = \sum_q Q_q \, \overline{q}(x) \gamma_\mu q(x), \tag{3.15}$$

where $q = u, d, s$ denotes quark flavor with $Q_u = 2/3$ and $Q_d = Q_s = -1/3$. The second identity in Eq. (3.14) follows from Lorentz invariance and from the fact that the only momentum vector that is available to parameterize it is q_μ and the only tensor, which gives a non-zero contribution is $g_{\mu\nu}$.

In calculating the contribution of Fig. 3.1 (b) one must integrate over two momenta, including the one that flows through the vacuum polarization loop. That means that for quarks running in the loop one must integrate over all distances to get a complete hadronic contribution. One should be careful when evaluating quark contributions to a_μ, as perturbative QCD might not be a reliable tool for estimating total quark contribution, especially in the low momentum region. There, more sophisticated methods must be devised to deal with the hadronic vacuum polarization (HVP) contribution.

While there are several methods that have been proposed to provide an estimate of those contributions, we should mention two. One involves methods of lattice QCD and amounts to the direct calculation of $\Pi(Q^2)$ in discretized space-time. A thorough description of this method goes beyond the scope of this book, but a reader interested in learning more about lattice QCD methods could find a collection of review articles suggested at the end of this chapter in Section 3.4.

A more phenomenological way to approach the problem involves using experimental data from e^+e^- annihilation (or hadronic τ-lepton decay) to evaluate vacuum polarization contribution due to hadrons.

As follows from the analyticity of vacuum polarization function, $\Pi(q^2)$ satisfies a dispersion relation [147],

$$\Pi(q^2) = \frac{1}{\pi}\int_{s_0}^{\infty} \frac{ds}{s-q^2}\text{Im}\Pi(s), \tag{3.16}$$

where $s_0 = 4m_\pi^2$ represents the threshold for the lowest mass state production, which in this case is a two pion ($\pi\pi$) state, and $\text{Im}\Pi(s)$ is the imaginary part which $\Pi(s)$ develops for $s > s_0$. We shall write this dispersion relation in its "once-subtracted form", i.e. by subtracting $\Pi(q^2 = 0)$ from both sides of Eq. (3.16) to make sure that we do not have to deal with (ultraviolet) divergent integrals,

$$\Pi(q^2) = \frac{q^2}{\pi}\int_{s_0}^{\infty} \frac{ds}{s(s-q^2)}\text{Im}\Pi(s), \tag{3.17}$$

and set $\Pi(0) = 0$, as required by derivative couplings of pions in the chiral limit.

The key observation is that the imaginary (dispersive) part of $\Pi(s)$ can be related to a total cross-section for e^+e^- annihilation into hadrons,

$$\text{Im}\Pi(s) = \frac{1}{12\pi}R(s) = \frac{1}{12\pi}\frac{\sigma(e^+e^- \to hadrons)}{\sigma(e^+e^- \to \mu^+\mu^-)}, \tag{3.18}$$

where $R(s)$ represents the experimentally measured R-ratio of the cross-section for $e^+e^- \to$ hadrons. Then it is convenient to rewrite the HVP as a convolution,

$$a_\mu^{hvp} = \frac{1}{3}\left(\frac{\alpha}{\pi}\right)^2 \int_{4m_\pi^2}^{\infty} \frac{ds}{s}K(s)R(s), \tag{3.19}$$

where the QED kernel $K(s)$ is given by

$$K(s) = \int_0^1 dx \frac{x^2(1-x)m_\mu^2}{x^2 m_\mu^2 + (1-x)s}. \tag{3.20}$$

Needless to say that for other lepton flavors ℓ the corresponding expression can be found by substituting m_μ with m_ℓ.

Hadronic ligh-by-light scattering contribution

Another important, but difficult-to-compute contribution is depicted in Fig. 3.2. This contribution is usually classified according to what kind of particles are running in the loop connected to four bosons. The contributions of leptons are normally included in the three-loop QED result, while quarks' contribution is usually separated into a separate entity called (hadronic) *light-by-light scattering.*

In general, the contribution of light-by-light (LBL) scattering is parameterized by a four-index tensor, $\Pi^{\mu\nu\alpha\beta}(p_1, p_2, p_3)$. It appears from the diagram Fig. 3.2 when one considers the

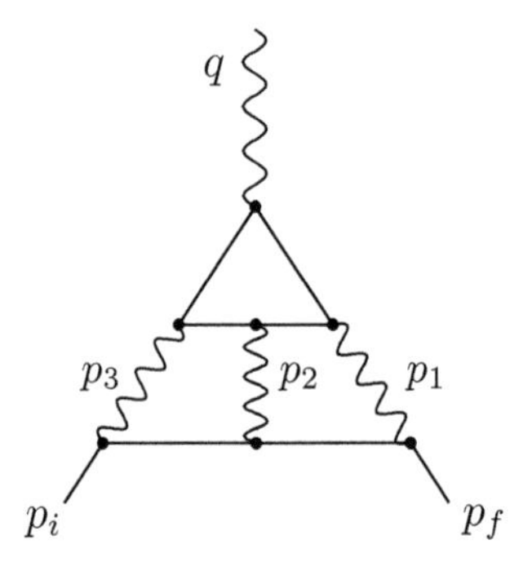

FIGURE 3.2 Example of a light-by-light scattering contribution. It is understood that permutations of photons' momenta p_i are also included. The "triangle" contains quark contributions, so it is usually parameterized by a tensor $\Pi^{\mu\nu\alpha\beta}(p_1, p_2, p_3)$, which is computed using various methods (see text).

matrix element of Eq. (3.3),

$$\begin{aligned}\overline{u}(p_i)\Gamma^\mu u(p_f) &= \int \frac{d^4p_1}{(2\pi)^4}\frac{d^4p_2}{(2\pi)^4}\frac{(-i)^3}{p_1^2p_2^2(p_1+p_2-q)^2}\frac{i}{(p_f-p_1)^2-m_f^2}\frac{i}{(p_f-p_1-p_2)^2-m_f^2} \\ &\times (-ie)^3\,\overline{u}(p_f)\gamma_\kappa\left(\not{p}_f-\not{p}_1+m\right)\gamma_\alpha\left(\not{p}_f-\not{p}_1-\not{p}_2+m\right)\gamma_\beta u(p_i) \\ &\times (-ie)^3\,\Pi^{\mu\kappa\alpha\beta}(p_1,p_2,q-p_1-p_2)\end{aligned} \tag{3.21}$$

Note that the quark current for the hadronic light-by-light scattering contribution denoted in Eq. (3.21) is defined in Eq. (3.15). The fourth-rank tensor $\Pi^{\mu\nu\alpha\beta}(p_1, p_2, p_3)$ is defined as a matrix element of the time-ordered product of four quark currents,

$$\Pi^{\mu\kappa\alpha\beta}(p_1,p_2,p_3) = \int d^4x_1d^4x_2d^4x_3e^{i(p_1\cdot x_1+p_2\cdot x_2+p_3\cdot x_3)}\langle 0|T\left[J^\mu(x_1)J^\kappa(x_2)J^\alpha(x_3)J^\beta(0)\right]|0\rangle, \tag{3.22}$$

where $q = p_f - p_i = p_1 + p_2 + p_3$. Such object would generically depend on 138 Lorentz structures, out of which only 32 would actually contribute to $(g-2)$ calculation. Since $\Pi^{\mu\nu\alpha\beta}(p_1, p_2, p_3)$ contains four photon legs, gauge invariance conditions on each index

$$q_\mu\Pi^{\mu\kappa\alpha\beta}(q,p_1,p_2,p_3) = p_{1\kappa}\Pi^{\mu\kappa\alpha\beta}(q,p_1,p_2,p_3) = \ldots = 0 \tag{3.23}$$

would leave out only 43 gauge-invariant structures with Bose symmetry relating some of those. Lorentz invariance also requires that those structures would only depend on scalars, either four-momenta squared or scalar products of q, p_1, p_2, and p_3.

We can clarify the contribution of light-by-light scattering to a_f by applying the projection technique introduced in Sect. 3.2.1. First, let us note that taking a derivative with respect to q^μ of the first gauge invariance condition in Eq. (3.23) leads to the relation

$$\Pi^{\mu\kappa\alpha\beta}(q,p_1,p_2,p_3) = -q_\nu\frac{\partial}{\partial q^\mu}\Pi^{\mu\nu\alpha\beta}(q,p_1,p_2,p_3). \tag{3.24}$$

This allows to rewrite the Eq. (3.21) such that

$$\overline{u}(p_i)\Gamma^\mu u(p_f) = q_\nu\overline{u}(p_i)T^{\mu\nu}u(p_f), \tag{3.25}$$

where $T^{\mu\nu}$ is given by the Eq. (3.22) with $\Pi^{\mu\kappa\alpha\beta}$ substituted by $-\frac{\partial}{\partial q^\mu}\Pi^{\mu\nu\alpha\beta}$. Also, since Eq. (3.25) is already proportional to q we can set it to zero in $\Pi^{\mu\nu\alpha\beta}$ after the derivative is taken.

We can now find the most general expression for the tensor $M_{\mu\nu}$. There are two momenta that can be used in parameterizing it, P and q, just like in Eq. (3.4). We should note, however, that since the matrix element with $M_{\mu\nu}$ is dotted into q^ν, only P should enter the parameterization. Then,

$$T^{\mu\nu} = A\ g^{\mu\nu} + B\left(\gamma^\mu\gamma^\nu - \gamma^\nu\gamma^\mu\right) + C\ P^\mu\gamma^\nu + D\ P^\nu\gamma^\mu + E\ P^\mu P^\nu, \tag{3.26}$$

where A, B, C, D, and E are form factors that only depend on Lorentz-invariant quantities. Now, since $P \cdot q = 0$, terms proportional to D and E do not contribute to the magnetic moment. Vector current conservation relation $q^\mu \overline{u}\gamma_\mu \overline{u} = 0$ implies that the term proportional C does not contribute as well. Finally, as can be seen from Eq. (3.25), $A = 0$ by the requirement of gauge invariance. Then,

$$a_f = \mathrm{Tr}\left[H_{\mu\nu}T^{\nu\mu}\right]. \tag{3.27}$$

This formula allows us to compute the anomalous magnetic moment.

New Physics and SM EFT

Since anomalous magnetic moments of leptons arise from radiative corrections, they can be affected by possible contributions from virtual BSM particles. There are many models, but we could use SM EFT Lagrangian with entries from Table 2.4 to parameterize all of them.

The Lagrangian would take the form

$$\begin{aligned}\mathcal{L}_{(\mathrm{g}-2)} &= \frac{1}{2\Lambda^2}\left[\alpha_B g_1 Q_{eB} + \alpha_W g_2 Q_{eW}\right] \\ &= \alpha_B \frac{g_1}{2\Lambda^2}\ \overline{L}_f \sigma^{\mu\nu} H \ell_f B_{\mu\nu} + \alpha_W \frac{g_2}{2\Lambda^2}\ \overline{L}_f \sigma^{\mu\nu}\tau^I H \ell_f W^I_{\mu\nu},\end{aligned} \tag{3.28}$$

where we slightly changed our notation by renaming the Wilson coefficients of Q_{eW} and Q_{eB} operators as $\alpha_W g_2/2$ and $\alpha_B g_1/2$, respectively.

Experimental methods of detection

Experimentally, the anomalous magnetic moment of a lepton can be measured by studying particle's response to the applied magnetic field, in particular, by tracking the direction of its spin, i.e. polarization. The applied methods, however, differ drastically depending on the flavor of the lepton. This is because, contrary to the electron, both muons and taus have finite lifetimes with tau lifetime being significantly shorter than that of the muon.

Electrons. A variety of experimental methods can be used to measure an electron's anomalous magnetic moment a_e. One of the most successful methods in terms of precision is the use of a Penning trap (PT). A Penning trap is a device that uses a combination of electric and magnetic fields to confine particles inside them. In particular, a cylindrical Penning trap employs an axial magnetic field to confine particles in the radial direction and a quadrupole electric field to confine them in the axial ones. In the cylindrical PT the electric field is produced by a combination of three electrodes: two end-cap ones are charged with the same sign of charge as the trapped particle, while the middle electrode is charged opposite to it. In the experimental setup, the electrodes can be shaped as hyperboloids of rotation*. Since those are the equipotential surfaces of the quadrupole potential, this allows one to generate pure quadrupole electrostatic potential inside the trap. Such a combination of the magnetic and electric fields causes charged particles (electrons) to move in the radial plane with a motion that traces out an epitrochoid. An electron can be trapped indefinitely in such a combination of fields. The physics of motion of a single electron inside the Penning trap is studied reasonably well [31].

The motion of the electrons inside the PT can be described by the solution of an equation of motion for a charged particle in the quadrupole electrostatic potential $V \sim V(2z^2 - r^2)$ with

*In reality, those shapes are difficult to manufacture, so combinations of several ($n > 3$) rings are used.

applied magnetic field in the z-direction. It is composed out of three types of motion: (a) axial, represented by oscillation between the endcaps, (modified) cyclotron motion, represented by particle orbiting around a magnetic field line, and (c) magnetron, represented by particle's orbit around trap's center. Three frequencies associated with those motions represent frequencies of three fully decoupled harmonic single-particle oscillators and can be measured. If an inhomogeneous magnetic "bottle field" is applied to the system, a measurement of the magnetic moment of the trapped electron becomes possible.

Muons. The method, most suitable for measuring a_μ, due to a finite lifetime of a muon, is to place (highly polarized) muons in a storage ring and observe its decay products [81]. The polarization of the muon can be tracked via its main decay mode, $\mu^+ \to e^+\nu_\mu\bar{\nu}_e$ which has a differential decay rate proportional to $(1 + A(E)\cos\theta)$ in the muon rest frame. Here θ is the angle between muon spin and decay positron momentum vector and $A(E)$ is a positron energy-dependent asymmetry factor. Positrons are therefore emitted preferentially along the muon spin direction. The gist of the method consists of the measurement of the difference ω_a between spin precession frequency ω_s and the muon cyclotron frequency ω_c for muons moving in a magnetic field of the storage ring. The angular frequency of the cyclotron motion is given by

$$\omega_c = \frac{eqB}{m\gamma}, \tag{3.29}$$

where $\gamma = (1 - v^2)^{-1/2}$ is the Lorenz factor. The muon spin precession term is given by

$$\omega_s = g\frac{eqB}{2m} + (1-\gamma)\frac{eqB}{m\gamma}, \tag{3.30}$$

with the difference, $\omega_a = \omega_s - \omega_c$, being proportional to the anomalous magnetic moment,

$$\omega_a = a_\mu \frac{eqB}{m}. \tag{3.31}$$

In reality, the angular frequency of such precession is also affected by the electric fields present in the experiment, both stray and, possibly, from the electrostatic quadrupoles that can be employed in focusing the muon beam. Thus, the frequency becomes

$$\vec{\omega}_{\rm net} = -\frac{q}{m}\left[a_\mu \vec{B} - \left(a_\mu - \frac{1}{\gamma^2 - 1}\right)\vec{\beta}\times\vec{E} + \frac{\eta}{2}\left(\vec{\beta}\times\vec{B} + \vec{E}\right)\right] = \vec{\omega}_a + \vec{\omega}_E + \vec{\omega}_{\rm edm} \tag{3.32}$$

where the anomalous magnetic moment is obtained from the first term in the brackets proportional to the magnetic field $\vec{B}$ in Eq. (3.32). Note that $q = -1$ for the muons and $+1$ for antimuons, and β is the muon velocity vector. Choosing the so-called muon "magic" momentum, $p_{\rm magic} = 3.094$ GeV, which is equivalent to $\gamma = 29.3$, eliminates the electric field dependent contribution to ω_a.

For $g > 2$ the anomalous spin precession causes the muon spin to advance the muon momentum vector as the muons circulate the ring. In the case of $\theta = 0$ the positrons are Lorentz boosted to higher energies in the laboratory frame, while for $\theta = \pi$ the Lorentz boost reduces the positron energy in the laboratory. Thus, a calorimeter that detects the inwards spiraling positrons will see a count rate modulation of higher and lower energy particles with the anomalous spin precession frequency ω_a.

In the end, the quality of experimental determination of a_μ is determined by how many muons can be stored in the storage ring, how uniform the applied magnetic field is, and how well the precession frequency can be determined.

3.2.2 Electric dipole moments

Another interesting flavor-conserving observable is related to the one we have just discussed. Similar to Eqs. (3.2) and (3.3), one can introduce an *electric* dipole moment (EDM) of a spin-1/2 particle. As we saw in Eq. (3.9), it is related to the imaginary part of a coupling of the effective Lagrangian of Eq. (3.8).

Similar to the anomalous magnetic moment, it might be convenient to define EDM in terms of a particle's electromagnetic form-factor. But contrary to Eq. (3.8), we now need to include parts of it that do not vanish under the requirement of CP-invariance. A matrix element of the current j^μ can now be written as

$$\langle p_i | j^\mu | p_f \rangle = \overline{u}(p_i) \Gamma^\mu (P, q) u(p_f), \tag{3.33}$$

where the generalized vertex Γ^μ now includes new parts,

$$\begin{aligned} \Gamma^\mu &= F_1(q^2)\gamma^\mu + \frac{iF_2(q^2)}{2m}\sigma^{\mu\nu}q_\nu \\ &+ F_A(q^2)\left(\gamma^\mu\gamma_5 q^2 - 2m\gamma_5 q^\mu\right) + \frac{F_3(q^2)}{2m}\sigma^{\mu\nu}\gamma_5 q_\nu. \end{aligned} \tag{3.34}$$

The EDM can be now defined as

$$d_f = -\frac{F_3(0)}{2m}. \tag{3.35}$$

Also, similar to Eq. (3.2), EDM can be defined in the nonrelativistic limit of the Lagrangian of Eq. (3.8),

$$\mathcal{H}_E = -\vec{d} \cdot \vec{E} = -d\, \vec{\sigma} \cdot \vec{E}. \tag{3.36}$$

This Hamiltonian describes the interaction of a particle with a dipole moment $\vec{d}$ with an electric field $\vec{E}$. Just like a particle's magnetic moment, it must lie along the direction of its spin.

It is interesting to see that despite looking quite similar to Eq. (3.2), the Hamiltonian of Eq. (3.36) transforms very differently under fundamental P and T symmetries. Since $\vec{E}$ is a polar vector, it changes sign under parity transformation P but stays the same under time reversal T. Since, as we discussed in Sec. 3.2.1, $\vec{\sigma}$ has exactly the opposite properties under P and T, it must be that $\mathcal{H}_E$ is *not invariant* under separate P and T transformations, but is invariant under combined PT parity transformation! Since it must also be invariant under CPT transformation, it means that $\mathcal{H}_E$ is invariant under the charge conjugation C but is *not invariant* under a combined CP-transformation!

This is what constitutes a crucial difference between the dipole magnetic and electric moments: the EDM is only nonzero in the presence of CP-violating interactions. In the minimal Standard Model the contribution to CP-violating amplitudes must come from the quark sector and be proportional to Jarlskog invariant J,

$$Im\left[V_{ij}V_{kl}V_{il}^*V_{kj}^*\right] = J\sum_{m,n=1}^{3}\epsilon_{ikm}\epsilon_{jln}, \quad \text{with } J \approx A^2\lambda^6\eta\left(1-\frac{\lambda^2}{2}\right), \tag{3.37}$$

where we employed Wolfenstein parameterization for the CKM matrix V_{ik}. It can then be shown that the non-zero standard model's contribution to d first appears at four-loop order! If neutrinos are Majorana particles, a larger EDM (around 10^{-33} e·cm) could be possible in the Standard Model.

It also means that only CP-violating pieces of possible New Physics interactions could be probed by studying the EDM of various particles. In many particular models of NP, where other CP-violating mechanisms besides CKM matrix are present, EDMs can be generated at a lower loop level than in the SM, which makes them very sensitive tests of those models [155].

EDM: New Physics and SM EFT

It is straightforward to obtain the expression for the EDM via a tree-level matching with SMEFT Lagrangian. There are two operators from Table 2.4 that do the trick, which are again $\mathcal{O}_{eW}$ and

$\mathcal{O}_{eB}$,

$$\mathcal{L}_{\rm EDM} = \frac{1}{\Lambda^2}\left[C_{eW}Q_{eB} + C_{eB}Q_{eW}\right] = \frac{C_{eW}}{\Lambda^2}\overline{L}_f\sigma^{\mu\nu}\tau^I H\ell_f W^I_{\mu\nu} + \frac{C_{eB}}{\Lambda^2}\overline{L}_f\sigma^{\mu\nu}H\ell_f B_{\mu\nu}. \tag{3.38}$$

In the broken phase, the Higgs field in Eq. (3.38) acquires vacuum expectation value as in Eq. (2.17). Separating out W^0_μ and B_μ contributions and rotating to the physical basis of Z_μ and a photon A_μ as in Eq. (2.10), we get an effective Lagrangian of Eq. (3.8). Taking the imaginary part of the coefficient gives

$$d_\ell(\mu) = \frac{\sqrt{2}v}{\Lambda^2}\mathrm{Im}\left[s_{\rm W}C_{\ell W}(\mu) - c_{\rm W}C_{\ell B}(\mu)\right], \tag{3.39}$$

where $s_{\rm W}$ and $c_{\rm W}$ are the sine and cosine of the Weinberg angle defined in Eq. (2.10), and we spelled out scale dependence of the Wilson coefficients and thus of the EDM.

It must be pointed out that EDM measurements could be also sensitive to Wilson coefficients other than $C_{\ell W}$ and C_{eB}. This happens because other operators could mix with Q_{eW} and Q_{eB} at one or more loops. We shall illustrate how it happens at one loop, leaving the details – and higher-order calculations – to the special literature [145].

At one loop there are other dimension-6 operators that can mix with Q_{eW} and Q_{eB} to give a contribution to d_ℓ. Examining Tables 2.3-2.6 again, we see that since the operators Q_{eW} and Q_{eB} are of the $\psi^2 XH$ type, only operators ψ^4, X^3, and X^2H^2 can contribute.

There are several ψ^4 operators in Table 2.6. Following [145], it would be convenient to use Fierz relations to rewrite

$$Q^{(3)}_{lequ} = -8Q_{luqe} - 4Q^{(1)}_{lequ} \tag{3.40}$$

and use Q_{luqe} in the following analysis. Out of all ψ^4 operators, only Q_{luqe} contributes at the one-loop level to the anomalous dimensions of Q_{eW} and Q_{eB},

$$\mu\frac{d}{d\mu}\begin{pmatrix} C_{eB} \\ C_{eW}\end{pmatrix} = \frac{Y^u g_2}{16\pi^2}\begin{pmatrix} -t_{\rm W}N_c\left(Y_Q+Y_u\right)/2 \\ N_c/4\end{pmatrix} C_{luqe}, \tag{3.41}$$

where N_c is the number of colors, Y_f the hypercharge of the fermion f, $t_{\rm W} = s_{\rm W}/c_{\rm W}$, and Y^u are the Yukawa couplings. In the basis where the Yukawa matrix is diagonal, the renormalization of the imaginary part of C_{eW} and C_{eB} needed for d_ℓ (see Eq. (3.39)) comes from the imaginary part of C_{luqe}.

There are also three operators of the type X^2H^2 that can mix into Q_{eW} and Q_{eB}. They contribute as

$$\mu\frac{d}{d\mu}\begin{pmatrix} C_{eB} \\ C_{eW}\end{pmatrix} = -\frac{Y^e g_2}{16\pi^2}\begin{pmatrix} 0 & 2t_{\rm W}\left(Y_L+Y_e\right) & 3/4 \\ 1 & 0 & t_{\rm W}\left(Y_L+Y_e\right)\end{pmatrix}\begin{pmatrix} C_{W\widetilde{W}} \\ C_{B\widetilde{B}} \\ C_{W\widetilde{B}}\end{pmatrix}. \tag{3.42}$$

Finally, there are possible contributions from the operators of the class X^3 (see Table 2.3). Out of those, only $Q_{\widetilde{W}}$ can contribute to Q_{eW}. Its contribution turns out to be finite,

$$\mathrm{Im}\left[C_{eQ}\right] = \frac{3}{64\pi^2}Y^e g_2^2 C_{\widetilde{W}}. \tag{3.43}$$

Putting all these contributions together and solving the RG equations above we get a one loop contribution to lepton's EDM in SM EFT [145],

$$d_\ell \simeq -\frac{1}{16\pi^2}\frac{\sqrt{2}ve}{\Lambda^2}\left[Y^u\mathrm{Im}\ C_{luqe} - Y^e\left(2\tilde{\kappa}_{\gamma\gamma} + \frac{1-4s^2_{\rm W}}{s_{\rm W}c_{\rm W}}\tilde{\kappa}_{\gamma Z} - \frac{\Lambda^2}{2v^2}\delta\tilde{\kappa}_\gamma\right)\right]\log\frac{\Lambda^2}{m_h^2}, \tag{3.44}$$

where the parameters $\tilde{\kappa}_{\gamma\gamma}$, $\tilde{\kappa}_{\gamma Z}$, and $\delta\tilde{\kappa}_{\gamma}$ are defined as

$$\begin{aligned}
\tilde{\kappa}_{\gamma\gamma} &= c_{\mathrm{W}}^2 C_{B\widetilde{B}} + s_{\mathrm{W}}^2 C_{W\widetilde{W}} - s_{\mathrm{W}} c_{\mathrm{W}} C_{W\widetilde{B}}, \\
\tilde{\kappa}_{\gamma Z} &= s_{\mathrm{W}} c_{\mathrm{W}} \left(C_{W\widetilde{W}} - C_{B\widetilde{B}}\right) - \frac{1}{2}\left(c_{\mathrm{W}}^2 - s_{\mathrm{W}}^2\right) C_{W\widetilde{B}}, \\
\delta\tilde{\kappa}_{\gamma} &= \frac{1}{t_{\mathrm{W}}} \frac{v^2}{\Lambda^2} C_{W\widetilde{B}}.
\end{aligned} \tag{3.45}$$

Now, experimental constraints on EDMs could be used to probe Wilson coefficients of SM EFT operators as in Eq. (3.44).

EDM: experimental methods of detection

Experimentally, EDM should be searched by studying particle's response to the applied electric field. In this chapter, we shall only concentrate on studies of EDMs of charged leptons, largely that of an electron. EDMs of strongly interacting particles will be discussed in subsequent chapters.

Electrons. It should be obvious that studies of d_e in free electrons are not practical, as they would simply accelerate away in large electric fields. It would then make sense to study EDMs of electrons bound in neutral atoms and molecules, where the interaction of Eq. (3.36) would produce a shift in bound electron's energy linearly proportional to the local value of the electric field $\vec{E}$, i.e. linear Stark effect [19]. The following theorem, however, complicates the observation of EDMs in atoms.

THEOREM 3.1 (Schiff) *The nuclear dipole moment causes the atomic electrons to rearrange themselves so that they develop a dipole moment opposite that of the nucleus. In the limit of nonrelativistic electrons and a point nucleus, the electrons' dipole moment exactly cancels the nuclear moment, so that the net atomic dipole moment vanishes.*

PROOF 3.1 We present a simple classical hand-waiving argument to justify the claim [83]. Let us consider a slightly more generic case of an atom with the nuclear charge Ze and N_e electrons on its shells, i.e. an ion. If $N_e = Z$ then we are dealing with the neutral atom. The second Newton's law for the N_e electrons (each with the mass m_e), nuclei (with the mass M_N), and the ion in external electric field E_0 can be written as

$$\begin{aligned}
(M_N + N_e m_e) a_I &= (Z - N_e) e E_0, \\
M_N a_N &= Z e E_N, \\
m_e a_e &= e E_e,
\end{aligned} \tag{3.46}$$

where a_e, a_N, and $a_I$are the accelerations of the electron, nucleon, and the whole ion, respectively.

We also denoted the average electric field at the nucleus, E_N, and by E_e - the average electric field at one of the ion's electrons. If the system of all those particles is moving, it is moving together, so

$$a_I = a_e = a_N. \tag{3.47}$$

Eliminating the acceleration one arrives at the electric field at the electron

$$E_e = \frac{m_e}{M_N}(Z - N_e)E_0, \tag{3.48}$$

which means that the average electric field for the electrons is suppressed by a very small ratio m_e/M_N, which means that it has practically no effect on electron's EDM, as $\mathbf{d}_e \cdot \mathbf{E}_e \approx 0$, so no energy shifts related to electron's EDM can be observed.

Note that it also follows that nuclear EDM cannot be observed either, as the average electric field at the location of the nuclei is

$$E_N = \frac{Z - N_e}{Z} \frac{M_N}{M_N - N e m_e} E_0 \sim \left(1 - \frac{N_e}{Z}\right) E_0, \tag{3.49}$$

which means that $\mathbf{d}_N \cdot \mathbf{E}_N \sim d_N(1 - N_e/Z)E_0 \approx 0$. A complete proof is available in the classic paper by Schiff [167,170].

The ways to evade Shiff's theorem are spelled out in the conditions that we had to assume in order to prove the theorem: (1) nonrelativistic electrons and (2) a point nucleus. Both of those conditions are not applicable for heavy elements with large nuclear charge Z, which could prove to be good places to search for EDMs [155]. For example, effective electric field that is acting on electron's EDM inside a polarized ThO molecule is $E_e \sim P\alpha^2 Z^3 e/a_0 \sim 80$ GV/cm due to relativistic motion. Here the polarization P could be nearly 100% with $E_0 \sim 10$ V/cm [167].

There are several physical systems that can be used for the searches of EDMs. For instance, atoms or molecules with an unpaired electron spin, or paramagnetic systems, could be good candidates, as permanent EDMs of paramagnetic atoms and molecules arise primarily from the electron's EDM [39], as its contribution scales as Z^3.

Muons. Methods described above are not very suitable for measuring muon's EDM, as muonic atoms have very short lifetimes. It is, however, possible to use the experimental set up used for the measurement of the anomalous magnetic moment of the muon to constrain muon's EDM.

Notice that Eq. (3.32) that describes the set of frequencies measured in an experiment designed to measure muon's (g-2) depends on muon's EDM – see the term proportional to η, the EDM written in Bohr magnetons. As it can be seen from Eq. (3.32), muon's EDM couples to the external magnetic field in the experiment because Lorentz transformation to the muon's rest frame assures that in that frame the electric field is non-zero. Choosing the "magic momentum" eliminates the electric field present in muon's lab frame. Adding a radial magnetic field to cancel the muon's spin precession due to (g-2) so that spin precession due to muon's EDM can be observed.

3.3 Charged lepton flavor violation: background-free searches for New Physics

Flavor-changing neutral current (FCNC) interactions serve as a powerful probe of physics beyond the standard model (BSM). Since no operators generate FCNCs in the Standard Model at tree level, New Physics degrees of freedom can effectively compete with the SM particles running in the loop graphs, making their discovery possible. This is, of course, only true provided the BSM models include flavor-violating interactions.

The observation of charged lepton flavor violating (CLFV) transitions would provide especially clean probes of New Physics. This is because in the Standard Model with massive neutrinos the CLFV transitions are suppressed by the powers of m_ν^2/m_W^2, which renders the predictions for their transition rates vanishingly small, e.g. $\mathcal{B}(\mu \to e\gamma)_{\nu SM} \sim 10^{-54}$ [26]. Such decay rates are not experimentally observable in the near future.

This vice can be turned into a virtue. A variety of well-established models of New Physics predict significantly larger rates for CLFV transitions, which make them (theoretically) background-free modes for the discovery of NP. In what follows we study those processes.

3.3.1 Effective Lagrangians and CLFV

Any New Physics scenario which involves lepton flavor violating interactions can be matched to an effective Lagrangian, $\mathcal{L}_{\text{eff}}$, whose Wilson coefficients would be determined by the ultraviolet (UV) physics that becomes active at some scale Λ. Below the electroweak symmetry breaking scale, this Lagrangian must be invariant under unbroken $SU(3)_c \times U(1)_{\text{em}}$ groups. The effective operators would reflect degrees of freedom relevant at the scale at which a given process takes place. If we assume that no new light particles (such as "dark photons" or axions) exist in the low energy spectrum, those operators would be written entirely in terms of the SM degrees of freedom such as leptons: $\ell_i = \tau, \mu$, and e; and quarks: b, c, s, u, and d. We shall not consider neutrinos in this Chapter. We also assume that top quarks have been integrated out.

The effective Lagrangian describing FCNC in the lepton sector, $\mathcal{L}_{\text{eff}}$, can be divided into a dipole part, $\mathcal{L}_D$; a part that involves four-fermion interactions, $\mathcal{L}_{\ell q}$; and a gluonic part, $\mathcal{L}_G$.

$$\mathcal{L}_{\text{eff}} = \mathcal{L}_D + \mathcal{L}_{\ell q} + \mathcal{L}_G + \ldots. \tag{3.50}$$

Here the ellipses denote higher-order effective operators that are not relevant for the following discussion. The dipole part in Eq. (3.50) is usually written as

$$\mathcal{L}_D = -\frac{m_{\ell_1}}{\Lambda^2} \left[\left(C_{DR}^{\ell_1 \ell_2}\, \bar{\ell}_2 \sigma^{\mu\nu} P_L \ell_1 + C_{DL}^{\ell_1 \ell_2}\, \bar{\ell}_2 \sigma^{\mu\nu} P_R \ell_1 \right) F_{\mu\nu} + h.c. \right], \tag{3.51}$$

where $P_{\text{R,L}} = (1 \pm \gamma_5)/2$ is the right (left) chiral projection operator. The Wilson coefficients would, in general, be different for different leptons ℓ_i.

The four-fermion dimension-six Lagrangian takes the form:

$$\begin{aligned}
\mathcal{L}_{\text{eff}}^{\Delta L_\mu = 1} = & - \frac{1}{\Lambda^2} \sum_f \Big[\Big(C_{VR}^f\, \overline{\mu}_R \gamma^\alpha e_R + C_{VL}^f\, \overline{\mu}_L \gamma^\alpha e_L \Big) \overline{f} \gamma_\alpha f \\
& + \Big(C_{AR}^f\, \overline{\mu}_R \gamma^\alpha e_R + C_{AL}^q\, \overline{\mu}_L \gamma^\alpha e_L \Big) \overline{f} \gamma_\alpha \gamma_5 f \\
& + m_e m_f G_F \Big(C_{SR}^f\, \overline{\mu}_R e_L + C_{SL}^f\, \overline{\mu}_L e_R \Big) \overline{f} f \\
& + m_e m_f G_F \Big(C_{PR}^f\, \overline{\mu}_R e_L + C_{PL}^f\, \overline{\mu}_L e_R \Big) \overline{f} \gamma_5 f \\
& + m_e m_f G_F \Big(C_{TR}^f\, \overline{\mu}_R \sigma^{\alpha\beta} e_L + C_{TL}^f\, \overline{\mu}_L \sigma^{\alpha\beta} e_R \Big) \overline{f} \sigma_{\alpha\beta} f + h.c. \Big],
\end{aligned} \tag{3.52}$$

where f represents fermions that are not integrated out at the scale at which the experiment takes place. These could be quark, charged lepton, and/or neutrino fields. The subscripts on the Wilson coefficients are for the type of Lorentz structure: vector, axial-vector, scalar, pseudo-scalar, and tensor. The Wilson coefficients would in general be different for different flavor and type of fermions f.

We note that the tensor operators are often omitted when constraints on the Wilson coefficients in Eq. (3.52) are derived. We would like to point out that those are no less motivated than others in Eq. (3.52). For example, they would be induced from Fierz rearrangement of operators of the type $\mathcal{Q} \sim (\overline{q}\ell_2)\left(\overline{\ell}_1 q\right)$ that often appear in leptoquark models. Also, as we shall see later, the experimental constraints on those coefficients follow from studying vector meson decays, where the best information on LFV transitions in quarkonia is available.

Another form of the Lagrangian Eq. (3.52) has been historically used for leptonic decays of muons and taus, such as $\mu^+ \to e^+ \nu_e \bar{\nu}_\mu$. It contains ten complex parameters g_{XY}^Z. We present it here as it is still often used for experimental analyses.

$$\begin{aligned}
\mathcal{L}_{\ell_1\to\ell_2\nu_2\bar{\nu}_1} &= -\frac{4G_F}{\sqrt{2}}\Big[g^S_{RR}\left(\overline{\ell_2}_R\nu_{\ell_2 L}\right)\left(\overline{\nu_{\ell_1}}_L\ell_{1R}\right) + g^S_{RL}\left(\overline{\ell_2}_R\nu_{\ell_2 L}\right)\left(\overline{\nu_{\ell_1}}_R\ell_{1L}\right) \\
&+ g^S_{LR}\left(\overline{\ell_2}_L\nu_{\ell_2 R}\right)\left(\overline{\nu_{\ell_1}}_L\ell_{1R}\right) + g^S_{LL}\left(\overline{\ell_2}_L\nu_{\ell_2 R}\right)\left(\overline{\nu_{\ell_1}}_R\ell_{1L}\right) \\
&+ g^V_{RR}\left(\overline{\ell_2}_R\gamma^\alpha\nu_{\ell_2 R}\right)\left(\overline{\nu_{\ell_1}}_R\gamma_\alpha\ell_{1R}\right) + g^V_{RL}\left(\overline{\ell_2}_R\gamma^\alpha\nu_{\ell_2 R}\right)\left(\overline{\nu_{\ell_1}}_L\gamma_\alpha\ell_{1L}\right) \qquad (3.53) \\
&+ g^V_{LR}\left(\overline{\ell_2}_L\gamma^\alpha\nu_{\ell_2 L}\right)\left(\overline{\nu_{\ell_1}}_R\gamma_\alpha\ell_{1R}\right) + g^V_{LL}\left(\overline{\ell_2}_R\gamma^\alpha\nu_{\ell_2 R}\right)\left(\overline{\nu_{\ell_1}}_L\gamma_\alpha\ell_{1L}\right) \\
&+ \frac{g^T_{RL}}{2}\left(\overline{\ell_2}_R\sigma_{\alpha\beta}\nu_{\ell_2 L}\right)\left(\overline{\nu_{\ell_1}}_R\sigma^{\alpha\beta}\ell_{1L}\right) + \frac{g^T_{LR}}{2}\left(\overline{\ell_2}_L\sigma_{\alpha\beta}\nu_{\ell_2 R}\right)\left(\overline{\nu_{\ell_1}}_L\sigma^{\alpha\beta}\ell_{1R}\right) + h.c.\Big],
\end{aligned}$$

with the normalization condition on the couplings g^Z_{XY}

$$\begin{aligned}
&\frac{1}{4}\left(\left|g^S_{RR}\right|^2 + \left|g^S_{LL}\right|^2 + \left|g^S_{RL}\right|^2 + \left|g^S_{LR}\right|^2\right) \\
+ &\left(\left|g^V_{RR}\right|^2 + \left|g^V_{LL}\right|^2 + \left|g^V_{RL}\right|^2 + \left|g^V_{LR}\right|^2\right) + 3\left(\left|g^T_{RL}\right|^2 + \left|g^T_{LR}\right|^2\right) = 1, \qquad (3.54)
\end{aligned}$$

which follows from the fact that the Lagrangian of Eq. (3.53) describes the decay completely. In the Standard Model all $g^Z_{XY} = 0$, except for $g^V_{LL} = 1$, respecting the normalization of Eq. (3.54). Note that G_F is the Fermi constant. It is important to point out that the Lagrangian Eq. (3.53) is completely equivalent to the Lagrangian of Eq. (3.52). The relations among the coefficients can be obtained using Fierz identities in Section A.4 of the Appendix. The coefficients g^m_{ik} scale as m^2/Λ^2, where m is the scale relevant to the problem. For instance, it could be a mass of the decaying state. We leave it as a problem for the reader to find the relations between the coefficients g^Z_{XY} of Eq. (3.53) and Wilson coefficients of the Lagrangian of Eq. (3.52) and those of SM EFT (see the end of this chapter).

The dimension seven gluonic operators can be either generated by some high scale physics or by integrating out heavy quark degrees of freedom, which make them scale like $1/\Lambda^2$ [151],

$$\begin{aligned}
\mathcal{L}_G = -\frac{m_2 G_F}{\Lambda^2}\frac{\beta_L}{4\alpha_s}\Big[&\left(C^{\ell_1\ell_2}_{GR}\,\bar{\ell}_1 P_L \ell_2 + C^{\ell_1\ell_2}_{GL}\,\bar{\ell}_1 P_R \ell_2\right)G^a_{\mu\nu}G^{a\mu\nu} \\
&+ \left(C^{\ell_1\ell_2}_{\tilde{G}R}\,\bar{\ell}_1 P_L \ell_2 + C^{\ell_1\ell_2}_{\tilde{G}L}\,\bar{\ell}_1 P_R \ell_2\right)G^a_{\mu\nu}\widetilde{G}^{a\mu\nu} \quad + \quad h.c.\Big]. \qquad (3.55)
\end{aligned}$$

Here $\beta_L = -9\alpha_s^2/(2\pi)$ is defined for the number of light active flavors, L, relevant to the scale of the process, which we take $\mu \approx 2$ GeV. All Wilson coefficients should also be calculated at the same scale. Here $\widetilde{G}^{a\mu\nu} = (1/2)\epsilon^{\mu\nu\alpha\beta}G^a_{\alpha\beta}$ is a dual to the gluon field strength tensor.

The experimental constraints on the Wilson coefficients of effective operators in $\mathcal{L}_{\text{eff}}$ could be obtained from a variety of LFV decays. Deriving constraints on those Wilson coefficients usually involves an assumption that only one of the effective operators dominates the result. This does not usually happen in many particular UV completions of the LFV EFTs. Nevertheless, *single operator dominance* hypothesis is a useful theoretical assumption in placing constraints on the parameters of $\mathcal{L}_{\text{eff}}$. One has to remember that certain cancellations among contributions of various operators are possible and use the constraints obtained this way with a grain of salt.

3.3.2 Muon decays and Michel parameters

Charged muon decay into an electron and a pair of neutrino-antineutrino, $\mu^+ \to e^+\nu_e\bar{\nu}_\mu$, which is sometimes referred to as "Michel muon decay", is probably one of the best-studied processes in particle physics. It is well described by the Standard Model and contains no strongly interacting particles in its initial or final states. Moreover, higher-order SM contributions, such as QED or $1/m_W$ corrections, can be computed with the required accuracy. It is therefore a good candidate for studies of BSM extensions of the Standard Model. While we will concentrate on muon decay below, everything we say can be immediately applied to tau decays $\tau^+ \to \mu^+\nu_\mu\bar{\nu}_\tau$ or $\tau^+ \to e^+\nu_e\bar{\nu}_\tau$ as well. In the SM the muon width can be written as

$$\Gamma_\mu^{\rm SM} = \frac{G_F^2 m_\mu^5}{192\pi^3} F\left(\frac{m_e^2}{m_\mu^2}\right)\left(1+\frac{3}{5}\frac{m_\mu^2}{m_W^2}\right)\left[1+\frac{\alpha(m_\mu)}{2\pi}\left(\frac{25}{4}-\pi^2\right)\right], \tag{3.56}$$

where $F(x) = 1-8x+8x^3-x^4-12x^2\log x$. Note that, following [121,136], we included both one-loop electromagnetic corrections and $1/m_W^2$ effects coming from the next-to-leading corrections to the Fermi model. They are obtained from matching the effective operators to the full Standard Model. A two-loop result for the electromagnetic corrections can be found in [176]. A similar decay for any lepton $\ell_i \to \ell_f\nu\bar{\nu}$ can be obtained by trivial substitution of lepton masses.

A differential muon decay rate into an electron and a neutrino-antineutrino pair is also possible to compute. If the polarization of the final state lepton (electron) is not measured,

$$\frac{d^2\Gamma_{\rm SM}(\mu^\pm \to e^\pm\nu\bar{\nu})}{d\cos\theta_e dx} = \frac{G_F^2 m_\mu^5}{192\pi^3}x^2\left[(3-2x)\pm P_{(\mu)}(2x-1)\cos\theta_e\right], \tag{3.57}$$

where $x = 2m_\mu E_e/(m_\mu^2+m_e^2)$ is the rescaled electron's energy, and $\vec{P}_{(\mu)}$ is the muon's polarization vector. The angle θ_e is the angle between the muon polarization and the electron momentum. The plus (minus) sign in Eq. (3.57) corresponds to decays of a $\mu^+(\mu^-)$ muon state.

How would possible BSM contributions affect muon decay? The leading-order corrections should come from dimension-six four-fermions operators. The most general form of the differential decay rate for the polarized muon is given by [125]

$$\frac{d^2\Gamma(\mu^\pm \to e^\pm\nu\bar{\nu})}{d\cos\theta_e dx} = \frac{G_F^2 m_\mu}{4\pi^3}W_{e\mu}^4\sqrt{x^2-x_0^2}\left(F_{IS}(x)\pm P_{(\mu)}F_{AS}(x)\cos\theta_e\right)\left(1+\vec{P}_e(x,\theta_e)\cdot\hat{\vec{\zeta}}\right), \tag{3.58}$$

where $W_{e\mu} = (m_\mu^2+m_e^2)/(2m_\mu)$, and $x_0 = m_e/W_{e\mu}$. Here $\vec{P}_e(x,\theta_e)$ is the polarization vector of the electron, while $\hat{\vec{\zeta}}$ is a unit vector along the direction of the measurement of electron (positron) spin polarization. The functions $F_{IS}(x)$ and $F_{AS}(x)$ are defined as

$$\begin{aligned} F_{IS}(x) &= x(1-x)+\frac{2\rho}{9}\left(4x^2-3x-x_0^2\right)+\eta x_0(1-x) \\ F_{AS}(x) &= \frac{\xi}{3}\sqrt{x^2-x_0^2}\left[1-x+\frac{2\delta}{3}\left(4x-3+\left(\sqrt{1-x_0^2}-1\right)\right)\right], \end{aligned} \tag{3.59}$$

where ρ, η, ξ, and δ are the so-called Michel parameters [27] that parameterize BSM contribution to muon decay. In the Standard Model $\rho = \delta = 3/4$, $\eta = 0$, and $\xi = 1$. Michel parameters can be written in terms of the coefficients g_{XX}^L of the effective Lagrangian of Eq. (3.53),

$$\begin{aligned} \rho &= \frac{3}{16}\left[\left|g_{RR}^S\right|^2+\left|g_{LL}^S\right|^2+\left|g_{RL}^S-2g_{RL}^T\right|^2+\left|g_{LR}^S-2g_{LR}^T\right|^2+\frac{3}{4}\left(\left|g_{RR}^V\right|^2+\left|g_{LL}^V\right|^2\right)\right], \\ \eta &= \frac{1}{2}{\rm Re}\left[g_{RR}^V g_{LL}^{S*}+g_{LL}^V g_{RR}^{S*}+g_{RL}^V\left(g_{LR}^{S*}+6g_{LR}^{T*}\right)+g_{LR}^V\left(g_{RL}^{S*}+6g_{RL}^{T*}\right)\right], \\ \xi &= \frac{1}{4}\left(\left|g_{LL}^S\right|^2-\left|g_{RR}^S\right|^2\right)-\frac{1}{4}\left(\left|g_{LR}^S\right|^2-\left|g_{RL}^S\right|^2\right)+\left(\left|g_{LL}^V\right|^2-\left|g_{RR}^V\right|^2\right) \\ &+ 3\left(\left|g_{LR}^V\right|^2-\left|g_{RL}^V\right|^2\right)+5\left(\left|g_{LR}^T\right|^2-\left|g_{RL}^T\right|^2\right)+4{\rm Re}\left[g_{LR}^S g_{RL}^{T*}-g_{RL}^S g_{LR}^{T*}\right], \\ \xi\delta &= \frac{3}{16}\left(\left|g_{LL}^S\right|^2-\left|g_{RR}^S\right|^2+\left|g_{RL}^S-2g_{RL}^T\right|^2+\left|g_{LR}^S-2g_{LR}^T\right|^2\right)+\frac{3}{4}\left(\left|g_{LL}^V\right|^2-\left|g_{RR}^V\right|^2\right). \end{aligned} \tag{3.60}$$

The muon's decay width, including Michel parameters, can be written as

$$\begin{aligned}\Gamma_\mu &= \frac{G_F^2 m_\mu^5}{192\pi^3}\left[F\left(\frac{m_e^2}{m_\mu^2}\right) + 4\eta\frac{m_e}{m_\mu}G\left(\frac{m_e^2}{m_\mu^2}\right) - \frac{32}{3}\frac{m_e^2}{m_\mu^2}\left(\rho - \frac{3}{4}\right)\left(1 - \frac{m_e^4}{m_\mu^4}\right)\right] \\ &\times \left(1 + \frac{3}{5}\frac{m_\mu^2}{m_W^2}\right)\left[1 + \frac{\alpha(m_\mu)}{2\pi}\left(\frac{25}{4} - \pi^2\right)\right], \qquad (3.61)\end{aligned}$$

where $G(x) = 1 + 9x - 9x^2 - x^3 + 6x(1+x)x^2 \log x$. Careful experimental studies of polarized muon decay allows for excellent bounds on Michel parameters, and thus, possible New Physics interactions. For current experimental constraints, please see [1].

3.3.3 Flavor-violating lepton decays: $\mu \to e\gamma$ and its relatives

Now let us turn to the decays that are not allowed in the Standard Model at the tree level. While FCNC processes can be generated in the SM at a one-loop level, their rates are vanishingly small, as they are proportional to powers of neutrino masses. A set of processes with only leptons (and possibly photons) in the initial and final states is a popular way to study lepton flavor violation.

Radiative decays: $\ell_i \to \ell_f \gamma$

Two-body flavor-changing neutral current (FCNC) radiative decay of a lepton state $\ell_i \to \ell_f\gamma$ is one of the simplest to consider. Many experiments have searched for this decay, with the muon being the most common initial state.

Let us write the most general amplitude for such transition, $\ell_i(p) \to \ell_f(p')\gamma(q)$,

$$A_{\ell_1 \to \ell_2\gamma}(p,p') = i\overline{u}_{\ell_2}(p')V_\mu(q,P)u_{\ell_1}(p)\epsilon^{*\mu} = J_\mu^{\ell_2\to\ell_2}\epsilon^{*\mu}, \qquad (3.62)$$

where u_ℓ are spinors representing lepton states, and $q = p - p'$ is the photon's momentum. The most general form for the vertex $V_\mu(q,P)$ can be written as

$$\begin{aligned}V_\mu(q,P) &= \frac{\sigma_{\mu\nu}q^\nu}{m_{\ell_1}}\left(F_1 + F_2\gamma_5\right) + \gamma_\mu\left(F_3 + F_4\gamma_5\right) \\ &+ \frac{(p-p')_\mu}{m_{\ell_1}}\left(F_5 + F_6\gamma_5\right) + \frac{(p+p')_\mu}{m_{\ell_1}}\left(F_7 + F_8\gamma_5\right), \qquad (3.63)\end{aligned}$$

where F_i are some functions of Lorentz-invariant products of initial and final state momenta and masses. We can call those functions *form factors.* We introduced the factor of $1/m_{\ell_1}$ to make sure that all form factors have the same dimension. The choice of m_{ℓ_1} might seem rather arbitrary, but it is natural to choose the mass of the decaying state as normalization in calculating decay amplitudes. We shall see how the dominant scale changes once we match onto the operators of Eq. (3.51), and, ultimately, to SM EFT operators [156].

Not all of the form factors F_i are independent. Since experimentally both leptons and a photon are on their mass shells, we could use gauge invariance and the equations of motion to reduce the number of independent form factors. Doing just that, we set $J_\mu^{\ell_2\to\ell_2}q^\mu = 0$ and employ the equations of motion for spinors, $\left(\not{p} - m_{\ell_1}\right)u(p) = 0$ and $\overline{u}(p')\left(\not{p}' - m_{\ell_2}\right) = 0$, to obtain

$$\begin{aligned}J_\mu^{\ell_2\to\ell_2}q^\mu &= \frac{i}{m_{\ell_1}}\overline{u}_{\ell_2}(p')\left[-m_{\ell_1}^2\left(F_3 - F_4\gamma_5\right) - m_{\ell_2}m_{\ell_1}\left(F_3 + F_4\gamma_5\right)\right. \\ &+ \left. q^2\left(F_5 + F_6\gamma_5\right) + \left(m_{\ell_1}^2 - m_{\ell_2}^2\right)\left(F_7 + F_8\gamma_5\right)\right]u_{\ell_1}(p) = 0. \qquad (3.64)\end{aligned}$$

The terms proportional to F_1 and F_2 drop out due to the anti-symmetry of $\sigma_{\mu\nu}$. It follows from Eq. (3.64) that the form factors F_5 and F_6 do not contribute for the on-shell photons $q^2 = 0$, and

$$F_3 = F_4 = F_7 = F_8 = 0. \qquad (3.65)$$

Using the properties of the left- and right-handed projection operators,

$$A_{\ell_1 \to \ell_2 \gamma}(p, p') = \frac{i}{m_{\ell_1}} \overline{u}_{\ell_2}(p') \left[A_L P_L + A_R P_R \right] \sigma_{\mu\nu} q^\nu u_{\ell_1}(p) \epsilon^{*\mu}, \tag{3.66}$$

where $A_{L/R} = F_1 \mp F_2$. The unpolarized decay rate averaged over possible polarizations, is given by

$$\Gamma(\ell_1 \to \ell_2 \gamma) = \frac{m_{\ell_1}}{16\pi} \left(|A_L|^2 + |A_R|^2 \right). \tag{3.67}$$

In the Standard Model, the rate $\Gamma(\ell_1 \to \ell_2\gamma)$ is tiny. For example, for the muon decay $\mu \to e\gamma$ [26],

$$\mathcal{B}(\mu \to e\gamma) = \frac{3\alpha}{32\pi} \left| \sum_i U_{\mu i}^* U_{ei} \frac{m_i^2}{m_W^2} \right|^2 \tag{3.68}$$

where U_{ik} is the neutrino mixing matrix and m_i are the corresponding masses of neutrino mass eigenstates (see Chapter 5 for more information). Numerically, this predicts the corresponding branching ratio in the Standard Model to be $\mathcal{B}(\mu \to e\gamma)_{\nu SM} \sim 10^{-54}$. This makes it interesting as a SM background-free test of NP.

The electron and a muon must have opposite chiralities. Since only left-handed fields interact with the W-bosons, in the Standard Model a chirality flip can only arise from the mass terms of the external leptons. As a result, for muon decay $A_R \sim m_\mu$, while $A_L \sim m_e$, which means that $|A_L|^2 \ll |A_R|^2$, and is therefore negligible. This can in principle be checked by measuring the photon's polarization. One possible avenue where New Physics can affect such arguments is with the models where the chirality flip occurs inside the loop diagram.

It would also be instructive to determine A_L and A_R from matching the decay amplitude of Eq. (3.62) to the one obtained directly from the SM EFT Lagrangian of Section 2.6.2,

$$A_R = A_L^* = \sqrt{2}\, \frac{v m_i^2}{\Lambda^2} \left(c_W C_{eB}^{fi} - s_W C_{eW}^{fi} \right) \equiv \sqrt{2}\, \frac{v m_i^2}{\Lambda^2} C_\gamma^{fi}, \tag{3.69}$$

where s_W and c_W are the sine and cosine of the Weinberg angle. As we can see, the experimental bound on $\mathcal{B}(\mu \to e\gamma)$ directly constraints a combination of the SM EFT Wilson coefficients $C_{eB}^{e\mu}$ and $C_{eW}^{e\mu}$.

Fully leptonic decays decays: $\ell_i \to \ell_j \ell_k \ell_l$

Other interesting processes that probe flavor-changing neutral currents in leptons involve decays of a charged lepton into three lighter charged leptons, $\ell_i \to \ell_j \ell_k \ell_l$. Such decays can be generated by flavor-violating photon couplings discussed above. They can also be mediated by a tree-level exchange of Z or Goldstone bosons with flavor-violating couplings, or by dimension-six four-lepton operators. As we show below, a rich structure of lepton-flavor violating interactions can be probed in the transitions $\ell_i \to \ell_j \ell_k \ell_l$, which makes them very interesting experimentally.

These decays can also be computed in SM EFT. Following [56] we distinguish three groups of $\ell_i \to \ell_j \ell_k \ell_l$ decays, depending on the final state composition: (a) those with three leptons of the same flavor, such as $\mu \to 3e$, $\tau \to 3\mu$, and $\tau \to 3e$, (b) those with distinguishable leptons, such as $\tau^\pm \to e^\pm \mu^+ \mu^-$ and $\tau^\pm \to \mu^\pm e^+ e^-$, and (c) those with two leptons of the same flavor and charge, and one of different flavor and opposite charge, such as $\tau^\pm \to 2\mu^\pm e^\mp$ or $\tau \to 2e^\pm \mu^\mp$. Their decay amplitudes can be conveniently written as

$$A_{\ell_i \to \ell_j \ell_k \ell_l}(p_i, p_j, p_k, p_l) = A_0 + A_\gamma, \tag{3.70}$$

where the notation follows from the fact that in A_0 one can set all momenta of external leptons to zero, while A_γ is the contribution generated by a photon exchange. The first part of the amplitude, for $\ell_i(p_i) \to \ell_j(p_j)\ell_k(p_k)\ell_l(p_l)$, A_0, can be parameterized in terms of lepton spinors as

$$A_0 = \frac{1}{\Lambda^2} \sum_I C_I \left(\overline{u}(p_j) Q_I u(p_i)\right) \left(\overline{u}(p_k) Q'_I v(p_l)\right), \tag{3.71}$$

where the Wilson coefficients C_I are defined according to the Dirac matrices $Q_I \times Q'_I$:

$$\begin{aligned} C_{VXY} &\to \gamma_\mu P_X \times \gamma^\mu P_Y, \\ C_{SXY} &\to P_X \times P_Y, \\ C_{TX} &\to \sigma_{\mu\nu} \times \sigma^{\mu\nu} P_X, \end{aligned} \tag{3.72}$$

where $\{X, Y\} = \{L, R\}$ are the chiral projectors. Other spinor structures can be related to Eq. (3.71) by Fierz transformations (see Appendix A.4). The coefficients C_{VXY}, C_{SXY}, and C_{TX} from Eq. (3.71) will be matched to the Wilson coefficients of SM EFT operators.

The photonic (penguin) contribution A_γ is different for the different cases (a), (b) and (c) outlined above. In fact, $A_\gamma^{(c)} = 0$. The two other contributions are given by [56]

$$\begin{aligned} A_\gamma^{(a)} &= \frac{ev}{\Lambda^2 p_{ij}^2} \left(\overline{u}(p_j)\sigma_{\mu\nu}\left[C_{\gamma L}P_L + C_{\gamma R}P_R\right] p_{ij}^\nu u(p_i)\right) \left(\overline{u}(p_k)\gamma^\mu v(p_l)\right) - (p_j \leftrightarrow p_k), \\ A_\gamma^{(b)} &= \frac{ev}{\Lambda^2 p_{ij}^2} \left(\overline{u}(p_j)\sigma_{\mu\nu}\left[C_{\gamma L}P_L + C_{\gamma R}P_R\right] p_{ij}^\nu u(p_i)\right) \left(\overline{u}(p_k)\gamma^\mu v(p_l)\right), \end{aligned} \tag{3.73}$$

where $p_{ij} = (p_i - p_j)$, and v is the vacuum expectation value defined in Eq. (2.17). The decay width of ℓ_i can then be written as

$$\begin{aligned} \Gamma(\ell_i \to \ell_j \ell_k \ell_l) &= \frac{\kappa_c m_{\ell_1}^5}{32\,(192\pi^3)\Lambda^4} \Big[X_\gamma + 4\left(|C_{VLL}|^2 + |C_{VRR}|^2 + |C_{VLR}|^2 + |C_{VRL}|^2\right) \\ &+ |C_{SLL}|^2 + |C_{SRR}|^2 + |C_{SLR}|^2 + |C_{SRL}|^2 + +48\left(|C_{TL}|^2 + |C_{TR}|^2\right)\Big], \end{aligned} \tag{3.74}$$

where $\kappa_c = 1/2$ if two of the final state leptons are identical and $\kappa = 1$ otherwise. The photon penguin contribution must be obtained by expanding the amplitude in m_f/m_i up to $\mathcal{O}(m_f^2/m_i^2)$, where m_f parameterizes the scale of the final state lepton. The result is

$$\begin{aligned} X_\gamma^{(a)} &= -\frac{16ev}{m_i}\mathrm{Re}\left[C_{\gamma L}^*\left(2C_{VLL} + C_{VLR} - \frac{1}{2}C_{SLR}\right) + C_{\gamma R}^*\left(2C_{VRR} + C_{VRL} - \frac{1}{2}C_{SRL}\right)\right] \\ &+ \frac{64e^2v^2}{m_i^2}\left(\log\frac{m_i^2}{m_f^2} - \frac{11}{4}\right)\left(|C_{\gamma L}|^2 + |C_{\gamma R}|^2\right), \\ X_\gamma^{(a)} &= -\frac{16ev}{m_i}\mathrm{Re}\Big[C_{\gamma L}^*\left(C_{VLL} + C_{VLR}\right) + C_{\gamma R}^*\left(C_{VRR} + C_{VRL}\right)\Big] \\ &+ \frac{32e^2v^2}{m_i^2}\left(\log\frac{m_i^2}{m_f^2} - 3\right)\left(|C_{\gamma L}|^2 + |C_{\gamma R}|^2\right). \end{aligned} \tag{3.75}$$

Note that since $A_\gamma^{(c)} = 0$, the corresponding contribution to the width $X_\gamma^{(c)} = 0$ as well.

It is most interesting to match Wilson coefficients in Eq. (3.74) to Wilson coefficients on SM EFT operators defined in Section 2.6.2. The result of tree-level matching into SMEFT can be found in Table 3.1.

CLFV decays: experimental methods of detection

CLFV decays of muons or tau leptons are experimentally studied at particle accelerators. Searches for those transitions are highly dependent on the availability of large numbers of produced muons and taus. As muons have longer lifetimes and smaller masses, specialized high-intensity muon beams can be produced at particle accelerators [20, 42]. Colliding high-power

TABLE 3.1 Matching of SM EFT Wilson coefficients in leptonic LFV decays of leptons. Here $X = L$ or R.

	Class (a): $\ell_i \to 3\ell_j$	Class (b): $\ell_i \to \ell_j 2\ell_k$	Class (c): $\ell_i^\pm \to \ell_j^\mp \ell_k^\pm \ell_k^\pm$
C_{VLL}	$2\left[(2s_W^2-1)\left(C_{\ell H}^{(1)ji}+C_{\ell H}^{(3)ji}\right)+C_{\ell\ell}^{jikk}\right]$	$(2s_W^2-1)\left(C_{\ell H}^{(1)ji}+C_{\ell H}^{(3)ji}\right)+C_{\ell\ell}^{jijj}$	$2C_{\ell\ell}^{kikj}$
C_{VRR}	$2\left(2s_W^2 C_{eH}^{ji}+C_{ee}^{jijj}\right)$	$2s_W^2 C_{eH}^{ji}+C_{ee}^{jikk}$	$2C_{ee}^{kikj}$
C_{VLR}	$2s_W^2\left(C_{\ell H}^{(1)ji}+C_{\ell H}^{(3)ji}\right)+C_{\ell e}^{jijj}$	$2s_W^2\left(C_{\ell H}^{(1)ji}+C_{\ell H}^{(3)ji}\right)+C_{\ell e}^{jikk}$	$C_{\ell e}^{kikj}$
C_{VRL}	$(2s_W^2-1)C_{eH}^{ji}+C_{\ell e}^{jjji}$	$(2s_W^2-1)C_{eH}^{ji}+C_{\ell e}^{jkki}$	$C_{\ell e}^{kjki}$
C_{SLR}	$-2\left[(2s_W^2-1)C_{eH}^{ji}+C_{\ell e}^{jjji}\right]$	$-2C_{\ell e}^{jkki}$	$-2C_{\ell e}^{kjki}$
C_{SRL}	$-2\left[2s_W^2\left(C_{\ell H}^{(1)ji}+C_{\ell H}^{(3)ji}\right)+C_{\ell e}^{jijj}\right]$	$-2C_{\ell e}^{jikk}$	$-2C_{\ell e}^{kikj}$
C_{SXX}	0	0	0
C_{TX}	0	0	0
$C_{\gamma L}$	$\sqrt{2}C_\gamma^{ij*}$	$\sqrt{2}C_\gamma^{ij*}$	0
$C_{\gamma R}$	$\sqrt{2}C_\gamma^{ji}$	$\sqrt{2}C_\gamma^{ji}$	0

proton beams with a target produce pions that eventually decay into muons. The long lifetime of the muons makes high-intensity muon beams possible. Due to their shorter lifetimes and larger masses, the taus require higher energy collisions. Therefore, practical studies of tau decays are done either at general-purpose flavor or tau-charm factories. For either state, the decays products are detected by various combinations of calorimetric or position-sensitive detectors.

Even though the decay products can be detected with reasonably high efficiency, the signals are expected to be quite rare, so background rejection plays an important role. For the muons, there are two main sources of background events. One is coming from the radiative muon decay $\mu^+ \to e^+\gamma\nu_e\bar{\nu}_\mu$. Such a process becomes a background when the positron and the photon are emitted almost back-to-back. In this part of the phase space the pair of neutrinos carries off little energy, so the decay closely resembles $\mu \to e\gamma$. The other source of background is more "instrumentation-dependent" and is due to the possible coincidence of a positron from a Michel muon decay, $\mu^+ e^+\nu_e\bar{\nu}_\mu$, with a random high energy photon from other sources. Similarly, for $\mu \to 3e$ transitions an important background is coming from decays like $\mu^+ \to e^+e^-e^+\nu_e\bar{\nu}_\mu$, in the part of phase space where the pair of neutrinos carries off little energy. The art of experimental design is to reduce the size of such backgrounds.

3.3.4 Muonium physics: decays

A theoretically clean bound-state system to study BSM effects in the lepton sector is muonium M_μ, a QED bound state of a positively-charged muon and a negatively-charged electron, $|M_\mu\rangle \equiv |\mu^+e^-\rangle$*. The main decay channel for both states is driven by the weak decay of the muon. The average lifetime of a muonium state τ_{M_μ} is expected to be the same as that of the muon, $\tau_\mu = (2.1969811 \pm 0.0000022) \times 10^{-6}$ s, apart from the tiny effect due to time dilation, $(\tau_{M_\mu} - \tau_\mu)/\tau_\mu = \alpha^2 m_e^2/(2m_\mu^2) = 6 \times 10^{-10}$ [58]. Just like positronium or a Hydrogen atom, muonium could be produced in two spin configurations, a spin-one triplet state called *ortho-muonium*, and a spin-zero singlet state called *para-muonium*. We shall denote the para-muonium state as $\left|M_\mu^P\right\rangle$ and the ortho-muonium state as $\left|M_\mu^V\right\rangle$. If the spin of the state does not matter, we shall employ the notation $|M_\mu\rangle$. The description of the muonium physics in this section can be easily generalized to any *leptonium* state by obvious renaming of the fields.

Muonium and muonium decays

The lifetime of muonium is by far dominated by the decay of the muon, $M_\mu \to e^+e^-\nu_e\bar{\nu}_\mu$ [58]. Other exclusive decays, especially the two-body ones, are interesting because they can probe

*A bound state of two muons, a heavier analogue of positronium, is conventionally called a *true muonium*.

leptonic FCNCs which are sensitive to NP particles, as we discussed in Sec. 3.3.3. It is interesting to notice that the largest SM-dominated decay channel of muonium is into invisible final states, i.e. into neutrinos. We hasten to note that the decay rates depend on the spin state of the muonium: due to the left-handed nature of weak interactions, the decay of a spin-singlet para-muonium would be suppressed by the factors of neutrino mass, i.e. be essentially zero in the SM. Moreover, the invisible decay of the para-muonium in the SM would be dominated by decays into a four-neutrino final state [21].

The computation of the two-neutrino decays of a muonium state in the SM is straightforward. The Fermi model Lagrangian,

$$\begin{aligned} \mathcal{L}_F &= -\frac{G_F}{\sqrt{2}} \left(\overline{\nu}_e \gamma^\alpha (1-\gamma_5) e\right) \left(\overline{\mu}\gamma_\alpha (1-\gamma_5)\nu_\mu\right) + h.c., \\ & -\frac{G_F}{\sqrt{2}} \left(\overline{\nu}_e \gamma^\alpha (1-\gamma_5) \nu_\mu\right) \left(\overline{\mu}\gamma_\alpha (1-\gamma_5) e\right) + h.c., \end{aligned} \tag{3.76}$$

where we have used Fierz identities to place neutrinos into one current and charged leptons into another. The computation of the muonium decay rate follows from the matrix element,

$$A(M_\mu \to \nu_e \bar{\nu}_\mu) = -\frac{G_F}{\sqrt{2}} \overline{\nu}_e \gamma^\alpha (1-\gamma_5)\nu_\mu \left\langle 0 \right| \overline{\mu}\gamma_\alpha (1-\gamma_5) e \left| M_\mu \right\rangle \tag{3.77}$$

where the muon and an electron are not free particles but are bound inside of the muonium state. This fact significantly affects the result of the calculation, as the matrix elements would be different for different spin configurations of the muonium state. We can push the calculation forward if we introduce the following parameterization of said matrix elements,

$$\begin{aligned} \left\langle 0 \right| \overline{\mu}\gamma^\alpha \gamma^5 e \left| M_\mu^P \right\rangle &= i f_P p^\alpha, \\ \left\langle 0 \right| \overline{\mu}\gamma^\alpha e \left| M_\mu^V \right\rangle &= f_V M_M \epsilon^\alpha(p), \\ \left\langle 0 \right| \overline{\mu}\sigma^{\alpha\beta} e \left| M_\mu^V \right\rangle &= i f_T \left(\epsilon^\alpha p^\beta - \epsilon^\beta p^\alpha\right), \end{aligned} \tag{3.78}$$

where p^α is the para-muonium's four-momentum, and $\epsilon^\alpha(p)$ is the ortho-muonium's polarization vector. In Eq. (3.78) we also introduced a matrix element of the tensor current that does not enter SM Lagrangian of Eq. (3.76), but could, in principle, be generated by New Physics interactions.

The parameterizations of the matrix elements in Eq. (3.78) follows from the fact that the only four-vectors that matrix elements could depend on are the momentum and (possibly) polarization of the muonium state. We will discuss parameterizations of matrix elements in more detail in Sec. 4.5.1 of Chapter 4.

Since we are dealing with a QED bound state, we can employ the Bethe-Salpeter formalism to compute the matrix elements in Eq. (3.78). This calculation could be instructive for future discussions of quark matrix elements where similar techniques for the computation were employed in various quark models. In QED, a bound state ${}^{2S+1}L_J$ is given by [147]

$$\left|{}^{2S+1}L_J; M\right\rangle = \sqrt{2M} \int \frac{d^3\vec{p}}{(2\pi)^3} \sum_{S_z=-s}^{s} C^{JM}_{\ell m, SS_z} \widetilde{\varphi}_{\ell m}(\vec{p}) |S, S_z\rangle \tag{3.79}$$

where M is the mass of the bound state and $C^{JM}_{\ell m, SS_z}$ are the Clebsch-Gordan coefficients connecting angular momentum eigenstates with spin eigenstates $|SS_z\rangle$. The term

$$\widetilde{\varphi}_{\ell m}(\vec{p}) = \int d^3\vec{x} e^{-i\vec{p}\cdot\vec{x}} \varphi_{\ell m}(\vec{x}) \tag{3.80}$$

represents the momentum-space wave function of the bound state.

We should only consider the lightest, s-wave muonium states, so $\ell = m = 0$. In this case, only spin states of a muon and an electron would define quantum numbers of the muonium. Using the language of representation theory, $1/2 \times 1/2 = 0 + 1$, i.e. a product of two doublet states makes a singlet (para-muonium) and a triplet (ortho-muonium) states. Then, for the singlet state, the relevant Clebsch-Gordan coefficients are

$$C^{0,0}_{\frac{1}{2},\frac{1}{2};\frac{1}{2},-\frac{1}{2}} = -C^{0,0}_{\frac{1}{2},\frac{1}{2};\frac{1}{2},\frac{1}{2}} = \frac{1}{\sqrt{2}}, \tag{3.81}$$

while for the triplet state

$$\begin{aligned} C^{1,+1}_{\frac{1}{2},\frac{1}{2};\frac{1}{2},\frac{1}{2}} &= 1, \quad C^{1,-1}_{\frac{1}{2},-\frac{1}{2};\frac{1}{2},-\frac{1}{2}} = 1, \\ C^{1,+1}_{\frac{1}{2},\frac{1}{2};\frac{1}{2},-\frac{1}{2}} &= \frac{1}{\sqrt{2}}, \quad C^{0,0}_{\frac{1}{2},\frac{1}{2};\frac{1}{2},\frac{1}{2}} = \frac{1}{\sqrt{2}}, \end{aligned} \tag{3.82}$$

As an example, let us compute the first of the matrix elements defined in Eq. (3.78); all other matrix elements can be computed in a similar way. The spin-singlet muonium state can be then written as

$$\begin{aligned} |^1S_0; 0\rangle &= \sqrt{2M} \int \frac{d^3\mathbf{p}}{(2\pi)^3} \widetilde{\varphi}_{00}(\mathbf{p}) \sqrt{\frac{2E_\mathbf{p}}{2m_\mu}} \sqrt{\frac{2E_{-\mathbf{p}}}{2m_e}} \\ &\times \frac{1}{\sqrt{2}} \left[a^\dagger \left(\frac{1}{2}, \mathbf{p}\right) b^\dagger \left(-\frac{1}{2}, -\mathbf{p}\right) - a^\dagger \left(-\frac{1}{2}, \mathbf{p}\right) b^\dagger \left(\frac{1}{2}, -\mathbf{p}\right) \right] |0\rangle. \end{aligned} \tag{3.83}$$

We expand each electron and muon field in the operator of Eq. (3.78) as

$$\psi(x) = \int \frac{d^3p}{(2\pi)^3} \frac{1}{\sqrt{2E_\mathbf{p}}} \sum_s \left(a^s_\mathbf{p} u^s(p) e^{-ipx} + b^{s\dagger}_\mathbf{p} v^s(p) e^{ipx} \right), \tag{3.84}$$

The matrix element of Eq. (3.78) can be written, albeit in a somewhat cumbersome form,

$$\begin{aligned} \langle 0 | \overline{\mu} \gamma_\mu \gamma_5 e | M_\mu \rangle &= \sqrt{\frac{2M_\mu}{4m_\mu m_e}} \frac{1}{\sqrt{2}} \int \frac{d^2\mathbf{p}}{(2\pi)^3} \widetilde{\varphi}(\mathbf{p}) \sum_{s,r} \int \frac{d^2\mathbf{k}_1}{(2\pi)^3 \sqrt{2E_{k_1}}} \frac{d^2\mathbf{k}_2}{(2\pi)^3 \sqrt{2E_{k_2}}} \\ &\times \overline{v}(s, \mathbf{k}_1) \gamma_\mu \gamma_5 u(r, \mathbf{k}_2)\, e^{-i(k_1+k_2)x} \langle 0 | b(s, \mathbf{k}_1) a(r, \mathbf{k}_2) \\ &\times \left[a^\dagger \left(\frac{1}{2}, \mathbf{p}\right) b^\dagger \left(-\frac{1}{2}, -\mathbf{p}\right) - a^\dagger \left(-\frac{1}{2}, \mathbf{p}\right) b^\dagger \left(\frac{1}{2}, -\mathbf{p}\right) \right] |0\rangle \end{aligned} \tag{3.85}$$

We will work in the non-relativistic approximation and neglect the momentum dependence of the spinors. The spinors are then defined as

$$\begin{aligned} u &= \sqrt{m_e} \begin{pmatrix} \xi \\ \xi \end{pmatrix}, \qquad v = \sqrt{m_e} \begin{pmatrix} \eta \\ -\eta \end{pmatrix}, \\ \overline{u} &= \sqrt{m_\mu} \left(\xi^\dagger, \xi^\dagger \right) \gamma^0, \qquad \overline{v} = \sqrt{m_\mu} \left(\eta^\dagger, -\eta^\dagger \right) \gamma^0, \end{aligned} \tag{3.86}$$

where ξ and η are the two-component spinors [147]. We will denote the spin-up and spin-down spinors as $\xi_\pm(\eta_\pm)$. We will also be employing the Weyl basis for the gamma matrices,

$$\gamma^0 = \begin{pmatrix} 0 & 1 \\ 1 & 0 \end{pmatrix}, \qquad \gamma^\alpha = \begin{pmatrix} 0 & \sigma^\alpha \\ \overline{\sigma}^\alpha & 0 \end{pmatrix}, \qquad \gamma^5 = \begin{pmatrix} -1 & 0 \\ 0 & 1 \end{pmatrix}, \tag{3.87}$$

where σ^α and $\overline{\sigma}^\alpha$ are defined as

$$\sigma^\alpha = (\mathbf{1}, \vec{\sigma}), \qquad \overline{\sigma}^\alpha = (\mathbf{1}, -\vec{\sigma}). \tag{3.88}$$

Note that $\vec{\sigma}$ is a vector comprised of the Pauli matrices, and $\mathbf{1}$ is the 2×2 identity matrix.

We can now compute the matrix element of Eq. (3.85). Using the anti-commutation relations $\{a(r,\mathbf{p}), a^\dagger(s,\mathbf{p}') = (2\pi)^3\delta^3(\mathbf{p}-\mathbf{p}')\delta^{rs}$ and $\{b(r,\mathbf{p}), b^\dagger(s,\mathbf{p}') = (2\pi)^3\delta^3(\mathbf{p}-\mathbf{p}')\delta^{rs}$, and the relation

$$\begin{aligned}
\overline{v}(r,\mathbf{k}_1)\Gamma u(s,\mathbf{k}_2) &= \text{Tr}\ [\Gamma_\mu u(s,\mathbf{k}_2)\overline{v}(r,\mathbf{k}_1)] \\
&= \frac{1}{\sqrt{2}}\sqrt{m_\mu m_e}\ \text{Tr}\ \left[(\sigma^\mu + \overline{\sigma}^\mu)\left(\xi_-\eta_+^\dagger - \xi_+\eta_-^\dagger\right)\right] \\
&= \frac{1}{\sqrt{2}}\sqrt{m_\mu m_e}\ \text{Tr}\ [(\sigma^\mu + \overline{\sigma}^\mu)],
\end{aligned} \tag{3.89}$$

where in this calculation $\Gamma = \gamma_\mu\gamma_5$, which we used in the second line of Eq. (3.89). As we can see, the combination of two-component spinors amounted to a unit matrix for the singlet $^1S_0$ state. The result will be different for the triplet $^3S_1$ state, requiring starting the computation anew.

This formulation is obviously cumbersome. Not all is lost, however! According to [147], there is a possible simplification: instead of picking the initial state as in Eq. (3.83), a simpler state can be chosen as

$$|M_\mu\rangle = \sqrt{\frac{2M_M}{2m_\mu 2m_e}}\int\frac{d^3p}{(2\pi)^3}\tilde{\psi}(p)\,|p,p'\rangle, \tag{3.90}$$

as long as we remember to project the spinors onto the singlet (spin-0) or the triplet (spin-1) states via the substitutions [147],

$$\xi\eta^\dagger = \frac{1}{\sqrt{2}}\mathbf{1}_{2\times 2} \tag{3.91}$$

for the spin-0 state and similarly

$$\xi\eta^\dagger = \frac{1}{\sqrt{2}}\vec{\epsilon}^{\,*}\cdot\vec{\sigma} \tag{3.92}$$

for the spin-1 state with three possible polarization states, $\vec{\epsilon}_1 = (0,0,1)$, $\vec{\epsilon}_2 = \frac{1}{\sqrt{2}}(1,i,0)$, and $\vec{\epsilon}_3 = \frac{1}{\sqrt{2}}(1,-i,0)$. It is convenient to introduce polarization four-vector, $\epsilon_\nu^* = (0,\vec{\epsilon^*})$.

Finally, we get for the matrix element of Eq. (3.85),

$$\begin{aligned}
\langle 0|\,\overline{\mu}\gamma_\mu\gamma_5 e\,|M_\mu\rangle &= \frac{\sqrt{M_\mu}}{2}\int\frac{d^2\mathbf{p}}{(2\pi)^3}\widetilde{\varphi}(\mathbf{p})e^{-i(E_p+E_{-p})x^0}\ \text{Tr}\left[(\sigma^\mu + \overline{\sigma}^\mu)\right] \\
&= \frac{\sqrt{M_\mu}}{2}\,|\varphi(0)|\ \ \text{Tr}\left[(\sigma^\mu + \overline{\sigma}^\mu)\right],
\end{aligned} \tag{3.93}$$

which implies that a decay constant f_M can be written in terms of the bound-state wave function,

$$f_M^2 = 4\frac{|\varphi(0)|^2}{M_M}, \tag{3.94}$$

which is the QED's version of the Van Royen-Weisskopf formula. For a Coulombic bound state, the wave function of the ground state is

$$\varphi(r) = \frac{1}{\sqrt{\pi a_{M_\mu}^3}}e^{-\frac{r}{a_{M_\mu}}}, \tag{3.95}$$

where $a_{\text{M}_\mu} = (\alpha m_{\text{red}})^{-1}$ is the muonium Bohr radius, α is the fine structure constant, and $m_{red} = m_e m_\mu/(m_e + m_\mu)$ is the reduced mass. Then,

$$|\varphi(0)|^2 = \frac{(m_{red}\alpha)^3}{\pi} = \frac{1}{\pi}(m_{red}\alpha)^3. \tag{3.96}$$

Before we proceed, we note that this method allows computations of more complicated matrix elements as well. For example, for the matrix element of the four-quark operator

$$\langle Q_1 \rangle = \langle \overline{M}_\mu | \left(\overline{\mu}\gamma_\alpha P_L e\right)\left(\overline{\mu}\gamma^\alpha P_L e\right) | M_\mu \rangle \tag{3.97}$$

for both pseudoscalar and vector muonium states one can obtain

$$\langle \bar{M}_\mu^P | Q_1 | M_\mu^P \rangle = 4M_\mu |\varphi(0)|^2, \qquad \langle \bar{M}_\mu^V | Q_1 | M_\mu^V \rangle = -12 M_\mu |\varphi(0)|^2. \tag{3.98}$$

These formulas will be useful for us in the next section, where we discuss muonium-antimuonium oscillations.

Invisible decays of muonium. Studies of FCNC decays of the muonium state can be a useful tool in searches for New Physics. While most of the FCNC decays of muonia have very small backgrounds from SM processes, its invisible decay has a flavor-conserving tree-level SM contribution $M_\mu \to \bar{\nu}_\mu \nu_e$ that dominates the invisible width*.

How would one experimentally study invisible decays of muonium states? One way to do that is to do it indirectly, by counting all visible muonium decay channels. If the NP contribution to $M_\mu \to$ invisible is appreciably large, it would contribute to the total muonium decay width Γ_M,

$$\Gamma_M \equiv \Gamma(M_\mu \to \text{all}) = \Gamma(\mu \to \text{all}) + \Gamma(M_\mu \to \text{invisible}) + \ldots \tag{3.99}$$

and hence decrease muonium lifetime τ_M. One complication that is encountered in this approach is related to the fact that the dominant component of the total muonium width, the muon's width Γ_μ would also be affected by the same NP physics as $M_\mu \to$ invisible via $\mu \to e+$invisible decays. This makes the separation of terms in Eq. (3.99) ambiguous. Studies of invisible widths of para- vs ortho-muonium would help to separate exotic NP contributions*.

It is possible to derive a two-neutrino decay of a muonium state in the Standard Model. Squaring up the amplitude from Eq. (3.77), and averaging over initial polarizations of the muonium state, we arrive at the result,

$$\Gamma(M_\mu \to \nu_e \bar{\nu}_\mu) = 48\pi \left(\frac{\alpha m_e}{m_\mu}\right)^3 \Gamma(\mu \to e \nu_e \bar{\nu}_\mu), \tag{3.100}$$

which numerically leads to the branching ratio $\mathcal{B}(M_\mu \to \nu_e \bar{\nu}_\mu) = 6.6 \times \times 10^{-12}$, which is a very small number.

3.3.5 Muonium-antimuonium oscillations

A $\Delta L = 2$ interaction could change the muonium state into the anti-muonium one, leading to the possibility of muonium-anti-muonium oscillations [78, 153]. As a variety of well-established models of New Physics contain $\Delta L = 2$ interaction terms, observation of muonium conversion into anti-muonium could then provide especially clean probes of New Physics in the leptonic sector. It would be useful to perform a model-independent computation of the oscillation parameters using techniques of the effective theory that includes all possible BSM models encoded in a few Wilson coefficients of effective operators. Also, employing effective field theory techniques for computation of the contributions that are non-local at the muon mass scale, we present those contributions in terms of the series of local operators expanded in inverse powers of m_μ [53].

---

*Strictly speaking, this is only true for the ortho-muonium (spin-1) state. See the following footnote.

*In the Standard Model, para-muonium decays to two neutrinos, as are decays to any fermion-antifermion pairs, are helicity-suppressed, which makes its invisible width to be dominated by numerically negligible four-neutrino decay mode. However, its decays to two-scalar final states are not suppressed.

The phenomenology of $M_\mu - \overline{M}_\mu$ oscillations is very similar to the phenomenology of meson-antimeson oscillations, which we will discuss in Chapter 4. There are, however, several important differences that we will emphasize below. One major difference is related to the fact that both ortho- and para-muonium can, in principle, oscillate. It is also interesting to note that muonium oscillations could occur both via the BSM (heavy) states, and via the SM light states. Since such states can go on their mass shells, these contributions would lead to a possibility of the lifetime difference in the $M_\mu - \overline{M}_\mu$ system.

If the New Physics Lagrangian includes $\Delta L = 2$ lepton-flavor violating interactions, the time development of muonium and anti-muonium states would be coupled, so it would be appropriate to consider their combined evolution,

$$|\psi(t)\rangle = \left(\begin{array}{c} a(t) \\ b(t) \end{array} \right) = a(t)|M_\mu\rangle + b(t)|\overline{M_\mu}\rangle. \tag{3.101}$$

The time evolution of $|\psi(t)\rangle$ evolution is governed by a Schrödinger equation,

$$i\frac{d}{dt}\begin{pmatrix} |M(t)\rangle \\ |\overline{M}(t)\rangle \end{pmatrix} = \left(m - i\frac{\Gamma}{2}\right)\begin{pmatrix} |M(t)\rangle \\ |\overline{M}(t)\rangle \end{pmatrix}. \tag{3.102}$$

CPT-invariance dictates that the masses and widths of muonium and anti-muonium are the same, $m_{11} = m_{22}$, $\Gamma_{11} = \Gamma_{22}$, while CP-invariance of the $\Delta L_\mu = 2$ interaction, which we assume for simplicity, dictates that

$$m_{12} = m_{21}^*, \qquad \Gamma_{12} = \Gamma_{21}^*. \tag{3.103}$$

As we can see, the Hamiltonian in Eq. (3.102) is not Hermitian, which is a consequence of the fact that the states can decay. The presence of off-diagonal pieces in the mass matrix signals that it needs to be diagonalized. The mass eigenstates $|M_{\mu_{1,2}}\rangle$ can be defined as

$$|M_{\mu_{1,2}}\rangle = \frac{1}{\sqrt{2}}\left[|M_\mu\rangle \mp |\overline{M}_\mu\rangle,\right] \tag{3.104}$$

where we neglected CP-violation and employed a convention where $CP|M_{\mu_\pm}\rangle = \mp|M_{\mu_\pm}\rangle$. The mass and the width differences of the mass eigenstates are

$$\Delta m \equiv M_1 - M_2, \qquad \Delta\Gamma \equiv \Gamma_2 - \Gamma_1. \tag{3.105}$$

where M_i (Γ_i) are the masses (widths) of the mass eigenstates $|M_{\mu_{1,2}}\rangle$. We defined Δm and $\Delta\Gamma$ to be either positive or negative, which is to be determined by experiment. It is often convenient to introduce dimensionless quantities,

$$x = \frac{\Delta m}{\Gamma}, \qquad y = \frac{\Delta\Gamma}{2\Gamma}, \tag{3.106}$$

where the average lifetime $\Gamma = (\Gamma_1 + \Gamma_2)/2$. It is important to note that while Γ is defined by the standard model decay rate of the muon, x and y are driven by the lepton-flavor violating interactions. It is then expected that both $x, y \ll 1$.

The time evolution of flavor eigenstates follows from Eq. (3.102),

$$\begin{aligned} |M(t)\rangle &= g_+(t)\,|M_\mu\rangle + g_-(t)\,|\overline{M}_\mu\rangle, \\ |\overline{M}(t)\rangle &= g_-(t)\,|M_\mu\rangle + g_+(t)\,|\overline{M}_\mu\rangle, \end{aligned} \tag{3.107}$$

where the coefficients $g_\pm(t)$ are defined as

$$g_\pm(t) = \frac{1}{2}e^{-\Gamma_1 t/2}e^{-im_1 t}\left[1 \pm e^{\Delta\Gamma t/2}e^{i\Delta m t}\right]. \tag{3.108}$$

As $x, y \ll 1$ we can expand Eq. (3.108) to get

$$\begin{aligned} g_+(t) &= e^{-\Gamma_1 t/2} e^{-i m_1 t} \left[1 + \frac{1}{8} (y - ix)^2 (\Gamma t)^2 \right], \\ g_-(t) &= \frac{1}{2} e^{-\Gamma_1 t/2} e^{-i m_1 t} (y - ix) (\Gamma t). \end{aligned} \quad (3.109)$$

Denoting an amplitude for the muonium decay into a final state f as $A_f = \langle f | \mathcal{H} | M_\mu \rangle$ and an amplitude for its decay into a CP-conjugated final state $\overline{f}$ as $A_{\bar{f}} = \langle \overline{f} | \mathcal{H} | M_\mu \rangle$, we can write the time-dependent decay rate of M_μ into the $\overline{f}$,

$$\Gamma(M_\mu \to \overline{f})(t) = \frac{1}{2} N_f \left| A_f \right|^2 e^{-\Gamma t} (\Gamma t)^2 R_M(x, y), \quad (3.110)$$

where N_f is a phase-space factor and $R_M(x, y)$ is the oscillation rate,

$$R_M(x, y) = \frac{1}{2} \left(x^2 + y^2 \right). \quad (3.111)$$

Integrating over time and normalizing to $\Gamma(M_\mu \to f)$ we get the probability of M_μ decaying as $\overline{M}_\mu$ at some time $t > 0$,

$$P(M_\mu \to \overline{M}_\mu) = \frac{\Gamma(M_\mu \to \overline{f})}{\Gamma(M_\mu \to f)} = R_M(x, y). \quad (3.112)$$

This equation gives the oscillation probability. Note that it depends on both the normalized mass x and the lifetime y differences. We will compute those parameters in the next section.

Effective operators and matrix elements

Muonium-anti-muonium oscillations could be an effective probe of flavor-violating New Physics in leptons. One of the issues is that at this point we do not know which particular model of New Physics will provide the correct ultraviolet (UV) extension for the Standard Model. However, since the muonium mass is most likely much smaller than the new particle masses, it is not necessary to know it. Any New Physics scenario which involves lepton flavor violating interactions can be matched to an effective Lagrangian, $\mathcal{L}_{\text{eff}}$, whose Wilson coefficients would be determined by the UV physics that becomes active at some scale Λ [106, 149],

$$\mathcal{L}_{\text{eff}} = -\frac{1}{\Lambda^2} \sum_i c_i(\mu) Q_i, \quad (3.113)$$

where the c_i's are the short distance Wilson coefficients. They encode all BSM model-specific information. Q_i's are the effective operators which reflect degrees of freedom relevant at the scale at which a given process takes place. If we assume that no new light particles (such as "dark photons" or axions) exist in the low energy spectrum, those operators would be written entirely in terms of the SM degrees of freedom. In the case at hand, all SM particles with masses larger than that of the muon should also be integrated out, leaving only muons, electrons, photons, and neutrinos.

It would be convenient for us to classify effective operators in Eq. (3.113) by their lepton quantum numbers. In particular, we can write the effective Lagrangian as

$$\mathcal{L}_{\text{eff}} = \mathcal{L}_{\text{eff}}^{\Delta L_\mu = 0} + \mathcal{L}_{\text{eff}}^{\Delta L_\mu = 1} + \mathcal{L}_{\text{eff}}^{\Delta L_\mu = 2} \quad (3.114)$$

The first term in this expansion contains both the Standard Model and the New Physics contributions. It then follows that the leading term in $\mathcal{L}_{\text{eff}}^{\Delta L_\mu = 0}$ is suppressed by powers of M_W, not the New Physics scale Λ. We should emphasize that only the operators that are local at the scale of the muonium mass are retained in Eq. (3.114).

The second term contains $\Delta L_\mu = 1$ operators. As we integrated out all heavy degrees of freedom, the operators of the lowest possible dimension that govern muonium oscillations must be of dimension six. The most general dimension six effective Lagrangian, $\mathcal{L}_{\text{eff}}^{\Delta L_\mu=1}$, has the form spelled out in Eq. (3.52). Note that the Lagrangian Eq. (3.52) also contains terms that do not follow from the dimension six operators in the SM EFT of Chapter 2, but could be effectively generated by nominally higher dimensional SM EFT operators [149]. This is taken into account by introducing mass and G_F factors emulating such suppression.

The last term in Eq. (3.114), $\mathcal{L}_{\text{eff}}^{\Delta L_\mu=2}$, represents the effective operators changing the lepton quantum number by two units. The leading contribution to muonium oscillations is given by the dimension six operators. The most general effective Lagrangian

$$\mathcal{L}_{\text{eff}}^{\Delta L_\mu=2} = -\frac{1}{\Lambda^2}\sum_i C_i^{\Delta L=2}(\mu) Q_i(\mu). \tag{3.115}$$

can be written with the operators written entirely in terms of the muon and electron degrees of freedom,

$$\begin{aligned} Q_1 &= (\overline{\mu}_L \gamma_\alpha e_L)(\overline{\mu}_L \gamma^\alpha e_L), \quad Q_2 = (\overline{\mu}_R \gamma_\alpha e_R)(\overline{\mu}_R \gamma^\alpha e_R), \\ Q_3 &= (\overline{\mu}_L \gamma_\alpha e_L)(\overline{\mu}_R \gamma^\alpha e_R), \quad Q_4 = (\overline{\mu}_L e_R)(\overline{\mu}_L e_R), \\ Q_5 &= (\overline{\mu}_R e_L)(\overline{\mu}_R e_L). \end{aligned} \tag{3.116}$$

We did not include operators that could be related to the presented ones via Fierz relations (see Appendix, Section A.4). There are other $\Delta L_\mu = 2$ local operators that are relevant for the computation of the lifetime difference y. They can be written as

$$Q_6 = (\overline{\mu}_L \gamma_\alpha e_L)\left(\overline{\nu_\mu}_L \gamma^\alpha \nu_{eL}\right), \quad Q_7 = (\overline{\mu}_R \gamma_\alpha e_R)\left(\overline{\nu_\mu}_L \gamma^\alpha \nu_{eL}\right), \tag{3.117}$$

where we only included SM EFT operators that contain left-handed neutrinos [149]. In order to see how these operators (and thus New Physics) contribute to the mixing parameters, it is instructive to consider off-diagonal terms in the mass matrix,

$$\left(m - \frac{i}{2}\Gamma\right)_{12} = \frac{1}{2M_M}\left\langle \overline{M}_\mu \left|\mathcal{H}_{\text{eff}}\right| M_\mu\right\rangle + \frac{1}{2M_M}\sum_n \frac{\left\langle \overline{M}_\mu \left|\mathcal{H}_{\text{eff}}\right| n\right\rangle \left\langle n \left|\mathcal{H}_{\text{eff}}\right| M_\mu\right\rangle}{M_M - E_n + i\epsilon}, \tag{3.118}$$

where the first term does not contain an imaginary part, so it contributes only to m_{12}, and, eventually, to the mass difference. The second term contains bi-local contributions connected by physical intermediate states. This term has both real and imaginary parts and thus contributes to both m_{12} and Γ_{12}.

We can rewrite Eq. (3.118) to extract the physical mixing parameters x and y of Eq. (3.106). For the mass difference,

$$x = \frac{1}{2M_M\Gamma}\text{Re}\left[2\langle \overline{M}_\mu \left|\mathcal{H}_{\text{eff}}\right| M_\mu\rangle + \langle \overline{M}_\mu| i \int d^4x \; \text{T}\left[\mathcal{H}_{\text{eff}}(x)\mathcal{H}_{\text{eff}}(0)\right] |M_\mu\rangle\right] \tag{3.119}$$

Assuming the LFV NP is present, the dominant local contribution to x comes from the last term in Eq. (3.114),

$$\langle \overline{M}_\mu | \mathcal{H}_{\text{eff}} | M_\mu\rangle = \langle \overline{M}_\mu | \mathcal{H}_{\text{eff}}^{\Delta L_\mu=2} | M_\mu\rangle \tag{3.120}$$

provided that only $Q_1 - Q_5$ operators are taken into account. It is easy to see that the relevant contributions are only suppressed by Λ^2. Other contributions, including the non-local double insertions of $\mathcal{L}_{\text{eff}}^{\Delta L_\mu=1}$, represented by the second term in Eq. (3.119), do contribute to the mass difference but are naively suppressed by Λ^4, i.e. of higher order in $1/\Lambda$.

Matrix elements of four-fermion operators. In order to obtain the mass difference contribution, we need to evaluate the matrix elements of the operators in Eqs. (3.116) and (3.117). Since we expect that both spin-0 singlet and spin-1 triplet muonium states would undergo oscillations, the oscillation parameters would in general be different, as the matrix elements would differ for those two cases.

One way to proceed is to employ a method of vacuum saturation or factorization. In this approach, the matrix elements can be written in terms of the muonium decay constant f_M. In the non-relativistic limit factorization gives the exact result for the QED matrix elements of the six-fermion operators. Let us describe how this approach works. Consider first the current-current matrix element

$$\langle \bar{M}_\mu^P | Q_1 | M_\mu^P \rangle = \langle \bar{M}_\mu^P | (\bar{\mu}_L \gamma_\alpha e_L)_1 (\bar{\mu}_L \gamma^\alpha e_L)_2 | M_\mu^P \rangle, \tag{3.121}$$

where we (temporarily) labeled the currents with subscripts 1 and 2 for the convenience of explanation. The vacuum saturation prescription tells us that the four-fermion matrix element can be computed by inserting a complete set of states $1 = \sum_I |I\rangle\langle I|$ in between the currents 1 and 2 and then discarding all contributions except for the vacuum one,

$$\begin{aligned}
\langle \bar{M}_\mu^P | (\bar{\mu}_L \gamma_\alpha e_L)_1 (\bar{\mu}_L \gamma^\alpha e_L)_2 | M_\mu^P \rangle &\to \sum_I \langle \bar{M}_\mu^P | (\bar{\mu}_L \gamma_\alpha e_L)_1 |I\rangle \langle I| (\bar{\mu}_L \gamma^\alpha e_L)_2 | M_\mu^P \rangle \\
&\to \langle \bar{M}_\mu^P | (\bar{\mu}_L \gamma_\alpha e_L)_1 |0\rangle \langle 0| (\bar{\mu}_L \gamma^\alpha e_L)_2 | M_\mu^P \rangle \quad (3.122) \\
&= \frac{1}{4} \langle \bar{M}_\mu^P | \bar{\mu} \gamma_\alpha \gamma_5 e |0\rangle \langle 0| \bar{\mu} \gamma^\alpha \gamma_5 e | M_\mu^P \rangle
\end{aligned}$$

We can now use the definitions of matrix elements between the vacuum and the M_μ state of Eq. (3.78) to express it in terms of the decay constants and some mass factors.

It is, however, clear that Eq. (3.122) does not encompass all possible contributions to the matrix element of a four-fermion operator. For example, since the muon field operator μ in the current 1 in Eq. (3.121) contains both creation and annihilation operators, it can not only act on the $\langle \bar{M}_\mu^P |$ state, but also on the $| M_\mu^P \rangle$, which is not possible after current separation in Eq. (3.122). The same is true for the electron field e in the same current 1. Moreover, the muon and electron fields can, in principle, act on different states: μ can act on $\langle \bar{M}_\mu^P |$, while e can act on $| M_\mu^P \rangle$, which is also not represented in Eq. (3.122)!

To account for this situation in the framework of vacuum saturation, it is convenient to apply Fierz rearrangement formulas (see Appendix A.4). For example, for the operator Q_1 under consideration,

$$(\bar{\mu}_L \gamma_\mu e_L)(\bar{\mu}'_L \gamma^\mu e'_L) = (\bar{\mu}_L \gamma_\mu e'_L)(\bar{\mu}'_L \gamma^\mu e_L), \tag{3.123}$$

where we again, temporarily, introduced primes to the second current. Fierz constructions for the operators other than those with left-left structure are more complicated and given in Appendix A.4. All points in the discussion above also apply to the fields in the current 2. Thus, a proper account of creations and annihilation operators in the factorization of currents would include the following

$$\begin{aligned}
\langle \bar{M}_\mu^P | Q_1 | M_\mu^P \rangle &= 2 \langle \bar{M}_\mu^P | \bar{\mu}_L \gamma_\alpha e_L |0\rangle \langle 0| \bar{\mu}_L \gamma^\alpha e_L | M_\mu^P \rangle \\
&+ 2 \langle \bar{M}_\mu^P | \bar{\mu}_L \gamma_\alpha e_L |0\rangle \langle 0| \bar{\mu}_L \gamma^\alpha e_L | M_\mu^P \rangle, \quad (3.124)
\end{aligned}$$

where the factor of two in the first line represents swapping the currents 1 and 2 and the second line gives the Fierzed contribution. Now, writing matrix elements in Eq. (3.124) using Eq. (3.78), we arrive at the following result,

$$\langle \bar{M}_\mu^P | Q_1 | M_\mu^P \rangle = \frac{2 \times 2}{4} (-i) i f_M^2 p_\mu p^\mu = f_M^2 M_M^2 B_1, \tag{3.125}$$

where we introduced a factor B_1 to account* for anything we did not take into account in this prescription.

How well does this procedure work? That is not such an easy question to answer. It is clear that the renormalization properties of a local four-fermion operator are different from those properties of a product of two currents. But it should at least work at some momentum scale. Then, we do not a priori know that all contributions of higher states are negligible. Thus, to address these questions, the right-hand part of Eq. (3.125) is multiplied by the factor B_1. We will drop this factor in muonium matrix elements because in the nonrelativistic limit we are employing here $B_1 = 1$, which can be checked by direct computation. Matrix elements of other operators can be computed in the same way and are presented below for both para- and ortho-muonium states. We will be using the same technique to parameterize four-quark matrix elements relevant for the computation of meson-antimeson oscillations.

Para-muonium. The matrix elements of the spin-singlet states can be obtained from Eq. (3.116) using the definitions of Eq. (3.78) [53],

$$\begin{aligned}
\langle \bar{M}_\mu^P | Q_1 | M_\mu^P \rangle &= f_M^2 M_M^2, \quad \langle \bar{M}_\mu^P | Q_2 | M_\mu^P \rangle = f_M^2 M_M^2, \\
\langle \bar{M}_\mu^P | Q_3 | M_\mu^P \rangle &= -\frac{3}{2} f_M^2 M_M^2, \quad \langle \bar{M}_\mu^P | Q_4 | M_\mu^P \rangle = -\frac{1}{4} f_M^2 M_M^2, \\
\langle \bar{M}_\mu^P | Q_5 | M_\mu^P \rangle &= -\frac{1}{4} f_M^2 M_M^2.
\end{aligned} \tag{3.126}$$

Combining the contributions from the different operators and using the definition from Eqs. (3.94), we obtain an expression for x_P for the para-muonium state,

$$x_P = \frac{4(m_{red}\alpha)^3}{\pi \Lambda^2 \Gamma} \left[C_1^{\Delta L=2} + C_2^{\Delta L=2} - \frac{3}{2} C_3^{\Delta L=2} - \frac{1}{4}\left(C_4^{\Delta L=2} + C_5^{\Delta L=2} \right) \right]. \tag{3.127}$$

This result is universal and holds true for any New Physics model that can be matched into a set of local $\Delta L = 2$ interactions.

Ortho-muonium. Using the same procedure, but computing the relevant matrix elements for the vector otrtho-muonium state, we obtain the matrix elements [53]

$$\begin{aligned}
\langle \bar{M}_\mu^V | Q_1 | M_\mu^V \rangle &= -3 f_M^2 M_M^2, \quad \langle \bar{M}_\mu^V | Q_2 | M_\mu^V \rangle = -3 f_M^2 M_M^2, \\
\langle \bar{M}_\mu^V | Q_3 | M_\mu^V \rangle &= -\frac{3}{2} f_M^2 M_M^2, \quad \langle \bar{M}_\mu^V | Q_4 | M_\mu^V \rangle = -\frac{3}{4} f_M^2 M_M^2, \\
\langle \bar{M}_\mu^V | Q_5 | M_\mu^V \rangle &= -\frac{3}{4} f_M^2 M_M^2.
\end{aligned} \tag{3.128}$$

Again, combining the contributions from the different operators, we obtain an expression for x_V for the ortho-muonium state,

$$x_V = -\frac{12(m_{red}\alpha)^3}{\pi \Lambda^2 \Gamma} \left[C_1^{\Delta L=2} + C_2^{\Delta L=2} + \frac{1}{2} C_3^{\Delta L=2} + \frac{1}{4}\left(C_4^{\Delta L=2} + C_5^{\Delta L=2} \right) \right]. \tag{3.129}$$

Again, this result is universal and holds true for any New Physics model that can be matched into a set of local $\Delta L = 2$ interactions.

It might be instructive to present an example of a BSM model that can be matched into the effective Lagrangian of Eq. (3.115) and can be constrained from Eqs. (3.126,3.128). Let us

*The compensating factors B_i, where i refers to the number of an operator in Eq. (3.116), are oftentimes called *bag factors*. The origin of this name lies in the late 1970s when such matrix elements were computed in the MIT-bag model.

consider a model which contains a doubly-charged Higgs boson. Such states often appear in the context of left-right models. A coupling of the doubly charged Higgs field Δ^{--} to the lepton fields can be written as

$$\mathcal{L}_R = g_{\ell\ell}\bar{\ell}_R\ell^c + H.c., \tag{3.130}$$

where $\ell^c = C\bar{\ell}^T$ is the charge-conjugated lepton state. Integrating out the Δ^{--} field, this Lagrangian leads to the following effective Hamiltonian,

$$\mathcal{H}_\Delta = \frac{g_{ee}g_{\mu\mu}}{2M_\Delta^2}\left(\overline{\mu}_R\gamma_\alpha e_R\right)\left(\overline{\mu}_R\gamma^\alpha e_R\right) + H.c., \tag{3.131}$$

below the scales associated with the doubly-charged Higgs field's mass M_Δ. Examining Eq. (3.131) we see that this Hamiltonian matches onto our operator Q_2 (see Eq. (3.116)) with the scale $\Lambda = M_\Delta$ and the corresponding Wilson coefficient $C_2^{\Delta L=2} = g_{ee}g_{\mu\mu}/2$.

The lifetime difference in the muonium system can be obtained from Eq. (3.118). It comes from the physical intermediate states, which is signified by the imaginary part in Eq. (3.118) and reads,

$$y = \frac{1}{\Gamma}\sum_n \rho_n \left\langle \overline{M}_\mu \left| \mathcal{H}_{\text{eff}} \right| n \right\rangle \left\langle n \left| \mathcal{H}_{\text{eff}} \right| M_\mu \right\rangle, \tag{3.132}$$

where ρ_n is a phase space function that corresponds to the intermediate state that is common for M_μ and $\overline{M}_\mu$. There are e^+e^-, $\gamma\gamma$, and $\nu\bar{\nu}$ intermediate states that can contribute to y*. Of those, the e^+e^- intermediate state corresponds to a $\Delta L = 1$ decay $M_\mu \to e^+e^-$, which implies that $\mathcal{H}_{\text{eff}} = \mathcal{H}_{\text{eff}}^{\Delta L_\mu = 1}$ in Eq. (3.132). According to Eq. (3.52), it appears that, quite generally, this contribution is suppressed by Λ^4, i.e. will be much smaller than x, irrespective of the values of the corresponding Wilson coefficients. Similarly, a possible $\gamma\gamma$ intermediate state is generated by higher-dimensional operators and therefore is further suppressed by either powers of Λ or the QED coupling α than other contributions.

The parametrically leading contribution to y comes from the $\nu\bar{\nu}$ intermediate state [53]. This common intermediate state can be reached by the Standard Model tree level decay $M_\mu \to \overline{\nu_\mu}\nu_e$ interfering with the $\Delta L = 2$ decay $\overline{M}_\mu \to \overline{\nu_\mu}\nu_e$. Such contribution is only suppressed by $\Lambda^2 M_W^2$. Writing y similarly to x in Eq. (3.119) as a correlation function of two Hamiltonians at different space-time points we obtain

$$\begin{aligned} y &= \frac{1}{2M_M\Gamma}\text{Im}\left[\langle\overline{M}_\mu|i\int d^4x\ \text{T}\left[\mathcal{H}_{\text{eff}}(x)\mathcal{H}_{\text{eff}}(0)\right]|M_\mu\rangle\right] \\ &= \frac{1}{M_M\Gamma}\text{Im}\left[\langle\overline{M}_\mu|i\int d^4x\ \text{T}\left[\mathcal{H}_{\text{eff}}^{\Delta L_\mu=2}(x)\mathcal{H}_{\text{eff}}^{\Delta L_\mu=0}(0)\right]|M_\mu\rangle\right], \end{aligned} \tag{3.133}$$

where the $\mathcal{H}_{\text{eff}}^{\Delta L_\mu=0} = -\mathcal{L}_{\text{eff}}^{\Delta L_\mu=0}$ is given by the Standard Model Lagrangian,

$$\mathcal{L}_{\text{eff}}^{\Delta L_\mu=0} = -\frac{4G_F}{\sqrt{2}}\left(\overline{\mu}_L\gamma_\alpha e_L\right)\left(\overline{\nu}_{eL}\gamma^\alpha\nu_{\mu L}\right), \tag{3.134}$$

and $\mathcal{H}_{\text{eff}}^{\Delta L_\mu=2}$ only contributes through the operators Q_6 and Q_7. Graphically, this procedure is illustrated in Fig. 3.3.

*Of course, intermediate states with four and more particles are also possible, but suppressed by a variety of factors, including phase space and electroweak couplings.

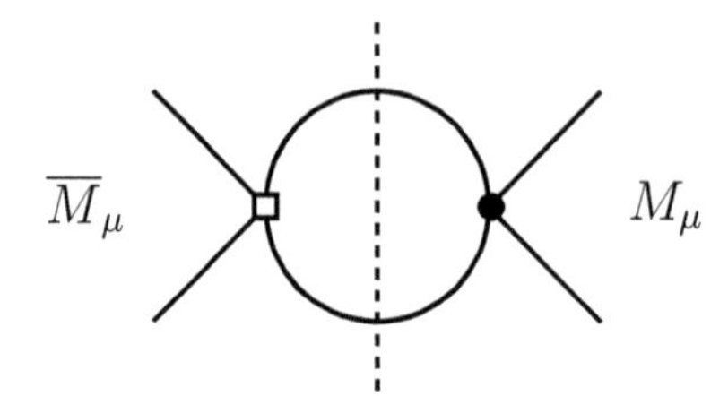

FIGURE 3.3 A contribution to y described in Eq. (3.133). A white square represents a vertex given by Eq. (3.117), while a black dot is given by the SM contribution of Eq. (3.134). A dotted line represents the imaginary part.

Since decaying muon injects a large momentum into the two-neutrino intermediate state, the integral in Eq. (3.133) is dominated by small distance contributions, compared to the scale set by $1/m_\mu$. We can compute the correlation function in Eq. (3.133) by employing a short distance operator product expansion, systematically expanding it in powers of $1/m_\mu$.

$$\begin{aligned} T &= i \int d^4x \, \mathrm{T}\left[\mathcal{H}_{\text{eff}}^{\Delta L_\mu=2}(x)\mathcal{H}_{\text{eff}}^{\Delta L_\mu=0}(0)\right] \\ &= i \int d^4x \, \mathrm{T}\left[(\overline{\mu}\Gamma_\alpha e)\left(\overline{\nu_\mu}_L\gamma^\alpha \nu_{eL}\right)(x)\left(\overline{\mu}\gamma_\beta P_L e\right)\left(\overline{\nu_e}_L\gamma^\beta \nu_{\mu L}\right)(0)\right], \end{aligned} \tag{3.135}$$

To simplify our discussion, let us only consider the case of Dirac neutrinos. The leading term is obtained by contracting the neutrino fields in Eq. (3.135) into propagators,

$$\begin{aligned} \overline{\nu_\mu}(x)\nu_\mu(0) &= iS_F(-x), \\ \nu_e(x)\,\overline{\nu_e}(0) &= iS_F(x), \end{aligned} \tag{3.136}$$

where $S_F(x)$ represents the propagator in coordinate representation.

Using Cutkoski rules (see Appendix, Section A.3) to compute the discontinuity, or imaginary part, of T and calculating the phase space integrals leads to

$$\text{Disc } T = \frac{G_F}{\sqrt{2}\Lambda^2}\frac{M_M^2}{3\pi}\left[C_6^{\Delta L=2}\left(Q_1+Q_5\right)+\frac{1}{2}C_7^{\Delta L=2}Q_3\right]. \tag{3.137}$$

We can now compute the lifetime difference y by using Eq. (3.133) and take the relevant matrix elements for the spin-singlet and the spin-triplet states of the muonium.

Para-muonium. The matrix elements of the spin-singlet state have been computed above and presented in Eq. (3.126). Computing the matrix elements in Eq. (3.133) using their definitions from Eqs. (3.94), we obtain an expression for the lifetime difference y_P for the para-muonium state,

$$y_P = \frac{G_F}{\sqrt{2}\Lambda^2}\frac{M_M^2}{\pi^2\Gamma}(m_{red}\alpha)^3\left(C_6^{\Delta L=2}-C_7^{\Delta L=2}\right). \tag{3.138}$$

It is interesting to note that if $C_6^{\Delta L=2} = C_7^{\Delta L=2}$ current conservation assures that no lifetime difference is generated at this order in $1/\Lambda$ for the para-muonium.

Ortho-muonium. Similarly, using the matrix elements for the spin-triplet state computed in Eq. (3.128), the expression to Eq. (3.137) leads to the lifetime difference

$$y_V = -\frac{G_F}{\sqrt{2}\Lambda^2}\frac{M_M^2}{\pi^2\Gamma}(m_{red}\alpha)^3\left(5C_6^{\Delta L=2}+C_7^{\Delta L=2}\right), \tag{3.139}$$

We emphasize that Eqs. (3.138) and (3.139) represent parametrically leading contributions to muonium lifetime difference, as they are only suppressed by two powers of Λ.

Experimental methods of detection

Studies of muonium decays and muonium-antimuonium oscillations require large quantities of muonia produced by a high-intensity source of muons. Muonia are usually produced following thermalization of a beam of positively charged muons in a variety of gases, liquids, and solids. It is interesting to point out that due to the fact that the dominant channel of muonium decay is a decay of a constituent muon $\mu^+ \to e^+\nu_e\bar{\nu}_\mu$, both muonium $M_\mu = \mu^+e^-$ and antimuonium $\overline{M}_\mu = \mu^-e^+$ have the same composition of the detectable particles in the final state. Indeed the decay channels of $M_\mu \to e^+e^-\nu_e\bar{\nu}_\mu$ and $\overline{M}_\mu \to e^+e^-\nu_\mu\bar{\nu}_e$ differ only by the flavor of neutrinos, which are not detected! This presents an interesting problem for detection of the antimuonium state, which is usually resolved kinematically: in the decays of $\overline{M}_\mu$ an electron is coming from the μ^- decay and therefore, usually has higher energy compared to the spectator positron, whose energy is typically around 13.5 eV. Often the electron and positron are detected by different detectors that are spatially separated, as in e.g. [183]. Major backgrounds include higher-order decays $M_\mu \to e^+e^-e^+e^-\nu_e\bar{\nu}_\mu$.

3.3.6 Lepton-flavor violating meson decays

In this section, we discuss the possibility of extracting Wilson coefficients of the effective Lagrangian in Eq. (3.50) for different ℓ_i from experimental data on quarkonium decays. Since our initial state is a strongly-interacting system, we will seek out the ways to isolate the effects of QCD in such a way that we can still say something useful about LFV.

We shall consider two- and three-body decays of the quarkonia of differing quantum numbers with quarks of various flavors such as $\Upsilon(nS) \to \ell_1\bar{\ell}_2$, $\Upsilon(nS) \to \gamma\ell_1\bar{\ell}_2$, etc. We will note that restricted kinematics of the two-body transitions would allow us to select operators with particular quantum numbers significantly reducing the reliance on the single operator dominance assumption.

We shall provide calculations of the relevant decay rates and establish possible constraints on Wilson coefficients of effective operators of the Lagrangian $\mathcal{L}_{\text{eff}}$ of Eq. (3.50). In the following sections we assume CP-conservation, which implies that all Wilson coefficients will be treated as real numbers. Finally, in studying branching ratios we assume that for a meson, M, the branching fraction $\mathcal{B}(M \to \ell_1\ell_2)$ is the average of $\mathcal{B}(M \to \bar{\ell}_1\ell_2)$ and $\mathcal{B}(M \to \ell_1\bar{\ell}_2)$, unless specified otherwise.

There is abundant experimental information on flavor off-diagonal leptonic decays of vector quarkonia, both from the ground and excited states. This information can be effectively converted to experimental bounds on the Wilson coefficients of vector and tensor operators, as well as on those of the dipole operators of Eq. (3.51). Those Wilson coefficients can then be related to the model parameters of explicit realizations of UV completions of effective Lagrangian in Eq. (3.50).

The most general expression for the $V \to \ell_1\bar{\ell}_2$ decay amplitude can be written as [111]

$$\mathcal{A}(V \to \ell_1\bar{\ell}_2) = \bar{u}(p_1, s_1)\left[A_V^{\ell_1\ell_2}\gamma_\mu + B_V^{\ell_1\ell_2}\gamma_\mu\gamma_5 + \frac{C_V^{\ell_1\ell_2}}{m_V}(p_2 - p_1)_\mu \right. \\ \left. + \frac{iD_V^{\ell_1\ell_2}}{m_V}(p_2 - p_1)_\mu\gamma_5\right] v(p_2, s_2)\ \epsilon^\mu(p). \qquad (3.140)$$

$A_V^{\ell_1\ell_2}$, $B_V^{\ell_1\ell_2}$, $C_V^{\ell_1\ell_2}$, and $D_V^{\ell_1\ell_2}$ are dimensionless constants that depend on the underlying Wilson coefficients of the effective Lagrangian of Eq. (3.50) as well as on hadronic effects associated with meson-to-vacuum matrix elements or decay constants.

The amplitude of Eq. (3.140) leads to the branching fraction, which is convenient to represent in terms of the ratio:

$$\frac{\mathcal{B}(V \to \ell_1 \bar{\ell}_2)}{\mathcal{B}(V \to e^+e^-)} = \left(\frac{m_V\left(1-y^2\right)}{4\pi\alpha f_V Q_q}\right)^2 \left[\left(\left|A_V^{\ell_1\ell_2}\right|^2 + \left|B_V^{\ell_1\ell_2}\right|^2\right) + \frac{1}{2}\left(1-2y^2\right)\left(\left|C_V^{\ell_1\ell_2}\right|^2 + \left|D_V^{\ell_1\ell_2}\right|^2\right)\right.$$
$$\left. + y\ \mathrm{Re}\left(A_V^{\ell_1\ell_2} C_V^{\ell_1\ell_2 *} + i B_V^{\ell_1\ell_2} D_V^{\ell_1\ell_2 *}\right)\right]. \tag{3.141}$$

We have neglected the mass of the lighter of the two leptons and set $y = m_2/m_V$. The form of the coefficients $A_V^{\ell_1\ell_2}$ to $D_V^{\ell_1\ell_2}$ depends on the initial state meson. For example, for $V = \Upsilon(nS)$ ($b\bar{b}$ states), $\psi(nS)$ ($c\bar{c}$ states), or ϕ ($s\bar{s}$ state), the coefficients are:

$$\begin{aligned}
A_V^{\ell_1\ell_2} &= \frac{f_V m_V}{\Lambda^2}\Big[\sqrt{4\pi\alpha}Q_q y^2\left(C_{DL}^{\ell_1\ell_2} + C_{DR}^{\ell_1\ell_2}\right) + \kappa_V\left(C_{VL}^{q\ell_1\ell_2} + C_{VR}^{q\ell_1\ell_2}\right) \\
&\quad + 2y^2\kappa_V\frac{f_V^T}{f_V}G_F m_V m_q\left(C_{TL}^{q\ell_1\ell_2} + C_{TR}^{q\ell_1\ell_2}\right)\Big], \\
B_V^{\ell_1\ell_2} &= \frac{f_V m_V}{\Lambda^2}\Big[-\sqrt{4\pi\alpha}Q_q y^2\left(C_{DL}^{\ell_1\ell_2} - C_{DR}^{\ell_1\ell_2}\right) - \kappa_V\left(C_{VL}^{q\ell_1\ell_2} - C_{VR}^{q\ell_1\ell_2}\right) \\
&\quad - 2y^2\kappa_V\frac{f_V^T}{f_V}G_F m_V m_q\left(C_{TL}^{q\ell_1\ell_2} - C_{TR}^{q\ell_1\ell_2}\right)\Big], \\
C_V^{\ell_1\ell_2} &= \frac{f_V m_V}{\Lambda^2}y\left[\sqrt{4\pi\alpha}Q_q\left(C_{DL}^{\ell_1\ell_2} + C_{DR}^{\ell_1\ell_2}\right) + 2\kappa_V\frac{f_V^T}{f_V}G_F m_V m_q\left(C_{TL}^{q\ell_1\ell_2} + C_{TR}^{q\ell_1\ell_2}\right)\right], \\
D_V^{\ell_1\ell_2} &= i\frac{f_V m_V}{\Lambda^2}y\left[-\sqrt{4\pi\alpha}Q_q\left(C_{DL}^{\ell_1\ell_2} - C_{DR}^{\ell_1\ell_2}\right) - 2\kappa_V\frac{f_V^T}{f_V}G_F m_V m_q\left(C_{TL}^{q\ell_1\ell_2} - C_{TR}^{q\ell_1\ell_2}\right)\right].
\end{aligned} \tag{3.142}$$

Here $Q_q = (2/3, -1/3)$ is the charge of the quark q and $\kappa_V = 1/2$ is a constant for pure $q\bar{q}$ states. It is a good approximation to drop terms proportional to y^2 in Eq. (3.142) for the heavy quarkonium states. Inspecting the ratio in Eq. (3.141), one immediately infers that the best constraints could be placed on the four-fermion coefficients, $C_{VL}^{q\ell_1\ell_2}$ and $C_{VR}^{q\ell_1\ell_2}$, as no final state lepton mass suppression exists for those coefficients. Constraints on the dipole coefficients, $C_{DL}^{\ell_1\ell_2}(C_{DR}^{\ell_1\ell_2})$, are also possible in this case. The NP constraints here are complementary to those obtained from the lepton decay experiments, especially for $\ell = \tau$, obtained in the radiative $\tau \to \mu(e)\gamma$ decays.

The constraints on the Wilson coefficients of tensor operators $C_{TL}^{q\ell_1\ell_2}(C_{TR}^{q\ell_1\ell_2})$, in Eq. (3.142) also depend on the ratio of meson decay constants,

$$\begin{aligned}
\langle 0|\bar{q}\gamma^\mu q|V(p)\rangle &= f_V m_V \epsilon^\mu(p), \\
\langle 0|\bar{q}\sigma^{\mu\nu} q|V(p)\rangle &= i f_V^T\left(\epsilon^\mu p^\nu - p^\mu \epsilon^\nu\right),
\end{aligned} \tag{3.143}$$

where $\epsilon^\mu(p)$ is the V-meson polarization vector and p is its momentum. These definitions are very similar to the ones used in the previous section to parameterize ortho-muonium matrix elements. What is different is that the relation between the decay constant and the wave function at the origin becomes dependent on the details of a particular quark model. We will discuss parameterizations of matrix elements in more detail in Sec. 4.5.1 of Chapter 4.

The decay constants f_V are known both experimentally from leptonic decays and theoretically from lattice or QCD sum rule calculations for a variety of quarkonium states V. The tensor (transverse) decay constant, f_V^T, has been calculated for the charmonium J/ψ state with the result $f_{J/\psi}^T(2\text{ GeV}) = (410 \pm 10)$ MeV. In what follows we assume that $f_V^T = f_V$. Note that the ratio of Eq. (3.141) is largely independent of the values of the decay constants.

Choosing other initial states would make it possible to constrain other combinations of the Wilson coefficients in Eq. (3.50). This is important for the NP models where several LFV operators can contribute, especially in the case where no operator gives an *a priori* dominant contribution. For example, choosing V to be the ρ meson with $\rho \sim \left(u\bar{u} - d\bar{d}\right)/\sqrt{2}$ gives:

$$
\begin{aligned}
A_\rho^{e\mu} &= \frac{f_\rho m_\rho}{\Lambda^2} y^2 \sqrt{2\pi\alpha}\,(Q_u - Q_d)\left(C_{DL}^{\ell_1\ell_2} + C_{DR}^{\ell_1\ell_2}\right), \\
B_\rho^{e\mu} &= -\frac{f_\rho m_\rho}{\Lambda^2} y^2 \sqrt{2\pi\alpha}\,(Q_u - Q_d)\left(C_{DL}^{\ell_1\ell_2} - C_{DR}^{\ell_1\ell_2}\right), \\
C_\rho^{e\mu} &= \frac{f_\rho m_\rho}{\Lambda^2} y \sqrt{2\pi\alpha}\,(Q_u - Q_d)\left(C_{DL}^{\ell_1\ell_2} + C_{DR}^{\ell_1\ell_2}\right), \\
D_\rho^{e\mu} &= -i\frac{f_\rho m_\rho}{\Lambda^2} y \sqrt{2\pi\alpha}\,(Q_u - Q_d)\left(C_{DL}^{\ell_1\ell_2} - C_{DR}^{\ell_1\ell_2}\right).
\end{aligned}
\tag{3.144}
$$

Here we imposed isospin symmetry on the NP operators and their coefficients, which results in the cancellation of the four-fermion operator contribution. The restricted kinematics of the decay implies that only μe operators can be constrained. The corresponding results for $V = \omega \sim \left(u\bar{u} + d\bar{d}\right)/\sqrt{2}$ decay can be obtained from Eq. (3.142) by substituting $Q_q \to (Q_u + Q_d)/\sqrt{2}$ and using $\kappa_\omega = 1/\sqrt{2}$. Again, the restricted kinematics of the decay implies that only μe operators interacting with up and down quarks can be constrained. Since we imposed isospin symmetry, it is convenient to use $m_q = (m_u + m_d)/2$.

LFV meson decays: experimental methods of detection

LFV decays of quarkonium states can be best studied at the flavor factories. The e^+e^- machines can be tuned to produce a particular 1^{--} heavy quarkonium state, while hadronic machines have a larger quarkonium production cross-section. Vector states are abundantly produced in e^+e^- annihilation, so these decays provide a powerful tool to study LFV transitions at flavor factories.

The resonant two-body radiative transitions of vector state $V \to \gamma(M \to \ell_1\bar{\ell}_2)$ could be used to study two-body decays considered above, provided the corresponding branching ratios for the radiative decays are large enough. Since vector states are abundantly produced in e^+e^- annihilation, these decays could provide a powerful tool to study LFV transitions at flavor factories by measuring

$$
\mathcal{B}(V \to \gamma\ell_1\bar{\ell}_2) = \mathcal{B}(V \to \gamma M)\mathcal{B}(M \to \ell_1\bar{\ell}_2), \tag{3.145}
$$

where $M = \chi_{q0}$ can be used for the studies of scalar decays $\chi_{q0} \to \ell_1\bar{\ell}_2$, while $M = \eta_q$ is important for the pseudoscalar transitions $\eta_q \to \ell_1\bar{\ell}_2$. The corresponding radiative transitions have large rates, e.g. $\mathcal{B}(\psi(2S) \to \gamma\chi_{c0}(1P)) = 9.99 \pm 0.27\%$ [111].

3.3.7 Muon conversion in nuclei: a cascade of effective theories

Muon conversion on a nucleus $\mu^- + (A,Z) \to e^- + (A,Z)$ offers a sensitive probe of New Physics, and a nice possibility to study it experimentally. It also provides an example of an interesting interplay of particle and nuclear physics effects. Just as in the previously discussed LFV decays of mesons, this process can probe New Physics that couples to both quarks and leptons. Experimentally, this process is very efficient with limitations set up by the intensity of the muonic beam. Theoretically, this is a hard process to describe completely model-independently, as it receives contributions from a multitude of scales. While perturbative techniques can be applied at the New Physics scale Λ, complicated nuclear dynamics enters the stage at the low energy QCD scale $\Lambda_{\rm QCD}$.

The set of relevant effective operators driving this transition is extensive. The number of relevant operators, however, is reduced if one only considers *coherent* $\mu^- + (A,Z) \to e^- + (A,Z)$ transitions. Coherent conversion selects only several particular operators to be probed, much as discussed in Sec. (3.3.6), as the matrix elements $\langle A,Z|\bar{q}\gamma_5 q|A,Z\rangle$, $\langle A,Z|\bar{q}\gamma_\alpha\gamma_5 q|A,Z\rangle$, and

$\langle A,Z|\bar{q}\sigma_{\alpha\beta}q|A,Z\rangle$ vanish identically for such process. In addition to the four fermion operators, there are also important non-local contributions from the operators governing $\mu \to e\gamma$ transitions with the photon interacting with a nucleus. For additional information, see [59]. Let us now analyze the muon-to-electron conversion process.

As we discussed above, the effective Lagrangian at the electroweak scale is given by Eq. (3.50) with fermions f given by quark fields q. Performing renormalization group running to the nucleon scale $M \sim 1$ GeV, we must integrate out all heavy quarks with masses larger than M. This results in the removal of operators with explicit heavy quark fields and changes to the Wilson coefficients of the remaining ones [47, 151], e.g.

$$C_{GX} \to C_{GQX} = \sum_{Q=t,b,c} C_{SX}^{Q}\kappa_Q + C_{GX}\kappa \tag{3.146}$$

for $X = L, R$ and κ's are given by the ratio of coupling constants and QCD beta functions for different numbers of flavors [47, 151]. Since at this low scale the relevant degrees of freedom are the nucleons, leptons, and photons, it would be appropriate to match the effective Lagrangian of Eq. (3.50) to the one written in terms of those fields. This is done by the replacement of the operators [47]

$$\begin{aligned} m_q\bar{q}q &\to f_{SN}^{(q)} m_N \overline{\psi}_N \psi_N, \\ \bar{q}\gamma_\mu q &\to f_{VN}^{(q)} \overline{\psi}_N \gamma_\mu \psi_N, \\ \frac{\beta_L}{2g_s^3} GG &\to f_{GN} m_N \overline{\psi}_N \psi_N, \end{aligned} \tag{3.147}$$

where $N = p, n$ is a nucleon, and f_{XY}'s are various nucleon form factors at zero momentum transfer. A trace anomaly relation for the nucleon energy-momentum tensor θ_α^β,

$$\langle N|\theta_\alpha^\alpha|N\rangle = m_N \langle N|\overline{\psi}_N\psi_N|N\rangle, \tag{3.148}$$

implies a sum rule,

$$\sum_{q=u,d,s} f_{SN}^{(q)} + f_{GM} = 1, \tag{3.149}$$

which allows us to eliminate the gluonic operator contribution to the effective Lagrangian. We can now combine the dipole contribution Eq. (3.51) to the nucleon-level Lagrangian, which can be written as

$$\begin{aligned} \mathcal{L}_{\text{eff}}^N = & - \frac{1}{\Lambda^2} \sum_{N=p,n} \Big[\left(C_{DR}^{\ell_1\ell_2}\, \bar{\ell}_2 \sigma^{\mu\nu} P_L \ell_1 + C_{DL}^{\ell_1\ell_2}\, \bar{\ell}_2 \sigma^{\mu\nu} P_R \ell_1 \right) F_{\mu\nu} \\ & + \left(\widetilde{C}_{VR}^{(N)}\, \overline{\mu}_R \gamma^\alpha e_R + \widetilde{C}_{VL}^{(N)}\, \overline{\mu}_L \gamma^\alpha e_L \right) \overline{\psi}_N \gamma_\alpha \psi_N \\ & + m_\mu m_N G_F \left(\widetilde{C}_{SR}^{(N)}\, \overline{\mu}_R e_L + \widetilde{C}_{SL}^{(N)}\, \overline{\mu}_L e_R \right) \overline{\psi}_N \psi_N + h.c. \Big], \end{aligned} \tag{3.150}$$

where the effective couplings $\widetilde{C}_{XY}^{(N)}$ are defined in terms of the Wilson coefficients of the Lagrangian $\mathcal{L}_{\text{eff}}^{\Delta L_\mu=1}$ of Eq. (3.52) and the form factors of Eq (3.147) [47]. For the vector operators,

$$\begin{aligned} \widetilde{C}_{VR}^{(N)} &= \sum_{q=u,d,s} C_{VR}^{q} f_{VN}^{(q)}, \\ \widetilde{C}_{VL}^{(N)} &= \sum_{q=u,d,s} C_{VL}^{q} f_{VN}^{(q)}, \end{aligned} \tag{3.151}$$

where $N = p, n$ (so there are actually four equations above). The values of $f_{VN}^{(q)}$ can be obtained from vector current conservation,

$$f_{Vp}^{(u)} = 2, \quad f_{Vn}^{(u)} = 1, \quad f_{Vp}^{(d)} = 1,$$
$$f_{Vn}^{(d)} = 2, \quad f_{Vp}^{(s)} = 0, \quad f_{Vn}^{(s)} = 0. \tag{3.152}$$

The coefficients of scalar operators in Eq. (3.150) are somewhat more complicated,

$$\widetilde{C}_{SX}^{(N)} = \sum_{q=u,d,s} C_{SX}^{q} f_{SN}^{(q)} + C_{GQX}\left(1 - \sum_{q=u,d,s} f_{SN}^{(q)}\right), \tag{3.153}$$

where $X = L, R$, and $N = p, n$. The calculation of the scalar coefficients in Eq. (3.153) is non-trivial, but one can use chiral arguments [47] or low energy theorems of QCD [151] to compute them.

The last step in evaluating quark contribution is matching between the nucleon and nuclei. Since the experiments are usually performed with Al or Au nuclei, we need to compute nuclear matrix elements $|A, Z\rangle$ of the effective operators written in terms of nucleon fields of Eq. (3.150). The relevant matrix elements can be written in terms of proton and neutron nuclear densities $\rho^{(p)}$ and $\rho^{(n)}$, respectively. In the nonrelativistic approximation,

$$\begin{aligned} \langle A, Z|\overline{\psi}_p \psi_p|A, Z,\rangle &= \langle A, Z|\overline{\psi}_p \gamma^0 \psi_p|A, Z,\rangle = Z\rho^p, \\ \langle A, Z|\overline{\psi}_n \psi_n|A, Z,\rangle &= \langle A, Z|\overline{\psi}_n \gamma^0 \psi_n|A, Z,\rangle = (A - Z)\rho^n, \\ \langle A, Z|\overline{\psi}_N \gamma^i \psi_N|A, Z,\rangle &= 0, \end{aligned} \tag{3.154}$$

where Z is the atomic number and A is the mass number of nuclei. The most common way to obtain nuclear densities is to implement a particular nuclear model. For example, for spherical nuclei, a two-parameter Fermi (2pF) charge distribution $\rho^{(N)}(r)$ can be used. In the isospin limit, the proton and neutron densities are the same,

$$\rho^{(p)}(r) = \rho^{(n)}(r) \equiv \rho(r) \tag{3.155}$$

with the nuclear density represented by the function

$$\rho(r) = \frac{\rho_0}{1 + \exp[(r - c)/z]}, \tag{3.156}$$

where c is the half-density radius. Another popular parameterization is the Fourier-Bessel expansion (FB),

$$\rho(r) = \begin{cases} \sum_v a_v \sin(v\pi r R^{-1})/(v\pi r R^{-1}) & \text{for} \quad r \leq R \\ 0 & \text{for} \quad r > R \end{cases} \tag{3.157}$$

The nucleon densities are assumed spherically symmetric and normalized as

$$\int_0^\infty dr 4\pi r^2 \rho^{(N)}(r) = 1. \tag{3.158}$$

The model parameters for the nuclei $^{48}_{22}$Ti and $^{197}_{79}$Au relevant for μ-to-e conversion experiments are shown in Table 3.2. We now have almost all the ingredients to compute the muon-to-electron conversion rate. The last important ingredient is related to atomic dynamics!

The muon conversion is *not* a simple scattering process of a muon on a nucleus. As a beam of slow muons hits the target, the muons get captured by the atoms eventually cascading into the 1s orbital, which means that the plane wave approximation is not appropriate and a proper solution for the muon wave function in a Coulomb field of the atomic nucleus must be employed.

TABLE 3.2 Nucleon densities model parameters, and the overlap integrals in the units of $m_\mu^{5/2}$

Nucleus	Model	c, fm	z, fm	$S^{(p)}$	$S^{(n)}$
$^{48}_{22}$Ti	FB	–	–	0.0368	0.0435
$^{197}_{79}$Au	2pF	6.38	0.535	0.0614	0.0918

The final state electron can be treated as a plane wave for light nuclei only because the Coulomb distortion of the electron wave function is large for heavy targets. A complete treatment of QED effects and nuclear model calculation was done in [47,122], which we follow below.

To properly account for the QED bound state effects it might be convenient to employ muon and electron wave functions that are obtained from the solution of the Dirac equation in a $1/r$ Coulomb potential [160],

$$\left[-i\gamma_5\sigma_r\left(\frac{\partial}{\partial r}+\frac{1}{r}-\frac{1}{r}\beta\hat{K}\right)+V(r)+m_{Ri}\beta\right]\psi=W\psi, \tag{3.159}$$

where W is the energy and m_{Ri} is the reduced mass of the muon or electron. We will explain how to compute the potential $V(r)$ later (for the impatient readers, it is given in Eq. (3.173)). Other important parts of Eq. (3.159) include

$$\hat{K}=\begin{pmatrix}\boldsymbol{\sigma}\cdot\boldsymbol{l}+1 & 0\\ 0 & -(\boldsymbol{\sigma}\cdot\boldsymbol{l}+1)\end{pmatrix},\qquad \sigma_r=\begin{pmatrix}\boldsymbol{\sigma}\cdot\boldsymbol{r}+1 & 0\\ 0 & -(\boldsymbol{\sigma}\cdot\boldsymbol{r}+1)\end{pmatrix}. \tag{3.160}$$

Here $\boldsymbol{\sigma}$ are the Pauli matrices, and $\boldsymbol{l}=-i\mathbf{r}\times\nabla$ is the angular momentum operator. Please note that the Eq. (3.159) is conventionally written using Dirac matrices in the Dirac representation,

$$\gamma_5=\begin{pmatrix}0 & 1\\ 1 & 0\end{pmatrix},\qquad \beta=\begin{pmatrix}1 & 0\\ 0 & -1\end{pmatrix}. \tag{3.161}$$

The initial muon is the $1s$ state of the muonic atom with the binding energy E_b, and the final state electron is the eigenstate with the energy $m_\mu - E_b$. Since the operators $\hat{K}$ and the third component of total angular momentum $\hat{J}_z$ commute with the Dirac Hamiltonian, the Ehrenfest's theorem states that their eigenvalues, κ and μ, are the conserved quantum numbers that label the wave function $\psi=\psi^\mu_\kappa$. Thus, the relevant wave functions can be defined as

$$\psi^\mu_\kappa=\begin{pmatrix}g(r)\chi^\mu_\kappa(\theta,\phi)\\ if(r)\chi^\mu_{-\kappa}(\theta,\phi)\end{pmatrix}. \tag{3.162}$$

The angular parts of the wave function $\chi^\mu_{\pm\kappa}(\theta,\phi)$ are the eigenspinors of the $\boldsymbol{\sigma}\cdot\boldsymbol{l}+1$ and $\hat{J}_z$ operators,

$$(\boldsymbol{\sigma}\cdot\boldsymbol{l}+1)\,\chi^\mu_{\pm\kappa}(\theta,\phi)=-\kappa\chi^\mu_{\pm\kappa}(\theta,\phi),\quad\text{and}\quad \hat{J}_z\chi^\mu_{\pm\kappa}(\theta,\phi)=\mu\chi^\mu_{\pm\kappa}(\theta,\phi), \tag{3.163}$$

and are normalized as

$$\int_{-1}^{1}d\cos\theta\int_0^{2\pi}d\phi\;\chi^{\mu *}_\kappa\chi^{\mu'}_{\kappa'}=\delta_{\mu\mu'}\delta_{\kappa\kappa'}\,. \tag{3.164}$$

The radial wave functions can be obtained by the usual substitutions $u_1(r)=rg(r)$ and $u_2(r)=rf(r)$. With such substitutions, the reduced radial wave functions $u_i(r)$ satisfy the following form of the Dirac equation,

$$\frac{d}{dr}\begin{pmatrix}u_1\\ u_2\end{pmatrix}=\begin{pmatrix}-\kappa/r & W-V+m_i\\ -(W-V-m_i) & \kappa/r\end{pmatrix}\begin{pmatrix}u_1\\ u_2\end{pmatrix}, \tag{3.165}$$

where W and $V(r)$ are the energy and potential, respectively. The normalizations of the wave functions of Eq. (3.162) for the bound muon and the final state electron are obviously different.

The initial muon state corresponds to the $1s$ shell state of the atom with the quantum numbers of $\mu = \pm 1/2$ and $\kappa = -1$ and a normalization

$$\int d^3\mathbf{r}\ \psi_{1s}^{(\mu)*}(\mathbf{r})\psi_{1s}^{(\mu)}(\mathbf{r}) = 1\ . \tag{3.166}$$

The final state electron's energies W belongs to the continuum spectrum, which implies that it must be normalized to a delta-function,

$$\int d^3\mathbf{r}\ \psi_{\kappa,W}^{\mu(e)*}(\mathbf{r})\psi_{\kappa',W'}^{\mu'(e)}(\mathbf{r}) = 2\pi\delta_{\mu\mu'}\delta_{\kappa\kappa'}\delta(W-W')\ . \tag{3.167}$$

Computing matrix element for the process $\mu^- + (A,Z) \to e^- + (A,Z)$ involves calculating the overlap of the initial muon and final electron wave functions weighted by various combinations of nuclear densities. This couples the nuclear and atomic dynamics and makes the resulting integrals not very illuminating. Fortunately, their effect is reduced to several *overlap* parameters that can be computed numerically. The resulting conversion rate Γ_{conv} is [47]

$$\begin{aligned}\Gamma_{\text{conv}} = & \frac{m_\mu^5}{4\Lambda^4}\left|C_{DR}D + 4G_F m_\mu\left(m_p\widetilde{C}_{SR}^{(p)}S^{(p)} + m_n\widetilde{C}_{SR}^{(n)}S^{(n)}\right) + 4\widetilde{C}_{VR}^{(p)}V^{(p)} + 4\widetilde{C}_{VR}^{(n)}V^{(n)}\right|^2 \\ + & \frac{m_\mu^5}{4\Lambda^4}\left|C_{DL}D + 4G_F m_\mu\left(m_p\widetilde{C}_{SL}^{(p)}S^{(p)} + m_n\widetilde{C}_{SL}^{(n)}S^{(n)}\right) + 4\widetilde{C}_{VL}^{(p)}V^{(p)} + 4\widetilde{C}_{VL}^{(n)}V^{(n)}\right|^2,\end{aligned} \tag{3.168}$$

where the overlap parameters are defined as the integrals [122],

$$\begin{aligned} D &= \frac{4}{\sqrt{2}}m_\mu \int_0^\infty dr r^2[-E(r)](g_e^- f_\mu^- + f_e^- g_\mu^-)\ , \\ S^{(N)} &= \frac{1}{2\sqrt{2}}\int_0^\infty dr r^2 N_N \rho^{(N)}(g_e^- g_\mu^- - f_e^- f_\mu^-)\ , \\ V^{(N)} &= \frac{1}{2\sqrt{2}}\int_0^\infty dr r^2 N_N \rho^{(N)}(g_e^- g_\mu^- + f_e^- f_\mu^-)\ , \end{aligned} \tag{3.169}$$

for $N = p, n$. In Eq. (3.169) we defined $N_p = Z$ and $N_n = (A-Z)$. The functions $g_{e/\mu}^-$ are the radial components of the muon wavefunction

$$\psi_{1s}^{(\mu)}(r,\theta,\phi) = \begin{pmatrix} g_\mu^-(r)\chi_{-1}^{\pm 1/2}(\theta,\phi) \\ if_\mu^-(r)\chi_1^{\pm 1/2}(\theta,\phi)\end{pmatrix}, \tag{3.170}$$

and $\kappa = \pm 1$ electron wave functions,

$$\begin{aligned} \psi_{\kappa=-1,W}^{\mu=\pm 1/2(e)}(r,\theta,\phi) &= \begin{pmatrix} g_e^-(r)\chi_{-1}^{\pm 1/2}(\theta,\phi) \\ if_e^-(r)\chi_1^{\pm 1/2}(\theta,\phi)\end{pmatrix}, \\ \psi_{\kappa=1,W}^{\mu=\pm 1/2(e)}(r,\theta,\phi) &= \begin{pmatrix} g_e^+(r)\chi_1^{\pm 1/2}(\theta,\phi) \\ if_e^+(r)\chi_{-1}^{\pm 1/2}(\theta,\phi)\end{pmatrix}. \end{aligned} \tag{3.171}$$

Note that in the limit of $m_e \to 0$ the following relations hold: $g_e^+ = if_e^-$, and $if_e^+ = g_e^-$ [122].

Finally, with chosen proton and neutron densities, the electric field $E(r)$ in the first equation of Eqs. (3.169) is obtained by integrating the Maxwell equations,

$$E(r) = \frac{Ze}{r^2}\int_0^r r'^2\rho^{(p)}(r')dr'\ . \tag{3.172}$$

Also, the electric potential $V(r)$ of Eq. (3.159)

$$V(r) = -e \int_r^\infty E(r')dr' \ . \tag{3.173}$$

It is often convenient to define the branching ratio for μ-e conversion on a nucleus (A, Z)

$$B_{\mu e}^Z = \frac{\Gamma_{\text{conv}}(\mu^- + (A,Z) \to e^- + (A,Z))}{\Gamma_{\text{capt}}(\mu^-(A,Z))}, \tag{3.174}$$

where $\Gamma_{\text{capt}}(\mu^-(A,Z)$ is muon's capture $(\mu^- + (A,Z) \to \nu_\mu + (A, Z-1)))$ rate on the same nucleus.

$\mu \to e$ conversion: experimental methods of detection

Muon conversion experiments utilize intense muon beams to produce muonium atoms by stopping muons in a target (e.g. aluminum foils) with individual muons captured into atomic orbits. The signal is a monochromatic electron, approximately 105 MeV/c in the case of muonic aluminum, that is emitted with a time delay of the lifetime of the muonic atom [20]. This energy is essentially given by the muon mass, with small corrections due to binding and nuclear recoil. The coherence of the process essentially results in enhancement of the signal, as the produced electron recoils off the entire nucleus.

As muon is approximately 200 times heavier than the electron, there is an overlap of the muon orbital wave function and the nucleus, which increases the chance of an LFV interaction of the muon and the nucleus. Since we are interested in a very rare process, it is important to reduce possible SM backgrounds, which include muon decay in orbit $\mu^- + (A,Z) \to e^- + \nu_e + \bar{\nu}_\mu + (A,Z)$, among other phenomena [20].

3.4 Notes for further reading

A great reference to review basic QFT material and fill in any gaps is always [147]. Any other textbook, such as [163] or [186] would also be useful. There are very nice resources describing Standard Model techniques for both high and low-energy physics, e.g. [65, 144].

Experimental techniques related to flavor-conserving probes of New Physics can be found in a number of excellent reviews, including [20]. Related theoretical techniques are reviewed in [117]. There are several excellent reviews of EDMs [19, 72, 155]. Experimental measurements of EDMs in atoms and molecules are discussed in [39]. The physics of a single electron inside the Penning trap that is relevant for the electron's EDM measurement is well developed [31] (for a simplified description, see [67]).

Muon conversion is a multiscale process. Various aspects of the process are discussed in [123]. A good deal of review of atomic physics techniques needed for muon conversion calculations is explained in [123, 160]. A nice review of the whole field can be found in [157].

Problems for Chapter 3

1. **Problem 1.** Using gauge invariance and equations of motion, show that Eq. (3.3) is the most general way to write a matrix element of the vector current.
2. **Problem 2.** The first term in Eq. (3.30) corresponds to a familiar case of Larmor frequency that can be obtained in non-relativistic quantum mechanics. Find this frequency by considering the motion of a non-relativistic electron in a constant magnetic field. To do so, choose the potential to be

$$\vec{A}(\vec{r}) = \frac{1}{2}\vec{B} \times \vec{r}, \tag{3.175}$$

and choose the corresponding magnetic field $\vec{B} = B\vec{n}_z$ to point along the z axis. Show that the Hamiltonian describing the motion in the xy plane can be cast in a form representing a harmonic oscillator with the frequency given by Eq. (3.30).

3. **Problem 3.** Express the parameters g^Z_{XY} of the effective Lagrangian in Eq. (3.53) in terms of the Wilson coefficients C^f_{AB} of the effective Lagrangian in Eq. (3.52). With that knowledge at hand, rederive the general formula for the muon width in Eq. (3.61) to verify the expressions of Michel parameters given in Eq. (3.60).

4

New Physics searches with quarks

4.1 Introduction

Quarks are ubiquitous. They are the only particles in the Standard Model that participate in strong, weak, and electromagnetic interactions, which means that most BSM models include NP degrees of freedom interacting with quarks. Quarks provide us with a tool that is often used by researchers to search for New Physics.

Being the most available tool, however, does not mean that it is the easiest one to use. While the effective operators involving quark fields can be easily written just below the New Physics scale Λ and classified according to their dimensions, their effects are often probed at low energies.

There is no conceptual problem in relating the SM EFT operators at the high and low scales. In the spirit of the effective field theory approach, the operators can be matched at the electroweak scale and then run down, using renormalization group (RG) equations, to the scales relevant for experimental measurements. It is at those low scales where things become complicated, as non-perturbative QCD effects become dominant. As is well known, quarks are not the correct degrees of freedom at low energies, mesons and baryons are. In this regime we have to properly translate the quark language of NP interactions to the dialect of hadronic interactions. This is not a trivial problem. Depending on the value of the quark masses, various methods based on symmetries of low energy QCD have been developed, which help even in the most complicated situations involving non-leptonic decays.

Alternatively, we can devise other strategies to extract possible signals of New Physics despite considerable background from non-perturbative QCD effects. Clever relationships between observables, often based on symmetries of QCD, have been developed where non-perturbative effects cancel out. Finally, extensive checks of relations among experimental observables that only hold in the Standard Model have been performed. Violations of such relations, if observed, do not point to the exact model of NP, but indicate that NP contribution is present, prompting searches in other places. Here we consider searches of New Physics with quark systems of various degree of complexity.

In this chapter we will discuss both flavor-conserving and flavor-changing processes. For the later, we will take a path of increasing complexity: first, we will discuss a bridge between the topics discussed in the previous chapter and in this one, the semileptonic interactions. We will see that soft QCD creates difficulties in model-independent description of even simple current-type operators and discuss ways to overcome them in searches for New Physics. Then we will see how glimpses of BSM physics can be seen in non-leptonic transitions.

4.2 Flavor-conserving interactions: EDM

The possibilities for studies of EDMs and other related CP-violating observables with quarks are enhanced considerably compared to studies of leptonic EDMs discussed in Sect. 3.2.2. EDM-like operators appear in the SM EFT at dimension six,

$$\frac{1}{\Lambda^2}\overline{q}_L\sigma^{\mu\nu}Hq_RX_{\mu\nu} \to \frac{v}{\Lambda^2}\overline{q}_L\sigma^{\mu\nu}q_RX_{\mu\nu}, \tag{4.1}$$

where $X_{\mu\nu} = \{B_{\mu\nu}, W_{\mu\nu}, G_{\mu\nu}\}$. The first two lead to both quark and lepton EDMs.

The third option above leads to something we have not considered yet: CP-violating quark EDM. Moreover, since gluons themselves carry color charge, a chromo-EDM of a gluon can also be induced. The relevant operator, called the Weinberg operator, is

$$Q_w = \frac{w}{3!}f^{ABC}\epsilon^{\mu\nu\alpha\beta}G^A_{\alpha\beta}G^B_{\mu\rho}G^{C,\rho}_{\nu}. \tag{4.2}$$

Some of those operators mix under renormalization group, complicating placing bounds of their Wilson coefficients.

Quark EDMs could be searched for by looking for EDMs of protons and neutrons, and, ultimately, those of atomic nuclei. To study those experimentally, one must also overcome Schiff's theorem (see Theorem 3.1) by carefully choosing the proper nuclear system. Analyses of such experiments are complicated [72,155]. Here we shall discuss a simpler system, the neutron.

4.2.1 Strong CP problem and electric dipole moment of the neutron

As we discussed in Chapter 2, there appears to be a unique source of CP-violation in the electroweak sector of the Standard Model that is associated with the single complex phase of the CKM matrix. It might be interesting to ask if there are other, non-CKM sources of CP-violation present in the full SM Lagrangian.

There are several candidates for CP-odd terms that could be added to the SM Lagrangian. At dimension-four, the seemingly easiest example can be readily found in QED. A term

$$\mathcal{L}_{\theta_{\rm QED}} = \frac{\theta_{\rm QED}}{4}F_{\mu\nu}\widetilde{F}^{\mu\nu} = -\theta_{\rm QED}\mathbf{E}\cdot\mathbf{B} \tag{4.3}$$

seems to fit the bill! Indeed, as we discussed in Sect. 2.3.1, the electric field $\mathbf{E}$ is P-odd and T-even, while the magnetic field $\mathbf{B}$ is P-even and T-odd. This immediately implies that the term in Eq. (4.3) is both P-odd and T-odd. Since the CPT theorem is respected by local QFTs, including the Standard Model, this also means that this term is CP-odd. Success?

It turns out our success is short-lived, as we soon realize that the term in Eq. (4.3) is not physical. In fact, we can rewrite it as a total derivative,

$$\mathcal{L}_{\theta_{\rm QED}} = \frac{\theta_{\rm QED}}{4}\partial_\mu\epsilon^{\mu\nu\alpha\beta}F_{\nu\alpha}A_\beta = \theta_{\rm QED}\partial_\mu K^\mu. \tag{4.4}$$

A part of the action that corresponds to the Lagrangian of Eq. (4.3) can then be written as

$$S_{\theta_{\text{QED}}} = \int d^4x \mathcal{L}_{\theta_{\text{QED}}} = \theta_{\text{QED}} \int d^4x \partial_\mu K^\mu = \theta_{\text{QED}} \int_{\partial V} dS_\mu K^\mu, \tag{4.5}$$

where in the last part of this equation we used Gauss theorem to relate the volume integral to the surface one. For the infinite volume and QED fields vanishing at infinity $S_{\theta_{\text{QED}}} = 0$, which means that the term in Eq. (4.3) has no physical consequences and can be safely dropped from the SM Lagrangian, as it is always done.

A similar term can also be written in electroweak theory for the $SU(2)$ fields, $\mathcal{L}_{\theta_W} = \theta_W g_2^2/(32\pi^2) W_{\mu\nu}\widetilde{W}^{\mu\nu}$. It turns out, however, that it also has no observable consequences [6], as we can perform chiral rotations separately on the left- and the right-handed fermion fields, eliminating the "weak theta term" from the action.

It is interesting to note that both tricks described above do not work in QCD. Since gluons couple to the left-handed and the right-handed quarks the same way, we cannot eliminate the QCD theta term by separate chiral rotations. Also, because the QCD vacuum is non-trivial, the argument with fields vanishing at infinity is not applicable. Thus, constructing the QCD Lagrangian in a way similar to the QED one, we could again ask ourselves why the term

$$\mathcal{L}_\theta = \theta \frac{g^2}{32\pi^2} G_{\mu\nu}\widetilde{G}^{\mu\nu} \tag{4.6}$$

is not normally added to it despite being a marginal operator of dimension-four, just like other operators. It is of course interesting that this term is CP-odd, which means it might play a role in generating CP-violating effects required for generating matter-antimatter asymmetry of the Universe – provided it is actually there! This Lagrangian is often colloquially referred to as the *QCD θ-term*. Similarly to our previous discussion of the QED theta-term, we can write it as

$$\begin{aligned} \mathcal{L}_\theta &= \theta \partial_\mu K^\mu, \quad \text{with} \\ K^\mu &= \frac{1}{16\pi^2} \epsilon^{\mu\nu\alpha\beta} \left(G_\nu^A \partial_\alpha G_\beta^A + \frac{1}{3} f_{ABC} G_\nu^A G_\alpha^B G_\beta^C \right). \end{aligned} \tag{4.7}$$

However, in QCD, the integral over the full derivative in Eq. (4.7) does not vanish. This is contrary to what happens in QED and points towards the fact that QCD has a richer vacuum structure than QED [61, 84]. This means that we are stuck with the QCD θ-term.

The θ-term of Eq. (4.7) would exist in a pure-gauge QCD, even if no quarks are present. In a full theory, quarks are added to the SM with their masses generated by the Higgs mechanism. It is interesting to point out that diagonalization of mass matrices of Eq. (2.27), as described in Sect. 2.4, would, in principle, result in a diagonal mass matrix with complex eigenvalues, which is also CP-violating,

$$(M_q)_{ii} = m_i e^{i\alpha_i}. \tag{4.8}$$

The phases α_i can be removed from the mass matrix by chiral transformations of the quark fields,

$$q_i \to q_i' = e^{i(\alpha_i/2)\gamma_5} q_i, \quad \overline{q}_i \to \overline{q}_i' = \overline{q} e^{i(\alpha_i/2)\gamma_5}. \tag{4.9}$$

However, the symmetry given by the transformation of Eq. (4.9) is anomalous*,

$$\partial_\mu \left(\overline{q}\gamma^\mu \gamma_5 q\right) = \frac{g_s^2}{16\pi^2} G_{\mu\nu}^A \widetilde{G}^{A,\mu\nu}. \tag{4.10}$$

*This is manifested as non-invariance of the measure in the Feynman path integral for the QCD action under this variable change.

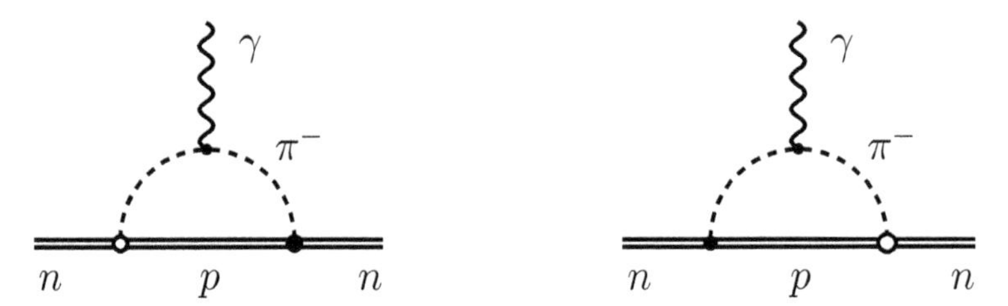

FIGURE 4.1 Diagrams representing the leading contribution to neutron EDM. Black dot represents a CP-conserving πNN vertex, while white circle represents CP-violating πNN vertex.

This means that we can trade the QCD θ-term for the complex quark masses. To be exact, the actual parameter that sets the magnitude of CP violation in QCD is a combination of the two,

$$\overline{\theta} = \theta - \arg\det M_q. \tag{4.11}$$

We shall often drop the bar over theta, but keep the relation in Eq. (4.11) in mind. As we shall show below, the QCD θ-term contributes to a number of observables, most notably, the neutron's electric dipole moment. As it is a well-constrained quantity, its non-observation puts severe constraints on the numerical value of θ,

$$\theta \ll 10^{-16}, \tag{4.12}$$

which begs the question of why it is not zero in the first place. Discussions on how to remove the θ-term from the QCD Lagrangian are very enlightening, with a number of scenarios of how to remove its physical consequences [61, 84]. One of such scenarios inspired the whole field of searches for axion-like particles.

We will not dwell on these scenarios in this section, although we will discuss their possible consequences later in the chapter. We will now discuss how to constrain the θ parameter from experimental data, most notably, from neutron's EDM. It would probably make sense to start our discussion with a description of how Eq. (4.6) affects neutron's EDM.

To estimate this contribution, we follow [55] to trade the θ term for the complex masses,

$$\mathcal{L}_{\text{CPV}} = \overline{\theta}\frac{g^2}{32\pi^2}G_{\mu\nu}\widetilde{G}^{\mu\nu} \to -\overline{\theta}\frac{im_u m_d}{m_u + m_d}\bar{q}\gamma_5 q, \tag{4.13}$$

which, in turn, leads to CP-violating pion-nucleon coupling $\bar{g}_{\pi NN}$ in

$$\mathcal{L}_{\pi NN} = \overline{N}\left(i\gamma_5 g_{\pi NN} + \bar{g}_{\pi NN}\right)\boldsymbol{\tau} N \cdot \boldsymbol{\pi}, \tag{4.14}$$

with $\bar{g}_{\pi NN}$ being related to θ in $|\bar{g}_{\pi NN}| \simeq 0.004|\theta|$ [55, 138], $\boldsymbol{\pi}$ representing the triplet of pions, and $\boldsymbol{\tau} \equiv \tau^a$ being the isospin matrices. Computing the diagrams in Fig. 4.1 and retaining leading (logarithmic) term brings us to the estimate of neutron's EDM [55],

$$d_n = \frac{e}{8\pi^2 m_n} g_{\pi NN}\bar{g}^{(0)}_{\pi NN} \log\frac{m_N^2}{m_\pi^2}, \tag{4.15}$$

which implies that $d_n \simeq 2 \times 10^{-3}\overline{\theta}$ e fm. Other estimates [154], done with different techniques, seem to confirm this result.

4.2.2 Experimental methods of detection: EDM

In order to extract neutron's EDM, one measures Larmor precession of the neutron spin in the presence of parallel and antiparallel magnetic and electric fields. The precession frequency for each of the two cases is given by

$$h\nu = 2\mu_n B \pm 2d_n E, \tag{4.16}$$

the addition or subtraction of the frequencies stemming from the precession of the magnetic moment around the magnetic field and the precession of the electric dipole moment around the electric field. From the difference of those two frequencies one readily obtains a measure of the neutron EDM:

$$d_n = \frac{h\Delta\nu}{4E}. \tag{4.17}$$

The biggest challenge of the experiment (and at the same time the source of the biggest systematic false effects) is to ensure that the magnetic field does not change during these two measurements. A great deal of progress in measuring neutron's EDM was achieved with ultracold neutrons in traps [93].

4.3 Flavor-conserving interactions: decays

Flavor-diagonal non-leptonic decays is a rather unlikely place to search for manifestations of New Physics. A reason for it is that flavor conserving non-leptonic decays occur due to strong interactions. In other words, the largest Standard Model background to such processes comes from the soft QCD effects, which are difficult to quantify model-independently. This fact, however, allows us to look for the signs of processes that break symmetries that are respected by QCD. For example, strong interactions conserve C, P, and combined CP symmetries. This means that any sign of breaking of those symmetries can be related to the presence of electroweak or New Physics interactions.

A direct consequence of the solution to the strong CP problem discussed in the previous section, the axion, suggests another way New Physics can be probed: by searching for flavorconserving decays into invisible states. Any possible light new states X, introduced in Chapter 2, that can couple to quarks could be produced in those decays. Processes like $J/\psi \to \gamma + X$ or $\eta^{(\prime)} \to \gamma + X$ are good examples of such transitions. The main Standard Model background to such processes would be flavor-conserving decays into the final states containing neutrinos.

Finally, if those light new states also couple to leptons, they can affect flavor-conserving transitions with leptons in the final states. New particle states can be seen as resonance structures in various spectra. For example, massive dark photon state could show up as peaks in the e^+e^- invariant mass spectrum in $\eta^{(\prime)} \to e^+e^-\gamma$ decays.

4.3.1 Flavor-conserving decays into invisible states

Invisible decays of quarkonium states might provide a way to see new neutral light states that are not part of the SM set-up. As we mentioned previously, examples of the processes that match the experimental signature of the decays to the states that do not leave traces in a detector are the SM processes with neutrinos. Let us discuss them first.

Standard Model backgrounds

The simplest process to compute (but not the simplest to observe) would be an invisible decay of a quarkonium state. For the sake of definiteness, let us discuss the decays of bottomonium ($b\bar{b}$) and charmonium ($c\bar{c}$) spin-one nS states [43,185]. In the SM the width of the invisible decay of a $b\bar{b}$ state into a two-neutrino final state would be given by

$$\Gamma(\Upsilon(nS) \to \nu\bar{\nu}) = \frac{N_\nu G_F^2}{48\pi\Gamma_{\Upsilon(nS)}} \left(1 - \frac{4}{3}\sin^2\theta_W\right)^2 f_{\Upsilon(nS)}^2 m_{\Upsilon(nS)}^3, \tag{4.18}$$

where $N_\nu = 3$ is the number of light neutrinos, G_F is the Fermi coupling, and θ_W is the Weinberg angle. The decay constant $f_{\Upsilon(nS)}$ can be extracted from experimental measurements of the $\Upsilon(nS)$ decays into the e^+e^- final state [1]. It is a QCD analogue of the muonium decay constant derived in Chapter 3 which parameterizes quark dynamics inside the Υ state. This decay rate is

$$\Gamma(\Upsilon(nS) \to e^+e^-) = \frac{4\pi\alpha^2 f_{\Upsilon(nS)}^2}{27 m_{\Upsilon(nS)}^3}, \tag{4.19}$$

so that the ratio of Eq. (4.18) and Eq. (4.19) is independent of the decay constant and thus the soft QCD dynamics. It might nevertheless be interesting to extract $f_{\Upsilon(nS)}$ from Eq. (4.19). Doing so we obtain $f_{\Upsilon(1S)} = (0.715 \pm 0.05)$ GeV for the $\Upsilon(1S)$ state. This implies that $\mathcal{B}(\Upsilon(nS) \to \nu\bar{\nu}) = (1.03 \pm 0.04) \times 10^{-5}$, which is a rather small number. Similar results for the decays of the charmonium states $\psi(nS)$ can be trivially obtained from Eq. (4.18) by appropriate substitution of masses and couplings.

It might be instructive to compare decays of charmonium states into neutrinos and charged lepton final states. For the $1S$ state charmonium, the J/ψ, the ratio is

$$\frac{\Gamma(J/\psi \to \nu\bar{\nu})}{\Gamma(J/\psi \to e^+e^-)} = \frac{9N_\nu G_F^2}{256\pi^2\alpha^2} M_{J/\psi}^4 \left(1 - \frac{8}{3}\sin^2\theta_W\right)^2 \simeq 4.54 \times 10^{-7}. \tag{4.20}$$

It is interesting to note that this result is about three orders of magnitude smaller than the corresponding decay of an Upsilon state. This implies that invisible decays of charmonium states provide a great opportunity to search for the glimpses of BSM physics, especially if it couples preferably to heavier states.

In BSM scenarios this decay rate might be enhanced either by new heavy particles modifying the interactions of neutrinos and heavy quarks, or by opening new decay channels into light DM states. Similarly, radiative decays of quarkonia with missing energy can also be computed. In fact, the calculation of $\Upsilon(nS) \to \nu\bar{\nu}\gamma$ branching ratio in the SM yields

$$\frac{dB(\Upsilon(nS) \to \nu\bar{\nu}\gamma)}{d\hat{s}} = \frac{N_\nu G_F^2}{162\pi} \frac{\alpha}{4\pi} \frac{f_{\Upsilon(nS)}^2 M_{\Upsilon(nS)}^3}{\Gamma_{\Upsilon(nS)}} (1 - \hat{s}^2) \tag{4.21}$$

and

$$\mathcal{B}(\Upsilon(nS) \to \nu\bar{\nu}\gamma) = \frac{N_\nu G_F^2}{243\pi} \frac{\alpha}{4\pi} \frac{f_{\Upsilon(nS)}^2 M_{\Upsilon(nS)}^3}{\Gamma_{\Upsilon(nS)}} \tag{4.22}$$

The SM branching ratios of all such decays are very small [185]. Thus, we may neglect neutrino-background effects when confronting theoretical predictions for quarkonia decays into invisible states with experimental data.

Decays into light New Physics states

There is a variety of light NP states that could be produced in quarkonia decays (see Sect. 2.7.2). Lets us consider the simplest example, a SM coupled to a complex scalar particle Φ with a Z_2 symmetry such that only a pair $\Phi\Phi^*$ can be produced. This models could be made as sophisticated as needed to satisfy current experimental constraints [185]. The simplest effective Lagrangian describing interactions of SM quarks q with Φ can be written as

$$\mathcal{L}_\Phi = -\frac{2}{\Lambda_\Phi^2} \sum_i C_i^{(q)} O_i, \tag{4.23}$$

where Λ_Φ is some heavy scale that signals where the ultraviolet completion of the model need to take over, and effective operators O_i are defined as

$$\begin{aligned} O_1 &= m_q\left(\bar{q}q\right)\left(\Phi^*\Phi\right), \qquad O_2 = i m_q\left(\bar{q}\gamma_5 q\right)\left(\Phi^*\Phi\right), \\ O_3 &= \left(\bar{q}\gamma^\mu q\right)\left(\Phi^* i \overleftrightarrow{\partial}_\mu \Phi\right), \quad O_4 = \left(\bar{q}\gamma^\mu\gamma_5 q\right)\left(\Phi^* i \overleftrightarrow{\partial}_\mu \Phi\right), \end{aligned} \tag{4.24}$$

where $\overleftrightarrow{\partial}_\mu = \left(\overrightarrow{\partial}_\mu - \overleftarrow{\partial}_\mu\right)/2$. We defined the operators O_i to be Hermitean, so that the Wilson coefficients $C_i^{(q)}$ are real. For an experimentally relevant decay of the J/ψ (or $\Upsilon(1S)$) states we can specify $q = c$ to write for the branching ratio $\mathcal{B}(J/\psi \to \Phi\Phi^*) = \Gamma(J/\psi \to \Phi\Phi^*)/\Gamma_{J/\psi}$,

$$\mathcal{B}(J/\psi \to \Phi\Phi^*) = \left[\frac{C_3^{(c)}}{\Lambda_\Phi^2}\right]^2 \frac{f_{J/\psi}^2 m_{J/\psi}^3}{48\pi\Gamma_{J/\psi}} \left(1 - \frac{4m_\Phi^2}{m_{J/\psi}^2}\right)^{3/2}, \tag{4.25}$$

where $f_{J/\psi}$ is a J/ψ decay constant (see below in Eq. (4.37)) and $\Gamma_{J/\psi}$ is its total width.

It is therefore possible to use invisible decays of quarkonium states to put constraints on various models of light DM, provided that DM states couple to quarks. Predictions for radiative decays with missing energy in light DM models are available in Ref. [80].

Other processes involving new invisible light final states are the ones that involve vector states, the "dark photons." From the computational point of view, transition amplitudes of particles that already decay into photon states can be easily converted to the decays into invisible final states. If dark photons couple to the visible sector via kinetic mixing of Eq. (2.103), it follows from Eq. (2.106) that a simple substitution $e \to \kappa e$ relates decays with visible photons to the decays to dark photons, up to a form factor that takes into account dark photon mass effects. For example, for the decays of the η,

$$\mathcal{B}(\eta \to V'\gamma) = 2\kappa^2 \mathcal{B}(\eta \to \gamma\gamma) \left|\bar{F}_{\eta\gamma^*\gamma^*}(m_{V'}^2, 0)\right|^2 \left(1 - \frac{m_{V'}^2}{m_\eta^2}\right)^3, \tag{4.26}$$

where $\bar{F}_{\eta\gamma^*\gamma^*}(m_{V'}^2, 0)$ is the form factor that takes into account kinematical dependence on the dark photon mass. The exact form of this form factor is known from the studies of η decays into the off-shell photon states that appear in such decays as $\eta \to e^+e^-\gamma$. Note that $\bar{F}_{\eta\gamma^*\gamma^*}(0,0) = 1$, as the decay constant cancels in the ratio in Eq. (4.26).

4.3.2 Experimental methods of detection

Decays to invisible final states

Due to their superb kinematic-reconstruction capabilities, decays of quarkonium states can be studied with general purpose detectors at flavor factories. Tau-charm factories can search for the invisible and radiative invisible decays of J/ψ and its higher excitations $\psi(nS)$ via decay chains $\psi(nS) \to \pi^+\pi^- J/\psi(\to \text{invisible})$ using $\pi^+\pi^-$ as a trigger. A similar strategy can be employed for the Υ states as well.

CP-violating strong decays

One of the manifestations of the QCD θ-term, as was pointed out in [55], is a possibility of CP-violating strong decays. Such decays are easy to construct: we just need to choose transitions with definite CP such that the CP quantum number of initial and final states did not match. An example is a decay $\eta \to \pi\pi$. Using the same trick of trading the θ-term for the complex masses, an effective Lagrangian for this transition can be constructed,

$$\mathcal{L}_{\eta\to\pi\pi} = -\frac{\theta m_\pi^2}{\sqrt{3}F_\pi} \frac{m_u m_d}{\left(m_u + m_d\right)^2} \pi^2 \eta. \tag{4.27}$$

TABLE 4.1 Four-fermion operators involving at least two quark fields. Operator configurations $(\overline{L}L)(\overline{L}L)$, $(\overline{R}R)(\overline{R}R)$, $(\overline{L}L)(\overline{R}R)$, and $(\overline{L}R)(\overline{R}L)$ are included.

$(\overline{L}L)(\overline{L}L)$		$(\overline{R}R)(\overline{R}R)$		$(\overline{L}L)(\overline{R}R)$	
$Q_{qq}^{(1)}$	$\left(\overline{Q}_p\gamma^\mu Q_r\right)\left(\overline{Q}_s\gamma^\mu Q_t\right)$	Q_{uu}	$(\overline{u}_p\gamma^\mu u_r)(\overline{u}_s\gamma^\mu u_t)$	Q_{lu}	$\left(\overline{L}_p\gamma^\mu L_r\right)(\overline{u}_s\gamma^\mu u_t)$
$Q_{qq}^{(3)}$	$\left(\overline{Q}_p\gamma^\mu\tau^I Q_r\right)\left(\overline{Q}_s\gamma^\mu\tau^I Q_t\right)$	Q_{dd}	$\left(\overline{d}_p\gamma^\mu d_r\right)\left(\overline{d}_s\gamma^\mu d_t\right)$	Q_{ld}	$\left(\overline{L}_p\gamma^\mu L_r\right)\left(\overline{d}_s\gamma^\mu d_t\right)$
$Q_{lq}^{(1)}$	$\left(\overline{L}_p\gamma^\mu L_r\right)\left(\overline{Q}_s\gamma^\mu Q_t\right)$	Q_{eu}	$(\overline{e}_p\gamma^\mu e_r)(\overline{u}_s\gamma^\mu u_t)$	Q_{qe}	$\left(\overline{Q}_p\gamma^\mu Q_r\right)(\overline{e}_s\gamma^\mu e_t)$
$Q_{lq}^{(3)}$	$\left(\overline{L}_p\gamma^\mu\tau^I L_r\right)\left(\overline{Q}_s\gamma^\mu\tau^I Q_t\right)$	Q_{ed}	$(\overline{e}_p\gamma^\mu e_r)\left(\overline{d}_s\gamma^\mu d_t\right)$	$Q_{qu}^{(1)}$	$\left(\overline{Q}_p\gamma^\mu Q_r\right)(\overline{u}_s\gamma^\mu u_t)$
		$Q_{ud}^{(1)}$	$(\overline{u}_p\gamma^\mu u_r)\left(\overline{d}_s\gamma^\mu d_t\right)$	$Q_{qu}^{(8)}$	$\left(\overline{q}_p\gamma^\mu T^A q_r\right)\left(\overline{u}_s\gamma^\mu T^A u_t\right)$
		$Q_{ud}^{(8)}$	$\left(\overline{u}_p\gamma^\mu T^A u_r\right)\left(\overline{d}_s\gamma^\mu T^A d_t\right)$	$Q_{qd}^{(1)}$	$\left(\overline{q}_p\gamma^\mu q_r\right)\left(\overline{d}_s\gamma^\mu d_t\right)$
				$Q_{qd}^{(8)}$	$\left(\overline{Q}_p\gamma^\mu T^A Q_r\right)\left(\overline{d}_s\gamma^\mu T^A d_t\right)$
$(\overline{L}R)(\overline{R}L)$		**B-violating**			
Q_{ledq}	$\left(\overline{L}_p^j e_r\right)\left(\overline{d}_s Q_t^j\right)$	Q_{duq}	$\epsilon^{\alpha\beta\gamma}\epsilon_{jk}\left[\left(d_p^\alpha\right)^T C u_r^\beta\right]\left[\left(Q_s^{\gamma j}\right)^T C L_t^k\right]$		
$Q_{quqd}^{(1)}$	$\left(\overline{Q}_p^j u_r\right)\epsilon_{jk}\left(\overline{Q}_s^k d_t\right)$	Q_{qqu}	$\epsilon^{\alpha\beta\gamma}\epsilon_{jk}\left[\left(Q_p^{\alpha j}\right)^T C Q_r^{\beta k}\right]\left[(u_s^\gamma)^T C e_t\right]$		
$Q_{quqd}^{(8)}$	$\left(\overline{Q}_p^j T^A u_r\right)\epsilon_{jk}\left(\overline{Q}_s^k T^A d_t\right)$	$Q_{qqq}^{(1)}$	$\epsilon^{\alpha\beta\gamma}\epsilon_{jk}\epsilon_{mn}\left[\left(Q_p^{\alpha j}\right)^T C Q_r^{\beta k}\right]\left[(Q_s^{\gamma m})^T C L_t^n\right]$		
$Q_{lequ}^{(1)}$	$\left(\overline{L}_p^j e_r\right)\epsilon_{jk}\left(\overline{Q}_s^k u_t\right)$	$Q_{qqq}^{(3)}$	$\epsilon^{\alpha\beta\gamma}\left(\tau^I\epsilon\right)_{jk}\left(\tau^I\epsilon\right)_{mn}\left[\left(Q_p^{\alpha j}\right)^T C Q_r^{\beta k}\right]\left[(Q_s^{\gamma m})^T C L_t^n\right]$		
$Q_{lequ}^{(3)}$	$\left(\overline{L}_p^j\sigma_{\mu\nu}e_r\right)\epsilon_{jk}\left(\overline{Q}_s^k\sigma^{\mu\nu}u_t\right)$	Q_{duu}	$\epsilon^{\alpha\beta\gamma}\left[\left(d_p^\alpha\right)^T C u_r^\beta\right]\left[(u_s^\gamma)^T C e_t\right]$		

It must be pointed out that since the Lagrangian of Eq. (4.27) depends on θ, an argument can be turned around: instead of using it to constrain θ, one can use current experimental constraints on theta from neutron EDM measurements to predict the CP-violating decay rates [189].

In addition to two-body decays, kinematical distributions in multibody transitions can be employed. For example, Dalitz plot analyses of $\eta \to \pi^+\pi^-\pi^0$ could serve as sensitive probes of CP-violation [85]. In particular, violation of "mirror" symmetry in the $\pi^+\pi^-\pi^0$ Dalitz plot could be a signal of C and CP violation. Such transitions could be studied both at the dedicated η/η' factories or in general purpose flavor experiments.

4.4 SM EFT and quark sector

To parameterize New Physics without specifying a particular model, effective operators can be written out that include quark and lepton degrees of freedom. The relevant dimension-6 operators are listed in Table 4.1 for convenience. They will be used later in this chapter to study various transitions among meson and baryon states.

4.5 Hadronic form-factors: from models to EFTs

Calculations of rare or semileptonic meson or baryon decays involve evaluation of matrix element of a current operator taken between the initial and final state hadrons. Such calculations are not trivial and cannot be performed easily, as quarks are interacting strongly. The first step in the evaluation of such matrix elements involves their kinematical parameterization following requirements of Lorentz covariance. The goal of this exercise is to reduce the nonperturbative quark dynamics to a set of scalar constants or form factors multiplied by momentum or polarization four-vectors of initial and/or final state hadrons.

Such exercise should be familiar to those readers who took a standard course of quantum field theory, as it is often employed in evaluation of vacuum polarization or vertex corrections functions. For example, a two-point function of QED or QCD can be written as

$$\Pi^{\mu\nu}(q) = i \int d^4x e^{iq\cdot x} \langle 0|T\left[j^\mu(x) j^\nu(0)\right]|0\rangle = \left(q^\mu q^\nu - g^{\mu\nu} q^2\right)\Pi(q^2) \tag{4.28}$$

which is a vacuum-to-vacuum matrix element of a time-ordered product of electromagnetic currents j. As there is only a single quantity with a Lorentz index that the integral can depend on, the transferred momentum q^μ, requirements of Lorentz covariance and current conservation unambiguously lead to the form displayed in Eq. (4.28). In that respect, the scalar function $\Pi(q^2)$ can be referred to as a form factor. It is often convenient to define $Q^2 = -q^2$.

4.5.1 Parameterization of hadronic matrix elements

Hadronic matrix elements of quark currents that are often encountered in computations of various decay and production rates can be viewed as Lorentz tensors with ranks determined by the number of Lorentz indices carried by the quark current. It might prove convenient to separate kinematics of those matrix elements by parameterizing them in terms of "global" vectors and spinors that characterize the hadronic system, such as momenta and polarization vectors of decaying and produced mesons, as well as spinors if baryons are participating in the transition. What is left are sets of Lorentz scalars that contain all dynamical information on quark hadronization. They are conventionally called *form factors* or decay constants, if the final state of a given matrix element is a vacuum. The form factors are computed using lattice formulations of QCD, QCD sum rules, or even quark models. We shall discuss such kinematical parameterizations below.

Decay constants

Let us start with the simplest matrix elements of quark currents that describe meson to vacuum transitions. Such matrix elements describe weak leptonic decays of meson states, so there are no strongly-interacting particles in the final state. Let us assume for simplicity that the quark current is between the quark states q_1 and q_2 that match the quark composition of the hadronic state.

Pseudoscalar states. Let us begin with the matrix elements of pseudoscalar state $|P\rangle$. Examples of such states include charged or neutral kaons K, charm D, or beauty B mesons. As the matrix elements of vector, scalar, or tensor current vanishes by parity conservation,

$$\langle 0|\overline{q}_1 \gamma^\mu q_2|P(p)\rangle = \langle 0|\overline{q}_1 q_2|P(p)\rangle = \langle 0|\overline{q}_1 \sigma^{\mu\nu} q_2|P(p)\rangle = 0, \tag{4.29}$$

the first non-trivial matrix element is that of the axial current, $j_A^\mu = \overline{q}_1\gamma^\mu\gamma_5 q_2$. As the axial current carries a Lorentz index μ, the matrix element must be proportional to a quantity that also carries a single Lorentz index. The only such quantity for the pseudoscalar meson is its momentum, p^μ. Thus, the matrix element must be proportional to p^μ. Since the axial current involves a γ_5-matrix, it is conventional to make the proportionality constant purely imaginary,

$$\langle 0|\overline{q}_1\gamma^\mu\gamma_5 q_2|P\rangle = i f_P p^\mu. \tag{4.30}$$

The single form-factor in Eq. (4.30), a constant f_P, is often referred to as the *decay constant**. It can be computed on the lattice, with quark models, or by employing other non-perturbative techniques.

---

*Sometimes, an alternative definition $F_P = f_P/\sqrt{2}$ is used. It is rather popular in kaon physics.

We already computed a similar matrix element in QED for a para-muonium state in Chapter 3, where it was related to a value of the wave function of the muonium state at the origin $|\psi(0)|$. Similar results can be obtained in (nonrelativistic) quark models, which provide an intuitive understanding of the decay constant as an amplitude of probability of quark and anti-quark finding each other inside the meson state.

A matrix element of the pseudoscalar current has even simpler parameterization. As the current has no Lorentz indices, the matrix element is simply a constant. Our intuition should tell us that this constant should also parameterize an amplitude of probability of quark and anti-quark finding each other inside the meson state, i.e. it should be related to f_P. It is indeed the case. Computing matrix element of the divergence of axial current,

$$\langle 0|i\partial_\mu j_A^\mu|P\rangle = -(m_{q_1}+m_{q_2})\langle 0|\overline{q}_1\gamma_5 q_2|P\rangle = -f_P m_P^2, \tag{4.31}$$

we can extract the value of the constant that parameterizes the matrix element of the pseudoscalar current,

$$\langle 0|\overline{q}_1\gamma_5 q_2|P\rangle = f_P \frac{m_P^2}{m_{q_1}+m_{q_2}}. \tag{4.32}$$

This is a standard way of parameterizing this matrix element. Note that the same result can be achieved by contracting the left and the right-handed parts of Eq. (4.30) with p^μ, and using the fact that $p^\mu = p_{q_1}^\mu + p_{q_1}^\mu$ with subsequent application of Dirac equation*.

The pseudoscalar decay constants can be computed in lattice formulations of QCD, in quark models or with QCD sum rules. Assuming that only Standard Model contributes to $P \to \ell\bar{\nu}_\ell$ leptonic decay of a pseudoscalar state, its decay width can be used to determine f_P from experimental data. For example, the width of the leptonic decay of a B-meson is given by

$$\Gamma(B^- \to \ell\bar{\nu}_\ell) = \frac{G_F^2}{8\pi} f_B^2 \left|V_{ub}\right|^2 m_\ell^2 m_B \left(1 - \frac{m_\ell^2}{m_B^2}\right). \tag{4.33}$$

The only non-perturbative parameter entering Eq. (4.33) is f_B, the B-meson decay constant, so it can be easily extracted if $\Gamma(B^- \to \ell\bar{\nu}_\ell)$ is measured experimentally. It is indeed possible, provided that helicity suppression factor m_ℓ^2 does not diminish the rate too much. This can be achieved by searching for the decays with muons or taus in the final state. Note that in the Standard Model the decay rate is proportional to a combination $f_B\left|V_{ub}\right|$, so it can be used to extract the value of the CKM matrix element V_{ub}, provided f_B is separately computed. We will discuss the sensitivity of leptonic decays to New Physics interactions later in this chapter.

Vector states. Analysis of the matrix elements with the vector initial state is quite similar. Parity conservation now requires that

$$\langle 0|\overline{q}_1\gamma^\mu\gamma_5 q_2|V(\epsilon,p)\rangle = \langle 0|\overline{q}_1\gamma_5 q_2|V(\epsilon,p)\rangle = \langle 0|\overline{q}_1\sigma^{\mu\nu}\gamma_5 q_2|V(\epsilon,p)\rangle = 0, \tag{4.34}$$

so it is the vector and the tensor quark currents that give non-zero matrix elements. There is, however, one major difference here: there are now two Lorentz vectors that can be employed to parameterize the matrix element: a polarization vector for the meson state, $\epsilon^\mu(p)$, and meson's momentum, p^μ. Thus, in general, one can write

$$\langle 0|\overline{q}_1\gamma^\mu q_2|V(\epsilon,p)\rangle = A m_V \epsilon^\mu + B p^\mu, \tag{4.35}$$

*Similar arguments can be used to relate matrix elements of the vector current to that of the scalar one for the scalar mesons. In this case the equation of motion is given by $i\partial_\mu\left(\overline{q}_1\gamma^\mu q_2\right) = (m_2 - m_1)\overline{q}_1 q_2$. This method can also be used to relate more complicated matrix elements.

where A and B are some constants, and we introduced a factor of m_V to ensure that both A and B have the same units. Multiplying Eq. (4.35) by the vector meson's momentum gives

$$0 = Am_V \ \epsilon \cdot p + Bm_V^2, \tag{4.36}$$

where the zero on the left hand side of Eq. (4.36) follows from vector current conservation. The first term on the right hand side of that equation is also zero due to the Lorentz condition $\epsilon \cdot p = 0$, resulting in the requirement that $B = 0$. The coefficient A is conventionally called f_V, the vector decay constant for a vector meson. Thus,

$$\langle 0|\overline{q}_1 \gamma^\mu q_2|V(\epsilon, p)\rangle = f_V m_V \epsilon^\mu. \tag{4.37}$$

Matrix element of the tensor current is parameterized in a similar way. One simply needs to remember that $\sigma^{\mu\nu}$ is anti-symmetric in $\mu \leftrightarrow \nu$ interchange, so the matrix element must be anti-symmetric too. There is only one possible anti-symmetric combination of p and ϵ,

$$\langle 0|\overline{q}_1 \sigma^{\mu\nu} q_2|V(\epsilon, p)\rangle = i\left(\epsilon^\mu p^\nu - p^\mu \epsilon^\nu\right) f_T. \tag{4.38}$$

Here f_T is the tensor decay constant, which can also be computed using various non-perturbative techniques. As there are no tensor currents in the SM Lagrangian, Eq. (4.38) parameterizes BSM operators.

Vector meson decay constants can also be extracted from experimental data. For the quarkonium states, such as J/ψ or Υ, leptonic decays $V \to \ell^+\ell^-$ can be used. The corresponding formula for a decay of the Υ state is given in Eq. (4.19).

There are no analogues of decay constants for the baryon states due to baryon number conservation. Matrix elements for baryon-number violating transitions are usually computed on a case-by-case basis.

Form factors for meson-to-meson transitions

The expressions for the matrix elements become more complicated if there are strongly interacting particles in both initial and final states. Just as in the cases considered previously, parameterizations of the matrix elements would depend on the same global four-vectors as meson momenta and polarizations. Let us consider matrix elements that are most often used in particle phenomenology.

Pseudoscalar to pseudoscalar transitions. Matrix elements describing transitions between two spin-less states can be parameterized in terms of two Lorentz vectors, the momenta of the initial p_1 and the final p_2 state particles. Again, parity forbids transitions driven by an axial or pseudo-tensor currents,

$$\langle P_2(p_2)|\overline{q}_1 \gamma^\mu \gamma_5 q_2|P_1(p_1)\rangle = \langle P_2(p_2)|\overline{q}_1 \sigma^{\mu\nu} \gamma_5 q_2|P_1(p_1)\rangle = 0. \tag{4.39}$$

Matrix elements of the vector current could in principle be parameterized with two form-factors, each multiplied by a momentum p_1^μ or p_2^μ. It is, however, conventional to choose linear combinations of those momenta, P^μ and q^μ,

$$\begin{aligned} P &= p_1 + p_2, \\ q &= p_1 - p_2. \end{aligned} \tag{4.40}$$

In terms of those two momenta the matrix element of the vector current is

$$\langle P_2(p_2)|\overline{q}_1 \gamma^\mu q_2|P_1(p_1)\rangle = f_+(q^2) P^\mu + f_-(q^2) q^\mu. \tag{4.41}$$

Note that the form factors $f_\pm$ depend on q^2, the square of the transferred momentum. This dependence will be discussed later in this section.

There is an alternative parameterization of this matrix element which is also commonly used in phenomenological analyses of semileptonic meson decays,

$$\langle P_2(p_2)|\overline{q}_1\gamma^\mu q_2|P_1(p_1)\rangle = F_+(q^2)\left(P^\mu - \frac{m_{P_1}^2 - m_{P_2}^2}{q^2}q^\mu\right) + F_0(q^2)\frac{m_{P_1}^2 - m_{P_2}^2}{q^2}q^\mu. \tag{4.42}$$

These parameterizations are completely equivalent. It is possible to relate the form factors in Eq. (4.42) to those defined in Eq. (4.41),

$$F_+(q^2) = f_+(q^2), \quad F_0(q^2) = f_+(q^2) + \frac{q^2}{m_{P_1}^2 - m_{P_2}^2}f_-(q^2). \tag{4.43}$$

To avid a singularity at $q^2 = 0$, and as it follows from Eq. (4.43), we need to require a relation between the normalizations of the form factors F_+ and F_0 in Eq. (4.42),

$$F_+(0) = F_0(0). \tag{4.44}$$

The form factors parameterize hadronic dynamics. They can be studied experimentally through the analyses of semileptonic differential decay rates $d\Gamma/dq^2$. It can be shown that in the transitions $P_1 \to P_2\ell\nu_\ell$ only one form factor would contribute if the mass of the final state lepton can be neglected. In this case, the differential decay rate can be written as

$$\frac{d\Gamma(P_1 \to P_2\ell\nu_\ell)}{dq^2} = \frac{G_F^2\left|V_{q_1q_2}\right|^2}{24\pi^3}\left|\mathbf{p}_{P_2}\right|^3\left|F_+(q^2)\right|^2, \tag{4.45}$$

where $|\mathbf{p}_{P_2}|$ is the magnitude of the P_2-meson's 3-momentum vector in the P_1-meson rest frame. As can be seen from Eq. (4.45), only a single form factor, $F_+(q^2)$, contributes. In Eq. (4.45) we introduced a factor $G_F^2\left|V_{q_1q_2}\right|^2$ which is common in semileptonic meson decays in the Standard Model.

Studies of q^2 dependence of the form factors, as well as their absolute normalizations, i.e. the values of $F_i(q^2 = 0)$, are quite non-trivial. It is possible to model-independently describe the form factors in some kinematical regions. For example, at large q^2 - or alternatively, at low recoil, the form factors for heavy to light transitions, such as $B \to \pi\ell\nu_\ell$ or $D \to \pi\ell\nu_\ell$ are dominated by the B^* or D^* poles. In this limit, the $B \to \pi$ form factor is

$$f_+^B(q^2) = \frac{f_{B^*}g_{B^*B\pi}}{2m_{B^*}(1 - q^2/m_{B^*}^2)}, \tag{4.46}$$

where f_{B^*} is the decay constant of the B^* meson, and the coupling $g_{B^*B\pi}$ is defined as

$$\langle B^-(p)\pi^+(q)|B^{*0}(p+q)\rangle = -g_{B^*B\pi}q\cdot\epsilon. \tag{4.47}$$

Here ϵ is the polarization vector of the B^*. Lattice QCD has a definite answer on the shapes of the form factors in some kinematical regions. Various phenomenological parameterizations of the form factors are also often used.

The most common and simplest model-dependent parametrization of $F_+(q^2)$ is a "single pole" shape, where the pole refers to the lowest mass vector resonance formed in the t-channel with quantum numbers of the quark current. For example, in the decay $D \to \pi e\bar{\nu}_e$ the dominant pole is the $D^\star$, a vector state with 1^- quantum numbers,

$$F_+^{\text{pole}}(q^2) = \frac{F_+(0)}{1 - \hat{q}^2}, \tag{4.48}$$

where $F_+(0)$ is the value of the form factor at zero momentum recoil that has to be fixed either from the lattice QCD or from other arguments, and $\hat{q}^2 = q^2/m_{D^*}^2$. While physical masses of the states $D^*(2010)$ (for $D \to \pi$ transition) or $D_s^*(2112)$ (for $D \to K$ transition) could be used, the mass m_{D*} is often taken as a fit parameter, as there is no reason to believe that the lowest-lying pole would saturate the form factor over the whole available kinematical range. More complicated shapes, with more effective poles, are also available,

$$F_+(q^2) = \frac{F_+(0)}{(1-\alpha)} \frac{1}{1 - q^2/m_V^2} + \sum_{k=1}^{N} \frac{\rho_k}{1 - \frac{1}{\gamma_k}\frac{q^2}{m_V^2}}, \tag{4.49}$$

where α determines the strength of the dominant pole, ρ_k gives the strength of the kth term in the expansion, and $\gamma_k = m_{V_k}^2/m_V^2$, with m_{V_k} representing masses of the higher mass states with vector quantum numbers. In principle, a form factor can be approximated to any desired accuracy by introducing a large number of effective poles. Keeping the number of terms in this expansion manageable, a popular parameterization due to Becirevic and Kaidalov (BK) [15] is often used, representing the $N = 1$ truncation of the expansion in Eq. (4.49),

$$F_+^{BK}(q^2) = \frac{F_+(0)}{(1-\hat{q}^2)(1 - a_{BK}\hat{q}^2)}, \tag{4.50}$$

where a_{BK} is a fit parameter. As with the case of a single pole shape in Eq. (4.48), a good fit to experimental distribution can be obtained if m_V is regarded as a fit parameter as well. While there is only one extra parameter between Eq. (4.48) and (4.50), the BK parameterization starts off with more parameters. A number of parameters is reduced by applying relationships obtained in the heavy-quark limit to decrease the number of parameters displayed in the above equation. A further extension of the BK parameterization was proposed by Ball and Zwicky (BZ) [12],

$$F_+^{BZ}(q^2) = \frac{F_+(0)}{1-\hat{q}^2}\left(1 + \frac{r_{BZ}\hat{q}^2}{1 - a_{BZ}\hat{q}^2}\right), \tag{4.51}$$

where r_{BZ} and a_{BZ} are the shape parameters. It must be pointed out that an alternative form of BZ parameterization exists,

$$F_+^{BZ}(q^2) = \frac{r_1}{1-\hat{q}^2} + \frac{r_2}{1 - a_{BZ}\hat{q}^2}, \tag{4.52}$$

can be related to the parameterization of Eq. (4.51) by the identification $F_+(0) = r_1 + r_2$ and $r_{BZ} = (a_{BZ} - 1)r_2/(r_1 + r_2)$. Note that a_{BZ} represents parameterization of the continuum states above D^* and therefore $a_{BZ} < 1$.

All form factor parameterizations discussed above represent physically-motivated ways to describe hadronic input. It is important to understand that the shape, or the functional form for the form factor induces a bias in the interpretation of results of an experimental studies. Model-independent arguments based on some general principles, such as unitarity, are needed. We will describe those below. Another question that might be asked is related to the uncertainty associated with the *choice* of a particular shape of the function that represents the form factor. This question may be addressed using bias-independent fits of the experimental form factors, for instance by using artificial neural networks [95].

The form factors described above are often employed to describe processes induced by the Standard Model operators. In principle, other effective operators can be can be generated by New Physics interactions. The simplest one is a scalar operator, $\overline{q}_1 q_2$, whose parameterization is simply a constant. It is conventional, again, to rewrite it in terms of $F_0(q^2)$ of Eq. (4.42),

$$\langle P_2(p_2)|\overline{q}_1 q_2|P_1(p_1)\rangle = \frac{m_{P_1}^2 - m_{P_2}^2}{m_{q_2} - m_{q_1}} F_0(q^2), \tag{4.53}$$

which is why the form factor $F_0(q^2)$ in Eq (4.42) is often referred to as a "scalar form factor".

Contrary to the case of parameterizations of the matrix elements describing leptonic decays of the pseudoscalar states, transitions between the pseudoscalar states governed by the tensor current do not vanish, which follows from the fact that an antisymmetric combination of two different four vectors can be constructed,

$$\langle P_2(p_2)|\bar{q}_1\sigma^{\mu\nu}q_2|P_1(p_1)\rangle = \frac{i}{m_{P_1}+m_{P_2}}\left(p_2^\mu p_1^\nu - p_1^\mu p_2^\nu\right)F_T(q^2). \tag{4.54}$$

The matrix element of Eq. (4.54) parameterizes a contribution from an operator that also does not appear in the Standard Model. If exists in BSM models, it can be constrained from the angular analysis of semileptonic transitions between P_1 and P_2, e.g. in semileptonic B or D decays. In principle, $F_T(q^2)$ is a new form factor, which needs to be separately computed. However, in the heavy quark limit (in both Heavy Quark Effective Theory and Soft-Collinear Effective Theory [149]) one finds that $F_T \approx F_+$.

Pseudoscalar to vector transitions. Transitions between a pseudoscalar and a vector state in general require four form factors for both vector and axial currents due to the number of polarization states of the vector particle that can be reached in this transition. In order to write the most general parameterization of those matrix elements we can employ the momentum of the initial p_1 state particle, as well as momentum p_2 and polarization vector $\epsilon^\mu(p_2)$ of the final state particle. From now on we shall not repeat the above discussion of how to properly use equations of motion, Lorentz condition, and other requirements to reduce the number of form factors. We will simply state the conventional form of the matrix element parameterization, and leave it as an exercise for the reader to arrive at the stated form.

The matrix element of the vector current is conventionally expressed in terms of a single form factor $V(q^2)$,

$$\langle V(p_2)|\bar{q}_1\gamma^\mu q_2|P(p_1)\rangle = \frac{2V(q^2)}{m_P+m_V}\epsilon^{\mu\nu\alpha\beta}\epsilon^*_\nu p_{2\alpha}p_{1\beta}. \tag{4.55}$$

The matrix element of the axial current depends on three form factors,

$$\begin{aligned}\langle V(p_2)|\bar{q}_1\gamma^\mu\gamma_5 q_2|P(p_1)\rangle &= i(m_P+m_V)\left(\epsilon^{*\mu}-\frac{\epsilon^*\cdot q}{q^2}q^\mu\right)A_1(q^2) \\ &- i\left(P^\mu-\frac{m_P^2-m_V^2}{q^2}q^\mu\right)\frac{\epsilon^*\cdot q}{m_P+m_V}A_2(q^2)+\frac{2im_V\epsilon^*\cdot q}{q^2}q^\mu A_0(q^2).\end{aligned} \tag{4.56}$$

In semileptonic decays $P \to V\ell\nu_\ell$ the decay rates are dominated by $V(q^2)$, $A_1(q^2)$, and $A_2(q^2)$ form factors if the masses of the final state leptons can be neglected.

Just like in the case of the pseudoscalar final state considered above, the q^2 dependence of the vector and axial form factors, as well as their absolute normalizations are non-trivial to obtain. A single pole parameterization of the form factors is popular,

$$V(A_i)_{\text{pole}}(q^2) = \frac{V(A_i)(0)}{1-\hat{q}^2}, \tag{4.57}$$

where $V(A_i)(0)$ are the values of the form factors at zero momentum recoil and $\hat{q}^2 = q^2/m_R^2$. The resonance R is a state with appropriate quantum numbers: the vector form factor $V(q^2)$ is dominated by vector resonances, the axial form factors $A_1(q^2)$ and $A_2(q^2)$ are dominated by axial resonant states, while the $A_0(q^2)$ form factor is dominated by the pseudoscalar resonances. Other parameterizations, such as BK or BZ shapes can be introduced as well with appropriate resonance mass substitutions.

The matrix element of the tensor current depends on three form factors,

$$\langle V(p_2)|\overline{q}_1\sigma^{\mu\nu}q_2|P(p_1)\rangle = i\epsilon^{*\alpha}\epsilon^{\mu\nu\rho\eta}\left\{-\left[\left(P_\rho - \frac{m_P^2-m_V^2}{q^2}q_\rho\right)g_{\alpha\sigma} + \frac{2}{q^2}p_{1\alpha}p_{1\rho}p_{2\sigma}\right]T_1(q^2)\right.$$
$$\left. + \left(\frac{2}{q^2}p_{1\alpha}p_{1\rho}p_{2\sigma} - \frac{m_P^2-m_V^2}{q^2}q_\rho g_{\alpha\sigma}\right)T_2(q^2) + \frac{2}{m_P^2-m_V^2}p_{1\alpha}p_{1\rho}p_{2\sigma}T_3(q^2)\right\}. \quad (4.58)$$

This matrix element is important in parameterizations of matrix elements of radiative decays of B or D mesons, such as $B \to K^*\gamma$ or $D \to \rho\gamma$.

Form factors for baryon-to-baryon transitions

There are several factors that make dealing with baryon transitions a more complicated affair. Of course, baryons have half-integer spin, which means that transitions between baryons initiated by quark currents will be dependent on baryons' spinors and, in general, Dirac gamma matrices that are sandwiched between those spinors. Lets us now consider matrix elements of currents between baryons of $J^P = (1/2)^+$ spin parity. We will denote the initial state baryon as $\mathcal{B}$ and the final state one as $\mathcal{B}'$ and their 4-momenta p_1 and p_2, respectively.

First, consider matrix elements of the vector current. Just like in case of the mesons described above, we can write the most general Lorentz structure built out of P, q (as defined in Eq. (4.40)), and Dirac γ-matrices,

$$\begin{aligned}\langle \mathcal{B}'(p_2)|\overline{q}_1\gamma^\mu q_2|\mathcal{B}(p_1)\rangle &= \overline{u}'(p_2)\big[f_1(q^2)\gamma^\mu + f_2 i\sigma^{\mu\nu}q^\nu + f_3(q^2)q^\mu \\ &\quad + f_4(q^2)i\sigma^{\mu\nu}P^\nu + f_4(q^2)P^\mu\big]u(p_1),\end{aligned} \quad (4.59)$$

where $u'(p_2)(u(p_1))$ is the spinor of the final (initial) state baryon. Using equations of motion we can remove dependence on P and write a conventional parameterization of this form factor,

$$\langle \mathcal{B}'(p_2)|\overline{q}_1\gamma^\mu q_2|\mathcal{B}(p_1)\rangle = \overline{u}'(p_2)\left[f_1(q^2)\gamma^\mu + f_2 i\sigma^{\mu\nu}q^\nu + f_3(q^2)q^\mu\right]u(p_1). \quad (4.60)$$

For the matrix element of the axial current,

$$\langle \mathcal{B}'(p_2)|\overline{q}_1\gamma^\mu\gamma_5 q_2|\mathcal{B}(p_1)\rangle = \overline{u}'(p_2)\left[F_1(q^2)\gamma^\mu + F_2 i\sigma^{\mu\nu}q^\nu + F_3(q^2)q^\mu\right]\gamma_5 u(p_1). \quad (4.61)$$

The form factors in Eq. (4.60) and (4.61) describe hadronic dynamics of semileptonic transitions between baryons, both in the SM, as $(V - A)$, or in BSM models, as $(V + A)$. Other effective operators might also include scalar or pseudoscalar quark currents.

For those, just like for the mesonic currents, we can contract both sides of Eqs. (4.60) and (4.61) with the momentum q^μ to get matrix elements for the scalar

$$\langle \mathcal{B}'(p_2)|\overline{q}_1 q_2|\mathcal{B}(p_1)\rangle = \left[f_1(q^2)\frac{m_\mathcal{B} - m_{\mathcal{B}'}}{m_{q_2}+m_{q_1}} + f_3(q^2)\frac{q^2}{m_{q_2}+m_{q_1}}\right]\overline{u}'(p_2)u(p_1), \quad (4.62)$$

and for the pseudoscalar currents,

$$\langle \mathcal{B}'(p_2)|\overline{q}_1\gamma_5 q_2|\mathcal{B}(p_1)\rangle = \left[F_1(q^2)\frac{m_\mathcal{B} + m_{\mathcal{B}'}}{m_{q_2}-m_{q_1}} + F_3(q^2)\frac{q^2}{m_{q_2}-m_{q_1}}\right]\overline{u}'(p_2)\gamma_5 u(p_1). \quad (4.63)$$

It is also possible to write out matrix elements of the tensor and pseudotensor currents. Using techniques described above it is possible to show that the most general expression describing transitions due to the tensor current is

$$\begin{aligned}\langle \mathcal{B}'(p_2)|\overline{q}_1\sigma^{\mu\nu}q_2|\mathcal{B}(p_1)\rangle &= \overline{u}'(p_2)\big[g_1(q^2)\sigma^{\mu\nu} - ig_2\gamma^{[\mu}P^{\nu]} \\ &- \quad ig_3(q^2)\gamma^{[\mu}q^{\mu]} - ig_4(p^2)P^{[\mu}q^{\nu]}\big]u(p_1),\end{aligned} \quad (4.64)$$

where we introduced a short-hand notation $A^{[\mu}B^{\nu]} = A^\mu B^\nu - A^\nu B^\mu$. Matrix elements of the pseudotensor current can be obtained using the relation given in Eq. (2.81).

4.5.2 Calculations of hadronic form factors

Form factors for various meson and baryon transitions described above represent genuine nonperturbative quantities whose computation is a non-trivial task. There are many different methods with varying degrees of reliability that can be applied, from various quark models [137] to QCD sum rules [50], and lattice QCD. A complete description of those methods goes beyond the scope of this book – in fact, it deserves its own book! We hope that the interested reader will follow provided references to see how much work goes into proper computations of hadronic form factors.

4.5.3 Analyticity constraints and a z-expansion

While different models of q^2-dependence of various form factors described in Sect. 4.5.1 provide good fits to experimental data, a separate, model-independent determination of such dependence is needed. We are, in fact, required to know such q^2 dependence to extract the values of the CKM matrix elements $V_{q_1 q_2}$ or detect the presence of non-SM physics. Let us discuss that dwelling on the example of the $F_{+,0}(q^2)$ functions. Our discussion will closely follow [113].

Model-independent studies of the form factors are possible by employing very general properties of the vertex functions, such as unitarity, analyticity, and crossing symmetry. In particular, analytic continuation of the function $F_{+,0}(q^2) \to F_{+,0}(t)$ to the complex values of t allows application of powerful theorems of complex analysis. Since the further discussion will be similar for F_+ and F_0, we will drop the index $+$ or 0 and spell it out only where it is needed.

Assuming that such analytic continuation is possible, the function $F(t)$ can be continued to a complex t plane. It is analytic throughout the plane with the exception of (possible) poles that correspond to the excited states with quantum numbers of the current in t-channel, and a branch cut along the real positive axis starting at the point $t = t_+$, which corresponds to the kinematical point where real $P_1 P_2$ states can be produced, $t_\pm = (m_{P_1} \pm m_{P_2})^2$.

The technique that we follow below is oftentimes referred to as the "z-expansion". While it is often applied to bound form factors, it can also be used to study the convergence of any perturbative expansion. Let us perform a transformation of the complex t-plane, such that all of it is mapped onto the interior of the unit circle in the plane of a new complex variable z,

$$z(t, t_0) = \frac{\sqrt{t_+ - t} - \sqrt{t_+ - t_0}}{\sqrt{t_+ - t} + \sqrt{t_+ - t_0}}. \tag{4.65}$$

One can check the that all values of t correspond to values of $|z| \leq 1$. A transformation like in Eq. (4.65) has a parameter t_0 that allows to choose which point in the t-plane maps onto $z = 0$. It is often chosen in such a way that it minimizes the maximum value of $|z|$, i.e.

$$t_0 = t_+ \left(1 - \sqrt{1 - \frac{t_-}{t_+}}\right). \tag{4.66}$$

Now, since the singularities of $F(z)$ are mapped onto a boundary $|z| = 1$, the form factor is analytic inside the unit circle in z. Thus, it is possible to expand it around any point inside $|z| \leq 1$. In particular, we can write

$$F(t) = \frac{1}{P(t)\phi(t, t_0)} \sum_{k=0}^{\infty} a_k(t_0) z^k(t, t_0), \tag{4.67}$$

where $P(t)$ and $\phi(t, t_0)$ are the functions that remove possible subthreshold poles and other singularities. A good example of an application of the function $P(t)$ is the case of the $B \to \pi$ form factor, where a vector state B^* appears to the left of the beginning of the branch cut t_+. Since we know the position of the pole associated with this state, the function $P(t)$ can be chosen such that it has a simple zero at $t = m_{B^*}^2$. If we also require that $|P| = 1$ along the cut, the function $P(t)$ can be chosen as

$$P_{F_+}^{B\to\pi}(t) = z(t, m_{B^*}^2). \tag{4.68}$$

Note that for a very similar transition form factor in semileptonic D-decays $D \to \pi$, the first vector excited state D^* appears to the right of the beginning of the cut, i.e. on a different physical sheet. Thus, there is no need to remove the pole from our analysis and the Blaschke factor can be chosen as $P_{F_+}^{D\to\pi}(t) = 1$. We shall discuss the choice of $\phi(t, t_0)$ momentarily.

What is the reason for such transformation? We did not do anything special but expressed a non-perturbative function $F(t)$ in terms of the infinite number of unknown parameters a_k. What is the rationale for such action? The claim is that the z-expansion of Eq. (4.67) converges much faster than the expansion in t, so only a few first terms would be needed to bound the form factor $F(t)$.

In fact, one can also bound the coefficients a_k by imposing constraints on their sum $\sum_k a_k^2$. To do so, let us point out that Eq. (4.67) is nothing but an expansion of the complex function $F(t)P(t)\phi(t, t_0)$ in a Laurent series about a point $z = 0$. For such expansion the coefficients a_k are given by the line integral

$$a_k = \frac{1}{2\pi i} \oint_C \frac{dz}{z} F(t)P(t)\phi(t, t_0). \tag{4.69}$$

A bound can be set by considering the norm,

$$\sum_{k=0}^{\infty} a_k^2 = \frac{1}{2\pi i} \oint_C \frac{dz}{z} |P\phi F|^2 = \int_{t_+}^{\infty} \frac{dt}{t - t_0} \sqrt{\frac{t_+ - t_0}{t - t_+}} |P\phi F|^2, \tag{4.70}$$

where the last relation follows from Eq. (4.65). This norm can be evaluated using the form factors for the related process of $P_1 P_2$ production, which follows from crossing symmetry. The idea is to bound the norm of Eq. (4.70) such that

$$\sum_{k=0}^{\infty} a_k^2 \leq \chi(Q^2), \tag{4.71}$$

where $\chi(Q^2)$ is some function that we are going to determine.

We now have to discuss a choice of the function $\phi(t, t_0)$. A default choice for this function is determined from arguments based on unitarity. That is, ϕ is chosen in such a way that the norm of Eq. (4.70) is identified as a rate for some (perturbatively) calculable inclusive process [113].

Let us use an example of a pseudoscalar-to-pseudoscalar form factor, which is driven by a vector current $j_V^\mu = \overline{q}_1 \gamma^\mu q_2$ in Eq. (4.42). Consider the correlation function

$$\Pi^{\mu\nu}(q) = i \int d^4x e^{iq\cdot x} \langle 0|T \left[j_V^\mu(x) j_V^{\nu\dagger}(0) \right] |0\rangle = \left(q^\mu q^\nu - g^{\mu\nu} q^2 \right) \Pi_1(q^2) + q^\mu q^\nu \Pi_0(q^2). \tag{4.72}$$

The functions $\Pi_1(q^2)$ and $\Pi_0(q^2)$ are such that the unsubtracted dispersion relations can be written for

$$\begin{aligned} \chi_{F_+}(Q^2) &= \frac{1}{2} \frac{\partial^2}{\partial(q^2)^2} \left[q^2 \Pi_1(q^2) \right] = \frac{1}{\pi} \int_0^\infty dt \, \frac{t \, \mathrm{Im}\Pi_1(t)}{(t + Q^2)^3}, \\ \chi_{F_0}(Q^2) &= \frac{\partial^2}{\partial(q^2)^2} \left[q^2 \Pi_0(q^2) \right] = \frac{1}{\pi} \int_0^\infty dt \, \frac{t \, \mathrm{Im}\Pi_0(t)}{(t + Q^2)^3}. \end{aligned} \tag{4.73}$$

Introducing an isospin factor η to account for different possible pseudoscalar states, we note that for $t > t_+$, i.e the threshold for production of P_1 and P_2 states from the vacuum under action of the current j_V^μ, such (exclusive) contribution would be always same or less than the inclusive production, i.e. production of all possible states. Inserting P_1P_2 intermediate state in Eq. (4.72), we can write the following inequalities for the imaginary parts of $\Pi_i(t)$,

$$\begin{aligned} \frac{\eta}{48\pi t^3}\left[(t-t_+)(t-t_-)\right]^{3/2}|F_+(t)|^2 &\leq \text{Im}\,\Pi_1(t), \\ \frac{\eta}{16\pi}\frac{t_+t_-}{t^3}\left[(t-t_+)(t-t_-)\right]^{1/2}|F_0(t)|^2 &\leq \text{Im}\,\Pi_0(t). \end{aligned} \tag{4.74}$$

This shows that an upper bound on the norm of Eq. (4.70) can be obtained if $\phi(t,t_0)$ is chosen such that

$$\begin{aligned} \phi_{F_+}(t,t_0) &= \sqrt{\frac{\eta}{48\pi}\frac{t_+-t}{(t_+-t_0)^{1/4}}}\left(\frac{z(t,0)}{t}\right)\left(\frac{z(t,-Q^2)}{t+Q^2}\right)^{3/2} \\ &\times \left(\frac{z(t,t_0)}{t_0-t}\right)^{-1/2}\left(\frac{z(t,t_-)}{t_--t}\right)^{-3/4}, \\ \phi_{F_0}(t,t_0) &= \sqrt{\frac{\eta t_+t_-}{48\pi}\frac{\sqrt{t_+-t}}{(t_+-t_0)^{1/4}}}\left(\frac{z(t,0)}{t}\right)\left(\frac{z(t,-Q^2)}{t+Q^2}\right) \\ &\times \left(\frac{z(t,t_0)}{t_0-t}\right)^{-1/2}\left(\frac{z(t,t_-)}{t_--t}\right)^{-1/4}, \end{aligned} \tag{4.75}$$

for the F_+ and F_0 form factors, respectively. Since the expansion of Eq. (4.67) converges fast, it is sufficient to retain only a few terms in the expansion. Retaining only terms up to z^2 we get

$$F(t) = \frac{1}{P(t)\phi(t,t_0)}\left[a_0 + a_1 z + a_2 z^2\right]. \tag{4.76}$$

It now can be used to constrain semileptonic form factors.

The method described in this section is quite flexible and can be applied to put model-independent constraints on other observables. For example, analyticity considerations based on z-expansion methods constrain electromagnetic and weak form factors of nucleons, which leads to predictions of such quantities as proton's charge and magnetic radii [73, 114].

4.6 Decays of heavy flavors into leptons

A considerable effort has been put into the understanding of non-relativistic heavy quark bound states, as the presence of several well-separated energy scales allows for a simplified description of non-perturbative QCD dynamics and puts additional restrictions on form factors involved in the parameterization of quark transitions. Let us thus discuss a heavy-light bound state, such as a B- or a D-meson.

Such a state is characterized by two essential energy scales: the heavy quark mass, m_Q, and the scale of non-perturbative physics, $\Lambda_{QCD} \sim 400$ MeV. All light degrees of freedom are relativistic and have energy and momentum of order Λ_{QCD}, which also characterizes the bound state energy. It then appears that the scale m_Q is not relevant for a description of a bound state and can be integrated out to build an effective theory of heavy mesons. This effective theory, which goes under the name of *Heavy Quark Effective Theory* or HQET [135, 149]. It allows for a systematic description of heavy quark systems in the limit $m_Q \to \infty$ and inclusion of $1/m_Q$ corrections to observables with any desired accuracy.

Bound state dynamics does simplify in the heavy quark limit. For example, spin effects of the heavy quark are proportional to $1/m_Q$. This means that at the leading order in $1/m_Q$ expansion heavy quark spin is irrelevant for bound state dynamics of heavy-light states, implying that the masses of pseudoscalar B and vector B^* mesons are degenerate in the heavy quark limit. This allows us to think about them as members of the same multiplet, similarly to the isospin doublet of protons and neutrons.

4.6.1 Taming strong QCD: heavy quark symmetry

The fact that heavy quark's spin decouples in the heavy quark limit has interesting consequences for the matrix elements of quark currents. In fact, it affects both the currents and the meson or baryon states. Let us first see how we can simplify the Lagrangian in the heavy quark limit. Then we will build the heavy quark states.

To build a heavy state, we would need to recall that it contains one heavy quark and one light antiquark, so it must transform as a triplet under $SU(3)_V$ chiral symmetry transformation. It will be convenient to first construct the states in the meson rest frame where $v = (1, \vec{0})$ and then boost it to the arbitrary fame labeled by its velocity v. It is interesting to point out that this means that we are describing D (or $c\bar{q}$) and $\bar{B}$ (or $b\bar{q}$) states (notice the bar!). The reasons for such naming convention are purely historical.

Effective operators and heavy quark limit

Let us show that the QCD Lagrangian that describes dynamics of heavy quarks $Q = \{b, c\}$ simplifies in the limit $m_Q \to \infty$. In this section we will mainly be talking about heavy quarks in meson or baryon states, where all other quarks are light, $m_q \sim \Lambda_{QCD}$. Such systems are sometimes called “hydrogen atoms of QCD" due to their relative simplicity.

Consider a field $\psi_Q(x)$ that describes a heavy quark. In a heavy-light bound state, such as a B-meson, its mass m_Q represents the highest scale in the problem. All other scales in the problem, such as the momentum of the light quark, are roughly represented by Λ_{QCD}, which is a few hundred MeV. This fact dictates the power counting for the system. This power counting is quite easy: we simply count inverse powers of inverse heavy quark mass, which grows with each operator's dimension. For instance, the heavy fermion's kinetic energy, $K = p_Q^2/(2m_Q)$, is now given by a power-suppressed operator. In the limit of an infinitely heavy mass, the heavy quark simply serves as a source of gluonic fields, while itself remaining (nearly) static. In general, in a frame where the heavy quark moves with nonrelativistic velocity v, the heavy quark momentum can be written as

$$p_Q = m_Q v + k \tag{4.77}$$

where k is the heavy quark's *residual momentum*, $|k| \ll m_Q$, and $v^2 = 1$. In this chapter it would be easier for us to use the standard parameterization of Dirac matrices. In this parameterization the static solution is simply given by the upper two components of the Dirac bi-spinor $\psi_Q(x)$.

Let us derive a Lagrangian for the heavy quark field in the $m_Q \to \infty$ limit. It can be derived from the usual Dirac Lagrangian that describes the motion of all quarks,

$$\mathcal{L} = \overline{\psi}_Q(x) \left(i \not{D} - m_Q \right) \psi_Q(x). \tag{4.78}$$

In order to take a nonrelativistic limit, it would be convenient separate out a large mechanical part of the field $e^{-im_Q v\cdot x}$, which is simply a solution of a free Dirac equation in the limit of static quark,

$$\begin{aligned} \psi_Q(x) = e^{-im_Q v\cdot x} \widetilde{\psi}_Q(x) &= e^{-im_Q v\cdot x} \left[\mathcal{P}_+^v \widetilde{\psi}_Q(x) + \mathcal{P}_-^v \widetilde{\psi}_Q(x) \right] \\ &\equiv e^{-im_Q v\cdot x} \left[h_v(x) + H_v(x) \right], \end{aligned} \tag{4.79}$$

where we labeled the fields h_v and H_v by their velocity and introduced projection operators

$$\mathcal{P}_\pm^v = \frac{1 \pm \not{v}}{2}. \tag{4.80}$$

It is easy to check that in the frame where the heavy quark is static, i.e. its velocity is $v^\mu = (1, \vec{0})$, $\mathcal{P}^v_\pm$ project out the upper (lower) two components of the bi-spinor $\psi_Q(x)$. It is easy to check by direct computation that

$$\mathcal{P}^v_+ + \mathcal{P}^v_- = 1 \quad \text{and} \quad \not{v}\mathcal{P}^v_\pm = \pm \mathcal{P}^v_\pm. \tag{4.81}$$

Now, because of Eq. (4.81), the fields h_v and H_v satisfy the relations

$$\not{v} h_v = h_v, \quad \text{and} \quad \not{v} H_v = -H_v. \tag{4.82}$$

We can now insert the field ψ_Q written in terms of h_v and H_v of Eq. (4.79) into the Dirac Lagrangian of Eq. (4.78). The derivative acting on the exponential brings down factors of $-im_Q v_\mu$ so we obtain

$$\mathcal{L} = \overline{h}_v i\not{D} h_v + \overline{H}_v i\not{D} H_v + 2m_Q \overline{H}_v H_v + \overline{h}_v i\not{D} H_v + \overline{H}_v i\not{D} h_v \tag{4.83}$$

So far we have not done much besides factoring out a part of the fermion field for which we supposedly know the solution of the equations of motion. It, however, does look like we now have two fields with different properties: a massless field h_v, and the "heavy" field H_v with the mass $2m_Q$. In the limit $m_Q \to \infty$ the field H_v describes infinitely heavy particles. It is quite similar to SM EFT, where all possible heavy NP particles are integrated out. Likewise, we can integrate H_v out of our theory.

Before we do so, let us simplify Eq. (4.83) a bit. We can use Eq. (4.82) to eliminate the gamma matrices in the first two terms. As a result,

$$\begin{aligned} \mathcal{L} &= \overline{h}_v iv \cdot D h_v - \overline{H}_v \left[iv \cdot D - 2m_Q\right] H_v \\ &+ \overline{h}_v i\not{D}_\perp H_v + \overline{H}_v i\not{D}_\perp h_v, \end{aligned} \tag{4.84}$$

where we introduced the notion of a "perpendicular" component of the derivative $D_\perp$. It can be defined for any 4-vector a^μ as

$$a^\mu_\perp = a^\mu - (a \cdot v) v^\mu \ . \tag{4.85}$$

as a result of a condition $a_\perp \cdot v = 0$. Note that in the rest frame $v^\mu = (1, \vec{0})$ the perp component $a^\mu_\perp$ is just a 3-vector $\vec{a}$.

Let us now eliminate H_v by diagonalizing the Lagrangian of Eq. (4.84). This can be done by deriving the equation of motion from Eq. (4.84) using, once again, Eq. (4.79).

$$\left(i\not{D} - m_Q\right) \psi_Q(x) \ \Rightarrow \ i\not{D} h_v + \left(i\not{D} - 2m_Q\right) H_v = 0. \tag{4.86}$$

We can, once again, eliminate the gamma matrices from Eq. (4.86) by applying projectors $\mathcal{P}^v_\pm$ from the left,

$$\left(iv \cdot D + 2m_Q\right) H_v = i\not{D}_\perp h_v \quad \text{or} \quad H_v = \left(2m_Q + iv \cdot D\right)^{-1} i\not{D}_\perp h_v, \tag{4.87}$$

where $\left(2m_Q + iv \cdot D\right)^{-1}$ is the operator inverse to $2m_Q + iv \cdot D$, which we define in terms of the Taylor series. Recall that in deriving Eq. (4.84) we separated parts of the heavy quark's field that corresponded to large, $\mathcal{O}(m_Q)$ momenta. Any time a derivative is acting on h_v, it brings down a momentum that is of order $\mathcal{O}(\Lambda_{QCD})$, since that is the only other scale left. Thus, the expansion

$$\left(2m_Q + iv \cdot D\right)^{-1} = \frac{1}{2m_Q} \sum_{n=0}^{\infty} (-1)^n \left(\frac{iv \cdot D}{2m_Q}\right)^n \tag{4.88}$$

is convergent and we can use it to derive effective Lagrangians of increasing field dimension. Inserting Eq. (4.87) into Eq. (4.83), and using the fact that $h_v = \mathcal{P}^v_+ h_v$ and

$$\mathcal{P}^v_+ i\not{D}_\perp i\not{D}_\perp \mathcal{P}^v_+ = \mathcal{P}^v_+ \left[\left(i\not{D}_\perp\right)^2 + \frac{g_s}{2} \sigma_{\mu\nu} G^{\mu\nu}\right] \mathcal{P}^v_+, \tag{4.89}$$

as well as keeping terms up to $\mathcal{O}(1/m_Q^2)$, we obtain the following result [135, 149],

$$\mathcal{L}_{\text{eff}} = \overline{h}_v iv \cdot D h_v + \frac{1}{2m_Q}\overline{h}_v \left(i\not{D}_\perp\right)^2 h_v + C_g \frac{g_s}{4m_Q}\overline{h}_v \sigma_{\mu\nu} G^{\mu\nu} h_v + \mathcal{O}(1/m_Q^2). \tag{4.90}$$

This is the HQET Lagrangian to order $\mathcal{O}(1/m_Q^2)$. It is clear now that the relative importance of the operators in Eq. (4.90) is determined by counting powers of $1/m_Q$. The leading term, dimension-4 operator $\overline{h}_v iv \cdot D h_v$, is not suppressed by any powers of $1/m_Q$, so it survives in the limit $m_Q \to \infty$. Higher-order terms are represented by the operators of higher dimension. This implies that the power counting scheme of HQET is similar to that of SM EFT. Note that we introduced a (Wilson) coefficient $C_g = 1$ for the operator $\overline{h}_v \sigma_{\mu\nu} G^{\mu\nu} h_v$, as it will receive QCD corrections.

It is interesting to note that the leading-order Lagrangian

$$\mathcal{L}_{\text{eff}}^{\infty} = \overline{h}_v iv \cdot D h_v \tag{4.91}$$

does not have any Dirac matrices, even though it is used to describe fermions. This is a sign of the emergence of enlarged spin symmetry in the heavy quark limit. Let us see how it happens. Let us perform an infinitesimal spin rotation of the h_v field

$$h_v \to h_v' = \left(1 + i\vec{\alpha} \cdot \vec{S}\right) h_v \tag{4.92}$$

with the parameter α_i parameterizing the angle of infinitesimal spin rotation, and look at the difference between the original and "rotated" effective Lagrangians,

$$\delta\mathcal{L}_{\text{eff}}^{\infty} = \overline{h}_v' iv \cdot D h_v' - \overline{h}_v iv \cdot D h_v, \tag{4.93}$$

where $\vec{S}$ are the usual fermion spin operators,

$$S_i = \frac{1}{2}\begin{pmatrix} \sigma_i & \hat{0} \\ \hat{0} & \sigma_i \end{pmatrix}, \qquad [S_i, S_k] = i\epsilon_{ijk} S_k. \tag{4.94}$$

Since $\vec{S}$ commutes with γ^0, and $v \cdot D$ contains no Dirac matrices, we immediately conclude that

$$\delta\mathcal{L}_{\text{eff}}^{\infty} = 0. \tag{4.95}$$

Since the spin transformation belongs to $SU(2)$, we realize that the Lagrangian $\mathcal{L}_{\text{eff}}^{\infty}$ possesses additional $SU(2)$ spin symmetry that is not present in the original Lagrangian of Eq. (4.78).

There is also another symmetry present in the Lagrangian of Eq. (4.91). There are several flavors of heavy quarks present in nature. Since the QCD Lagrangian does not contain any flavor-changing interactions, and $\mathcal{L}_{\text{eff}}^{\infty}$ has no quark mass term, we can sum over all flavors of heavy quarks. If the number of heavy quark flavors is N_f, then the total symmetry group of leading order HQET would be $SU(2N_f)$. This makes HQET a powerful effective theory for calculations of physical properties of heavy quark transitions.

Let us now discuss the HQET Lagrangian with the leading heavy-quark symmetry breaking terms included that is given in Eq. (4.90). The physical meaning of those operators becomes transparent in the rest frame of the heavy quark. The first operator,

$$\mathcal{O}_{kin} = \frac{1}{2m_Q}\overline{h}_v \left(i\not{D}_\perp\right)^2 h_v \quad \Rightarrow \quad \frac{1}{2m_Q}\overline{h}_v \left(i\vec{D}\right)^2 h_v, \tag{4.96}$$

clearly represents the kinetic energy of heavy quark motion inside the hadron. The second operator,

$$\mathcal{O}_{mag} = \frac{g_s}{4m_Q}\overline{h}_v \sigma_{\mu\nu} G^{\mu\nu} h_v \quad \Rightarrow \quad -\frac{1}{m_Q}\overline{h}_v \vec{S} \cdot \vec{B} h_v \tag{4.97}$$

represents its interaction with chromomagnetic field B^i present inside the heavy hadron, which follows if we recall the definition of magnetic field in terms of space components of $G^A_{\mu\nu}$,

$$B^i = -\frac{1}{2}\epsilon^{ijk}G^{jk}. \tag{4.98}$$

There are immediate consequences of this result. If we consider B_q and B_q^* (or D_q and D_q^*) meson states as bound states of two spin-1/2 quarks, we would realize that the only difference between those pseudoscalar and vector mesons is their spin state: they are in a relative singlet and triplet states, respectively. Since the spin interaction in HQET is suppressed by a power of a heavy quark mass, those two states will be degenerate in the heavy quark limit. This is indeed what is seen experimentally: $m_{B^*} - m_B = 45.0 \pm 0.4$ MeV, while mass differences between B and higher states are several hundred MeV. Since those states are degenerate in the heavy quark limit, it would make sense to talk about a *heavy meson superfield* that includes both B_q and B_q^* states. This is similar to a *nucleon* state, which is a doublet of isospin.

Another observation that is worth mentioning in this chapter concerns *field redefinitions* in HQET. Field redefinitions exploit the freedom to change the fields h_v and $\bar{h}_v$ in such a way that the equations of motion at a given order in $1/m_Q$ do not change.

To see what we can do, let us derive the equations of motion for the fields h_v and $\bar{h}_v$ using Eq. (4.90),

$$iv \cdot D h_v = -\frac{1}{2m_Q}\left[\left(i\not{D}_\perp\right)^2 + C_g\frac{g_s}{2}\sigma_{\mu\nu}G^{\mu\nu}\right] h_v. \tag{4.99}$$

We can immediately see that the equation of motion obtained from Eq. (4.90) mixes up terms of different orders in $1/m_Q$: the left-hand side of this equation is $\mathcal{O}(1)$, but the right-hand one is of the higher order in $1/m_Q$. This means that one can always redefine the field h_v such that some of the terms in the Lagrangian are absorbed into the definition of h_v – the equations of motions will only change by an operator that contributes at higher orders in $1/m_Q$.

An example of this field redefinition technique can be seen in the construction of the Lagrangian of Eq. (4.90). In principle, terms of the type

$$\mathcal{L}' = \frac{1}{2m_Q}\bar{h}_v\left(iv \cdot D\right)^2 h_v \tag{4.100}$$

are absent. The reason this kind of operator is not included, even though it is allowed by all symmetries, is that a field redefinition

$$h_v \to \left[1 - \frac{\left(iv \cdot D\right)^2}{4m_Q}\right] h_v \tag{4.101}$$

can be used to remove the operator $\mathcal{L}'$ from the Lagrangian of Eq. (4.90). Similar field redefinitions can also be done at higher orders in $1/m_Q$. In practice, they are used to construct minimal bases of HQET operators at higher orders.

External states and heavy quark limit

As we pointed out earlier, heavy quark limit simplifies operations with both currents and external states. Oftentimes, it is convenient to use tensor formalism to write the matrix elements of various currents [135,149]. We shall review this formalism below, specifying, for the sake of concreteness, to beauty states. Generalization to the charm states consists of trivial relabeling of the fields.

For the heavy quark the spin part of the quantum field would be

$$\begin{aligned} \text{Spin up: } b_\alpha^{(\Uparrow)} &= \begin{pmatrix} 1 \\ 0 \\ 0 \\ 0 \end{pmatrix} \equiv \delta_{1\alpha}, \\ \text{Spin down: } b_\alpha^{(\Downarrow)} &= \begin{pmatrix} 0 \\ 1 \\ 0 \\ 0 \end{pmatrix} \equiv \delta_{2\alpha}. \end{aligned} \tag{4.102}$$

Similarly, the spin part of the light antiquark field can be chosen as $\bar{q}_\alpha^{(\Downarrow)} = -\delta_{3\alpha}$ and $\bar{q}_\alpha^{(\Uparrow)} = -\delta_{4\alpha}$. Let us now construct tensors that would represent heavy meson spin states. For the pseudoscalars

$$\bar{B}^q_{\alpha\beta} \propto b_\alpha^{(\Uparrow)} \bar{q}_\beta^{(\Downarrow)} + b_\alpha^{(\Downarrow)} \bar{q}_\beta^{(\Uparrow)} = -\begin{pmatrix} \hat{0} & \hat{1} \\ \hat{0} & \hat{0} \end{pmatrix} = -\frac{1+\gamma^0}{2}\gamma_5, \tag{4.103}$$

where $\hat{0}$ represents a 2×2 matrix of zeros. Note that the index q goes over the flavors of the light quarks $q = u, d$ and s. In the arbitrary frame, where the heavy quark is moving with velocity v, we simply replace

$$\frac{1+\gamma^0}{2} \rightarrow \frac{1+\not{v}}{2}. \tag{4.104}$$

This would result in a pseudoscalar state in coordinate representation,

$$H_q^{(ps)} = -\frac{1+\not{v}}{2}\bar{B}_q\gamma_5. \tag{4.105}$$

Similarly, we can construct spin wave function of the vector B^* state. This spin-triplet state should be represented by three polarization vectors, as we are dealing with a massive vector state. These polarization vectors can be defined as

$$\begin{aligned} \epsilon_B^{(\pm)} &= (0, 1, \pm i, 0), \\ \epsilon_B^0 &= (0, 0, 0, 1). \end{aligned} \tag{4.106}$$

Just like in the case of the pseudoscalar (singlet) state considered earlier, we discuss each polarization (helicity) state in the rest frame of the heavy field. This leads us to

$$\begin{aligned} B^{q,(+)}_{\alpha\beta} &\propto b_\alpha^{(\Uparrow)} \bar{q}_\beta^{(\Uparrow)} = \frac{1}{\sqrt{2}} \begin{pmatrix} \hat{0} & \sigma_1 + i\sigma_2 \\ \hat{0} & \hat{0} \end{pmatrix} = -\frac{1+\gamma^0}{2}\not{\epsilon}_B^{(+)}, \\ B^{q,(0)}_{\alpha\beta} &\propto b_\alpha^{(\Uparrow)} \bar{q}_\beta^{(\Downarrow)} - b_\alpha^{(\Downarrow)} \bar{q}_\beta^{(\Uparrow)} = \begin{pmatrix} \hat{0} & \sigma_3 \\ \hat{0} & \hat{0} \end{pmatrix} = -\frac{1+\gamma^0}{2}\not{\epsilon}_B^{0}, \\ B^{q,(-)}_{\alpha\beta} &\propto b_\alpha^{(\Downarrow)} \bar{q}_\beta^{(\Downarrow)} = \frac{1}{\sqrt{2}} \begin{pmatrix} \hat{0} & \sigma_1 - i\sigma_2 \\ \hat{0} & \hat{0} \end{pmatrix} = -\frac{1+\gamma^0}{2}\not{\epsilon}_B^{(-)}. \end{aligned} \tag{4.107}$$

Similarly to the case of pseudoscalar mesons, this result can be generalized to an arbitrary frame of reference,

$$H_q^{(vect)} = \frac{1+\not{v}}{2}\not{\epsilon}_B \tag{4.108}$$

for all three helicity states in the momentum representation. In the coordinate representation we would have to replace $\epsilon_B^\mu \rightarrow \bar{B}_q^{*\mu}$. Note that the heavy vector states satisfy the condition $v_\mu \bar{B}_q^{*\mu} = 0$.

We have argued before that in the heavy quark limit the pseudoscalar and vector states are degenerate. Thus, it is convenient to introduce a supermultiplet state that includes both of them, which in coordinate representation would read

$$H_q(v) = \frac{1+\not{v}}{2}\left(\bar{B}_q^{*\mu}\gamma_\mu - \bar{B}_q\gamma_5\right). \tag{4.109}$$

Since we are dealing with the fields in the heavy quark limit, the fields $\bar{B}_q$ and $\bar{B}_q^{*\mu}$ do not create antiparticles. We should also define a conjugated field $\overline{H}$,

$$\overline{H}_i(v) = \gamma^0 H_i^\dagger(v)\gamma^0 = \left(\bar{B}_q^{*\dagger\mu}\gamma_\mu + \bar{B}_q^\dagger\gamma_5\right)\frac{1+\not{v}}{2}. \tag{4.110}$$

An introduction of this *tensor representation* for the heavy meson field significantly simplifies the calculations of matrix elements of various operators between heavy meson states.

Let us provide a recipe for computation of matrix elements of quark currents. First, substitute the bra and ket-vectors with either Eq. (4.103) or (4.107). Second, substitute the operator with all possible tensors transforming appropriately under the spin symmetry. Put those two ingredients together and multiply by an unknown coefficient that would have to be determined from experimental data. The object that we are building must be invariant under heavy quark spin symmetry, which means that we must take a trace over spin indices.

Let us show how it works for the matrix element of the unit operator, i.e. normalization of the pseudoscalar and vector heavy meson states,

$$\begin{aligned}
\langle \bar{B}_q(v)|\bar{B}_q(v)\rangle &= a\ \mathrm{Tr}\left[B^{q\dagger}B^q\right] = a\ \mathrm{Tr}\left[\gamma_5\frac{1+\not{v}}{2}\frac{1+\not{v}}{2}\gamma_5\right] = 2a, \\
\langle \bar{B}_q^*(v)|\bar{B}_q^*(v)\rangle &= a\ \mathrm{Tr}\left[\not{\epsilon}_B^*\frac{1+\not{v}}{2}\frac{1+\not{v}}{2}\not{\epsilon}_B\right] = 2a.
\end{aligned} \tag{4.111}$$

Note that both of these states should have the same normalization. Also, we implicitly eliminated all other possible tensors built out of $\not{v}$ due to Eqs. (4.81). If we require to normalize the external states to one, $a = 1/2$.

Heavy quark symmetry. Isgur-Wise function

Another application of the recipe discussed above is a flavor-changing transition between heavy quark states that are driven by heavy-to-heavy current $\bar{Q}_1\Gamma Q_2$, where Γ could be any Dirac matrix, $\Gamma = 1, \gamma_5, \gamma^\mu, \ldots$. An example of such transition can be exclusive semileptonic weak decay $B \to D\ell\bar{\nu}$ or $B \to D^*\ell\bar{\nu}$. In the Standard Model these decays are governed by $(V - A)$ current structure, in which case $\Gamma = \gamma^\mu(1-\gamma_5)$.

Depending on the spin state of initial and final states, a proper description of this transition would involve several form factors. In fact, to completely describe both semileptonic B decays into D and D^* states one would need six of them, as we discussed in Sect. 4.5.1. As was shown in [116], in the heavy quark limit*, all those form factors are related to only *one* form factor whose normalization can be determined model-independently.

To show this, let's consider the matrix elements

$$\langle H(v')|\overline{h}_{v'}^{Q_2}\Gamma h_v^{Q_1}|H(v)\rangle, \quad \langle H^*(v')|\overline{h}_{v'}^{Q_2}\Gamma h_v^{Q_1}|H(v)\rangle \tag{4.112}$$

*Since there are two heavy quarks participating in this transition, there are actually several ways to take the heavy quark limit. For this problem we take $m_b, m_c \to \infty$ with $m_c/m_b \sim \mathcal{O}(1)$.

for all possible Γ. Here $h_v^{(Q_i)}$ represent effective heavy fields in HQET for Q_i quarks. The actions of the heavy fields on the states can be represented as a Dirac matrix

$$\sqrt{m_{H_{Q_1}} m_{H_{Q_2}}}\, \overline{H}_{Q_2}(v')\Gamma H_{Q_1}^{ps}(v), \tag{4.113}$$

where index ps means that we are only interested in a pseudoscalar component of the meson superfield. This matrix must be coupled to the 4×4 matrix of light degrees of freedom $\mathcal{M}$, which can be represented by

$$\mathcal{M} = A + B\not{v} + C\not{v}' + D\not{v}\not{v}'. \tag{4.114}$$

Now, all matrix elements can be written as

$$\frac{\langle H^{(*)}(v')|\overline{h}_{v'}^{Q_2}\Gamma h_v^{Q_1}|H(v)\rangle}{\sqrt{m_{H_{Q_1}} m_{H_{Q_2}}}} = \mathrm{Tr}\left[\overline{H}_{Q_2}(v')\Gamma H_{Q_1}^{ps}(v)\mathcal{M}\right]. \tag{4.115}$$

Using the properties of $H_{Q_i}(v_i)$ states $\not{v}H_{Q_i}(v_i) = H_{Q_i}(v_i)$ and $H_{Q_i}(v_i)\not{v} = -H_{Q_i}(v_i)$ and computing the traces we can show that the result of Eq. (4.115) would be the same if

$$\mathcal{M} = A + B + C + D = \xi(v \cdot v'). \tag{4.116}$$

Here $\xi(v \cdot v')$ is called Isgur-Wise function and it is implied that all constants are multiplied by a 4×4 unit matrix in Dirac space. Also,

$$w = v \cdot v' = \frac{m_{H_{Q_1}}^2 + m_{H_{Q_2}}^2 - q^2}{2m_{H_{Q_1}} m_{H_{Q_2}}}. \tag{4.117}$$

Please note [116] that $\xi(v \cdot v = 1) = 1$. Similarly, we can show that matrix elements of other heavy-to-heavy currents can also be related to the same Isgur-Wise function. For the $B \to D$ transitions we obtain [77],

$$\begin{aligned}
\frac{\langle D(v')|\overline{h}_{v'}^c h_v^b|B(v)\rangle}{\sqrt{m_B m_D}} &= \xi(w)\,(1+w)\,, \\
\frac{\langle D(v')|\overline{h}_{v'}^c \gamma^\mu h_v^b|B(v)\rangle}{\sqrt{m_B m_D}} &= \xi(w)\,(v+v')^\mu\,, \\
\frac{\langle D(v')|\overline{h}_{v'}^c \sigma_{\mu\nu} h_v^b|B(v)\rangle}{\sqrt{m_B m_D}} &= i\xi(w)\left(v'_\mu v_\nu - v_\mu v'_\nu\right),
\end{aligned} \tag{4.118}$$

while for the $B \to D^*$ transition

$$\begin{aligned}
\frac{\langle D^*(v')|\overline{h}_{v'}^c \gamma_5 h_v^b|B(v)\rangle}{\sqrt{m_B m_D}} &= \xi(w)\,(\epsilon^* \cdot v)\,, \\
\frac{\langle D^*(v')|\overline{h}_{v'}^c \gamma^\mu h_v^b|B(v)\rangle}{\sqrt{m_B m_D}} &= i\xi(w)\epsilon^{\mu\nu\alpha\beta}\epsilon_\nu^* v'_\alpha v_\beta, \\
\frac{\langle D^*(v')|\overline{h}_{v'}^c \gamma^\mu\gamma_5 h_v^b|B(v)\rangle}{\sqrt{m_B m_D}} &= i\xi(w)\left[\epsilon^{*\mu}(1+w) + v^\mu(v\cdot\epsilon^*)\right], \\
\frac{\langle D^*(v')|\overline{h}_{v'}^c \sigma^{\mu\nu} h_v^b|B(v)\rangle}{\sqrt{m_B m_D}} &= \xi(w)\epsilon^{\mu\nu\alpha\beta}\epsilon_\alpha^*\,(v+v')_\beta\,.
\end{aligned} \tag{4.119}$$

Note that tensor currents in Eq. (4.118) and Eq. (4.119) do not appear in the Standard Model Lagrangian.

4.6.2 New Physics in leptonic and semileptonic decays of heavy mesons

Let us consider semileptonic transitions underlined by a decay of a heavy quark $Q = \{b, c\}$ into another quark q. Depending on the situation, q could be a light ($q = u, d, s$) or a heavy ($q = c$) quark. The most general effective Lagrangian can be obtained by direct generalization of the leptonic Lagrangian of Eq. (3.53) and is conventionally written as

$$\begin{aligned}\mathcal{L}_{\text{eff}} = & - \frac{4G_F}{\sqrt{2}} V_{qQ} \left[(1 - \widetilde{g}^V_{LL}) \overline{\ell}_L \gamma_\mu \nu_L \overline{q}_L \gamma^\mu Q_L + \widetilde{g}^V_{LR} \overline{\ell}_L \gamma_\mu \nu_L \overline{q}_R \gamma^\mu Q_R \right. \\ & + \left. \widetilde{g}^S_{RL} \overline{\ell}_R \nu_L \overline{q}_R Q_L + \widetilde{g}^S_{RR} \overline{\ell}_R \nu_L \overline{q}_L Q_R + \widetilde{g}^T_{RL} \overline{\ell}_R \sigma_{\mu\nu} \nu_L \overline{q}_R \sigma^{\mu\nu} Q_L \right] + h.c., \quad (4.120)\end{aligned}$$

where $\widetilde{g}^Z_{XY}$ for $\{X, Y\} = \{L, R\}$ and $Z = \{V, S, T\}$ are the vector, scalar, and tensor-type New Physics couplings. They are all zero in the Standard Model and are related to the Wilson coefficients of SM EFT operators displayed in Table 4.1,

$$\begin{aligned}\widetilde{g}^V_{LL} &= \frac{C_L}{\Lambda^2} \frac{\sqrt{2}}{4G_F V_{qQ}}, \quad \widetilde{g}^V_{LR} = \frac{C_R}{\Lambda^2} \frac{\sqrt{2}}{4G_F V_{qQ}}, \\ \widetilde{g}^S_{RL} &= \frac{C^S_L}{\Lambda^2} \frac{\sqrt{2}}{4G_F V_{qQ}}, \quad \widetilde{g}^S_{RR} = \frac{C^S_R}{\Lambda^2} \frac{\sqrt{2}}{4G_F V_{qQ}}, \quad (4.121) \\ g^T_{RL} &= \frac{C_T}{\Lambda^2} \frac{\sqrt{2}}{4G_F V_{qQ}}, \quad (4.122)\end{aligned}$$

It is convenient to define the NP parameters this way, as the connection to the SM results is then obvious. The Lagrangian of Eq. (4.120) can be seen to contribute to leptonic and semileptonic decays mediated by the $Q \to q$ transition.

Leptonic decays

Leptonic decays $P_1 \to \ell \overline{\nu}_\ell$ are sensitive probes of NP interactions mediated by charged particles. They are also rather easy to compute, as there is only a single nonperturbative quantity that parameterizes matrix elements of quark currents. There is no need to switch to HQET to compute those matrix elements.

Taking a matrix element of Eq. (4.120) by employing the results from Sec. 4.5.1 we obtain,

$$\begin{aligned}\mathcal{B}(P_1 \to \ell \overline{\nu}_\ell) &= \frac{G_F^2 \left|V_{qQ}\right|^2}{8\pi\Gamma} f^2_{P_1} m^2_\ell M_{P_1} \left(1 - \frac{m^2_\ell}{M^2_{P_1}}\right)^2 \\ &\times \left| (1 + \widetilde{g}^V_{LL} - \widetilde{g}^V_{LR}) - \frac{M^2_{P_1}}{m_\ell(m_Q + m_q)} (\widetilde{g}^S_{RL} - \widetilde{g}^S_{RR}) \right|^2 . \quad (4.123)\end{aligned}$$

The Standard Model value $\mathcal{B}_{\text{SM}}$ can be recovered by setting $\widetilde{g}^Z_{XY}$ to zero. The tensor current do not contribute, as the relevant matrix element is zero (see Sec. 4.5.1).

The values of Wilson coefficients in particular models can be found by matching the model Lagrangian into SM EFT or directly to Eq. (4.120). As an example, we can do so explicitly for the models with an extended Higgs sector, which include new charged scalar states. In particular, two Higgs doublet models, including Minimal Supersymmetric SM (MSSM), could give contributions to leptonic and semileptonic transitions. There are, in general, different implementations of this extension of the SM discussed in Sect. 2.5. For example, the first doublet (Φ_1) could give mass to the up-type fermions and the second (Φ_2) to the down-type fermions. In this case,

$$\begin{aligned}\mathcal{B}(D^+ \to \ell^+ \nu_\ell) &= \mathcal{B}_{\text{SM}} \left(1 + \frac{m^2_D}{m^2_{H^\pm}}\right)^2 \quad (4.124) \\ \mathcal{B}(D^+_s \to \ell^+ \nu_\ell) &= \mathcal{B}_{\text{SM}} \left[1 + \frac{m^2_{D_s}}{m^2_{H^\pm}} \left(1 - \tan^2\beta \frac{m_s}{m_c}\right)\right]^2\end{aligned}$$

Note that the latter model introduces a correction to the SM expectations that may be considerable and negative at large $\tan^2\beta$. A limit can also be set on the mass of a charged Higgs, $m_{H^+} > 2.2\tan\beta$. This limit is similar to the one obtained from the measurement of $\mathcal{B}(B \to \tau\nu)$ decay.

New Physics in semileptonic decays

The Lagrangian in Eq. (4.123) would also affect semileptonic decays of heavy quarks [77]. Let us, for the sake of an example, consider $b \to c$ exclusive transitions and a NP model with manifest right-handed currents. This model can be obtained from Eq. (4.123) by setting all coefficients but $\widetilde{g}_{LR}^V$ to zero. While this particular realization might not be practically possible, there are parts of the complete model parameters space where this situation is approximately realized. The effective Lagrangian now takes the form

$$\mathcal{L}_{\text{eff}} = -\frac{4G_F}{\sqrt{2}} V_{cb} \left(\bar{\ell}_L \gamma^\mu \nu_{\ell L}\right) J_{h,\mu}, \tag{4.125}$$

where in the heavy quark limit

$$\begin{aligned} J_{h,\mu} &= \bar{c}_L \gamma_\mu b_L + \widetilde{g}_{LR}^V \, \bar{c}_R \gamma_\mu b_R \\ &\to \overline{h}_{v'}^c \gamma^\mu P_L h_v^b + \widetilde{g}_{LR}^V \, h_{v'}^c \gamma^\mu P_R h_v^b, \end{aligned} \tag{4.126}$$

where $P_{L/R}$ are the left- and right-handed projectors and we matched the QCD quark currents to the leading order HQET currents.

It is possible to study implications of the Lagrangian of Eq. (4.125) to exclusive semileptonic $B \to D^{(*)}\ell\bar{\nu}$ decays. Since quark currents of different helicities contribute, it would be useful to consider polarized D^* states. We shall write the corresponding distributions for the longitudinal and transverse polarizations of the D^*.

We can compute the matrix elements of the currents from Eq. (4.126) using Eqs. (4.118) and (4.119). This will lead us to the following expressions for the differential decay widths [77],

$$\begin{aligned} \frac{d\Gamma(B \to D\ell\bar{\nu})}{dw} &= \Gamma_0(w) \left|V_c b\right|^2 \mathcal{A}(w) \left|\xi(w)\right|^2, \\ \frac{d\Gamma(B \to D_T^*\ell\bar{\nu})}{dw} &= \Gamma_0^*(w) \left|V_c b\right|^2 \mathcal{B}^T(w) \left|\xi(w)\right|^2, \\ \frac{d\Gamma(B \to D_L^*\ell\bar{\nu})}{dw} &= \Gamma_0^*(w) \left|V_c b\right|^2 \mathcal{B}^L(w) \left|\xi(w)\right|^2, \end{aligned} \tag{4.127}$$

where we introduced

$$\Gamma_0^{(*)} = \frac{G_F^2 m_B^5}{48\pi^3} x_{(*)}^3 \sqrt{w-1}(w+1)^2, \tag{4.128}$$

with $x_{(*)} = m_{D^{(*)}}/m_B$ and the following expressions for the form factors $\mathcal{A}(w)$ and $\mathcal{B}^{L/T}(w)$,

$$\begin{aligned} \mathcal{A}(w) &= \frac{w-1}{w+1}\left(1+\widetilde{g}_{LR}^V\right)^2 (1+x)^2, \\ \mathcal{B}^T(w) &= 2\left(1-2x_* w + x_*^2\right)\left[\left(1-\widetilde{g}_{LR}^V\right)^2 + \frac{w-1}{w+1}\left(1+\widetilde{g}_{LR}^V\right)^2\right], \\ \mathcal{B}^L(w) &= \left(1-\widetilde{g}_{LR}^V\right)^2 (x_*-1)^2. \end{aligned} \tag{4.129}$$

The Standard Model results are recovered by setting $\widetilde{g}_{LR}^V = 0$. These results are obtained at the leading order in $1/m_Q$. The biggest advantage of the EFT approach to matrix element computation is the possibility of systematic improvement of the result by taking into account higher-order terms in $1/m_Q$ expansion.

As we can see, the effective Lagrangian techniques are well suited for searches for New Physics in heavy quark decays into the final states containing leptons. As with any EFT technique, the exact identification of a particular NP model is not possible, but the chiral structure of the effective operators certainly helps to pinpoint classes of models that can contribute to the considered processes.

4.6.3 Rare decays with charged leptons

Since flavor-changing neutral current transitions are only possible at one loop level in the Standard Model, they are usually thought to be a good probe of BSM physics. Therefore, transitions initiated by the quark-level processes $b \to s(d)\bar{\ell}\ell$ or $c \to u\bar{\ell}\ell$ are quite popular as tools for NP searches.

Even though down-type quark transitions $b \to s(d)\bar{\ell}\ell$ are generated at one loop, Glashow-Iliopoulos-Maiani (GIM) mechanism ensures that they are dominated by top-quark effects, making the amplitudes proportional to powers of top-quark mass, which is a large quantity. Thus, one needs to account for a non-vanishing SM contribution before NP contributions are identified. The techniques for computations of the SM effects are very mature; their description can be found in [33].

The situation is different for charm decay, where the SM contribution is smaller [90], so NP contributions could be more pronounced. As it turns out, however, poorly controlled long-distance QCD effects might spoil unambiguous identifications of NP effects [38]. There is hope, however, that those effects could be ultimately brought under theoretical control. Let us consider rare decays of both charmed and beauty states.

The simplest rare decay is a purely leptonic transition of a neutral meson P^0 into a lepton pair, $P_q^0 \to \ell^+\ell^-$. The most general $P_q^0 \to \ell^+\ell^-$ decay amplitude can be written

$$\mathcal{M}(P_q^0 \to \ell^+\ell^-) = \overline{u}(p_-, s_-)\left[A + \gamma_5 B\right] v(p_+, s_+), \tag{4.130}$$

where $u(p_-, s_-)$ and $v(p_+, s_+)$ are the charged leptons' spinors, and A and B are the amplitudes that depend on the short-distance Wilson coefficients and strong-interaction parameters such as P-meson decay constants.

In the beauty sector $P_q^0 = B_q^0$ this transition has a dominant SM contribution, as heavy top quark dominates the leading one-loop SM contribution to the Wilson coefficients [33]. In the charm sector $P_q^0 = D^0$ this transition has a very small SM contribution, so it can serve as a clean probe of NP amplitudes. Other rare decays such as $D \to \rho\gamma$ receive significant SM contributions, which are often difficult to calculate. Similarly to our previous discussion, all NP contributions to $c \to u\ell^+\ell^-$ transitions can be accounted for in an effective Lagrangian,

$$\mathcal{L}_{\rm NP}^{\rm rare} = -\frac{1}{\Lambda^2}\sum_{i=1}^{10} \widetilde{C}_i(\mu)\, \widetilde{Q}_i, \tag{4.131}$$

where $\widetilde{\mathrm{C}}_i$ are Wilson coefficients, $\widetilde{Q}_i$ are the effective operators, and Λ represents the energy scale of NP interactions that generate $\widetilde{Q}_i$'s. There are only ten operators with canonical dimension six,

$$\begin{aligned}
\widetilde{Q}_1 &= (\overline{\ell}_L\gamma_\mu\ell_L)(\overline{u}_L\gamma^\mu c_L)\,,\\
\widetilde{Q}_2 &= (\overline{\ell}_L\gamma_\mu\ell_L)(\overline{u}_R\gamma^\mu c_R)\,,\\
\widetilde{Q}_3 &= (\overline{\ell}_L\ell_R)\,(\overline{u}_R c_L)\,,\\
\widetilde{Q}_4 &= (\overline{\ell}_R\ell_L)(\overline{u}_R c_L)\,,\\
\widetilde{Q}_5 &= (\overline{\ell}_R\sigma_{\mu\nu}\ell_L)(\overline{u}_R\sigma^{\mu\nu} c_L)\,,
\end{aligned} \tag{4.132}$$

and five additional operators $\widetilde{Q}_6, \ldots, \widetilde{Q}_{10}$ obtained from those in Eq. (4.132) by interchanging $L \leftrightarrow R$, e.g. $\widetilde{Q}_6 = (\overline{\ell}_R\gamma_\mu\ell_R)(\overline{u}_R\gamma^\mu c_R)$, $\widetilde{Q}_7 = (\alpha/4)(\overline{\ell}_R\gamma_\mu\ell_R)(\overline{u}_L\gamma^\mu c_L)$, etc.

The effective Lagrangian of Eq. (4.131) is quite general, and thus it also contains the SM contribution. It is worth noting that matrix elements of several operators or their linear combinations vanish in the calculation of $\mathcal{B}(D^0 \to \ell^+\ell^-)$. For example, $\langle \ell^+\ell^-|\widetilde{Q}_5|D^0\rangle = \langle \ell^+\ell^-|\widetilde{Q}_{10}|D^0\rangle = 0$ vanish identically, while $\langle \ell^+\ell^-|Q_9|D^0\rangle \equiv (\alpha/4)\langle \ell^+\ell^-|(\widetilde{Q}_1 + \widetilde{Q}_7)|D^0\rangle = 0$ due to vector current conservation, etc.

NP contribution given by the operators of Eq. (4.131) affects the amplitudes A and B,

$$
\begin{aligned}
|A| &= \frac{f_D M_D^2}{4\Lambda^2 m_c}\left[\widetilde{C}_{3-8} + \widetilde{C}_{4-9}\right] \\
|B| &= \frac{f_D}{4\Lambda^2}\left[2m_\ell\left(\widetilde{C}_{1-2} + \widetilde{C}_{6-7}\right) + \frac{M_D^2}{m_c}\left(\widetilde{C}_{4-3} + \widetilde{C}_{9-8}\right)\right],
\end{aligned}
\tag{4.133}
$$

with $\widetilde{C}_{i-k} \equiv \widetilde{C}_i - \widetilde{C}_k$. The amplitude of Eq. (4.130) results in the following branching fractions for the lepton flavor-diagonal and off-diagonal decays,

$$
\begin{aligned}
\mathcal{B}(D^0 \to \ell^+\ell^-) &= \frac{M_D}{8\pi\Gamma_D}\sqrt{1-\frac{4m_\ell^2}{M_D^2}}\left[\left(1-\frac{4m_\ell^2}{M_D^2}\right)|A|^2 + |B|^2\right] \\
\mathcal{B}(D^0 \to \mu^+e^-) &= \frac{M_D}{8\pi\Gamma_D}\left(1-\frac{m_\mu^2}{M_D^2}\right)^2\left[|A|^2 + |B|^2\right].
\end{aligned}
\tag{4.134}
$$

In the latter expression, the electron mass is safely neglected. Any NP model that contributes to $D^0 \to \ell^+\ell^-$ can be constrained by bounds on the Wilson coefficients appearing in Eq. (4.133). We note that, because of helicity suppression, studies of $D^0 \to e^+e^-$ (and consequently analyses of lepton universality using this channel) are experimentally challenging (see, however [120]). Experimental limits on $\mathcal{B}(D^0 \to \mu^+e^-)$ give constraints on lepton-flavor-violating interactions via Eq. (4.134). Similar limits can also be obtained from two-body charmed quarkonium decays [111].

In studying NP contributions to rare decays in charm, it can be advantageous to study *correlations* of various processes, for example $D^0 - \overline{D^0}$ mixing and rare decays [90]. In general, one cannot predict the rare decay rate by knowing just the mixing rate, even if both it and $\mathcal{B}(D^0 \to \ell^+\ell^-)$ are dominated by a single operator contribution. It is, however, possible to do so for a restricted subset of NP models [90].

4.6.4 Flavor-changing decays into invisible states

High-luminosity e^+e^- flavor factories provide a great opportunity to search for rare processes that require high purity of the final states. In particular, searches for B or D-decays into final states that leave no traces in a detector, or *invisible* final states, are possible. This is so due to the fact that pairs of B or D-mesons are produced in a charge-correlated state.

Such invisible final states could be light NP particles discussed in Sect. 2.7, they might or might not be dark matter candidates. Nevertheless, searches for light, feebly interacting particles is a well-defined and worthy exercise. If those new particles are such that they leave the detector before interacting, their experimental signature (or the absence of thereof) is very similar to that of the SM neutrinos. Thus, the only irreducible SM background that has the same experimental signature is heavy meson decays into the final states containing only neutrinos. Transitions of a $B_q^0(D^0)$ meson into such final states are described by an effective Lagrangian,

$$
\begin{aligned}
\mathcal{L}_{eff} &= -\frac{4G_F}{\sqrt{2}}\frac{\alpha}{2\pi\sin^2\theta_W} \\
&\times \sum_{l=e,\mu,\tau}\sum_k \lambda_k X^l(x_k)\left(J_{Qq}^\mu\right)\left(\overline{\nu}_L^l\gamma_\mu\nu_L^l\right),
\end{aligned}
\tag{4.135}
$$

where $J^{\mu}_{Qq} = \overline{q}_L \gamma^{\mu} b_L$ for beauty, and $J^{\mu}_{Qq} = \overline{u}_L \gamma^{\mu} c_L$ for charm transitions, and we consider Dirac neutrinos. The functions $\lambda_k X^l(x_k)$ are combinations of the Cabbibo-Kobayashi-Maskawa factors and Inami-Lim functions. These functions are dominated by the top-quark contribution for $b \to q$ transitions, so

$$\sum_k \lambda_k X^l(x_k) = V^*_{tq} V_{tb} X(x_t), \tag{4.136}$$

where $x_t = m_t^2/M_W^2$ and

$$X(x_t) = \frac{x_t}{8}\left[\frac{x_t+2}{x_t-1} + \frac{3(x_t-2)}{(x_t-1)^2}\ln x_t\right] \tag{4.137}$$

Perturbative QCD corrections numerically change Eq. (4.137) by at most 10%, so therefore will be neglected. For charm $c \to u$ transitions we keep the contributions from both internal b and s-quarks,

$$\sum_k \lambda_k X^l(x_k) = V^*_{cs} V_{us} X^l(x_s) + V^*_{cb} V_{ub} X^l(x_b), \tag{4.138}$$

where $X^l(x_q) = \overline{D}(x_q, y_l)/2$ with $y_l = m_l^2/m_W^2$ are related to the Inami-Lim functions [115],

$$\begin{aligned} \overline{D}(x_q, y_l) &= \frac{1}{8}\frac{x_q y_l}{x_q - y_l}\left(\frac{y_l-4}{y_l-1}\right)^2 \ln y_l \\ &+ \frac{x_q}{8}\left[\frac{x_q}{y_l - x_q}\left(\frac{x_q-4}{x_q-1}\right)^2 + 1 + \frac{3}{(x_q-1)^2}\right]\ln x_q \\ &+ \frac{x_q}{4} - \frac{3}{8}\left(1 + 3\frac{1}{y_l-1}\right)\frac{x_q}{x_q-1}. \end{aligned} \tag{4.139}$$

Given this, one can easily estimate branching ratios for $B_q(D) \to \nu\overline{\nu}$ decays. One can immediately notice that the left-handed structure of the Lagrangian results in helicity suppression of these decays due to the fact that initial state is a spin-0 meson. The branching ratio is

$$\mathcal{B}(B_q \to \nu\overline{\nu}) = \frac{G_F^2 \alpha^2 f_{B_q}^2 M_{B_q}^3}{16\pi^3 \sin^4\theta_W \Gamma_{B_q}} |V_{tb} V^*_{tq}|^2 X(x_t)^2 x_\nu^2, \tag{4.140}$$

where $x_\nu = m_\nu/M_{B_q}$ and $\Gamma_{B_q} = 1/\tau_{B_q}$ is the total width of the B_q meson. We also summed over all possible neutrino states. Similar formula for charm decays can be obtained by trivial substitution.

As can be seen from Eq. (4.140), the branching ratio is exactly zero in the minimal standard model with massless neutrinos! The factor $x_\nu \ll 1$ is small for any neutral meson state. Assuming for neutrino masses that $m_\nu \sim \sum_i m_{\nu_i} < 1$ eV, where m_{ν_i} is the mass of one of the neutrinos, Eq. (4.140) yields the branching ratios of $\mathcal{B}_{th}(B^0_s \to \nu\overline{\nu}) \simeq 3\times 10^{-24}$, $\mathcal{B}_{th}(B^0_d \to \nu\overline{\nu}) \simeq 1\times 10^{-25}$, and $\mathcal{B}_{th}(D^0 \to \nu\overline{\nu}) \simeq 1\times 10^{-30}$ for B_s, B_d, and D^0 states, respectively. These extremely small branching ratios show that those transitions constitute almost background-free modes for searches for new light particles.

While correct, the above conclusion needs clarification. Experimentally, the $\nu\overline{\nu}$ final state does not constitute a good representation of the invisible width of $B^0_q(D^0)$ mesons in the Standard Model. Indeed, in the SM the final state that is not detectable in a flavor factory setup contains an arbitrary number of neutrino pairs,

$$\mathcal{B}\left(B_q \to \not{E}\right) = \mathcal{B}\left(B_q \to \nu\overline{\nu}\right) + \mathcal{B}\left(B_q \to \nu\overline{\nu}\nu\overline{\nu}\right) + \ldots. \tag{4.141}$$

Similar formula exists for D-decays [21]. As discussed above (see Eq. (4.140)), decay to the $\nu\overline{\nu}$ final state is helicity-suppressed. The four-neutrino final state, on the other hand, does not suffer from such suppression, so it is expected to have a considerably larger branching ratio. Naively,

$$\frac{\mathcal{B}(B_q \to \nu\bar{\nu}\nu\bar{\nu})}{\mathcal{B}(B_q \to \nu\bar{\nu})} \sim \frac{G_F^2 M_B^4}{16\pi^2 x_\nu^2} \gg 1. \tag{4.142}$$

Still, a result of a calculation shows [21] that the SM result for the invisible width of a heavy meson is still small, the largest branching ratio of the B_s decay is $\mathcal{O}(10^{-15})$. Therefore, $B(D) \to$ invisible is still a good channel to search for invisible decays into NP particles.

Branching fractions for the heavy meson states decaying into $\chi_s\overline{\chi}_s$ and $\chi_s\overline{\chi}_s\gamma$, where χ_s is a DM particle of spin s, can be calculated in the EFT framework. Since production of scalar χ_0 states avoids helicity suppression, these will be discussed here. Also, we will write the following expressions for a generic heavy quark state $H_q^0 = \{B_q^0, D^0\}$.

A generic effective Lagrangian for scalar DM interactions has a simple form [11]

$$\mathcal{L}_{\text{eff}} = -2\sum_i \frac{C_i}{\Lambda^2} O_i, \tag{4.143}$$

where C_i are the Wilson coefficients. The effective operators O_i contain a heavy quark $Q = \{b, c\}$ and a light quark q of the same electric charge as Q and can be written as

$$\begin{aligned}
O_1 &= m_Q(\overline{q}_R Q_L)(\chi_0^*\chi_0), \\
O_2 &= m_Q(\overline{q}_L Q_R)(\chi_0^*\chi_0), \\
O_3 &= (\overline{q}_L\gamma^\mu Q_L)(\chi_0^* \overleftrightarrow{\partial}_\mu \chi_0), \\
O_4 &= (\overline{q}_R\gamma^\mu Q_R)(\chi_0^* \overleftrightarrow{\partial}_\mu \chi_0),
\end{aligned} \tag{4.144}$$

where $\overleftrightarrow{\partial} = (\overrightarrow{\partial} - \overleftarrow{\partial})/2$ and the DM *anti-particle* $\overline{\chi}_0$ may or may not coincide with χ_0. The branching fraction for the two-body decay $P_q^0 \to \chi_0\chi_0$ is

$$\mathcal{B}(P_q^0 \to \chi_0\chi_0) = \frac{(C_1 - C_2)^2}{4\pi m_{P_q^0}\Gamma_{P_q^0}} \left[\frac{f_P m_{P_q^0}^2 m_Q}{\Lambda^2(m_Q + m_q)}\right]^2 \times \sqrt{1 - 4x_\chi^2}, \tag{4.145}$$

where $x_\chi = m_\chi/m_{H^0}$ is a rescaled DM mass. This rate is not helicity-suppressed, so it could allow us to study DM properties at an e^+e^- flavor factory.

Using the formalism above, the photon energy distribution and the decay width of the radiative transition $H_q^0 \to \chi_0\,\chi_0\,\gamma$ can be calculated [11],

$$\frac{d\Gamma}{dE_\gamma}(P_q^0 \to \chi_0\chi_0\gamma) = \frac{f_P^2\alpha C_3 C_4}{3\Lambda^4}\left(\frac{F_P}{4\pi}\right)^2 \frac{2m_{P_q^0}^2 E_\gamma (m_{H_q^0}(1 - 4x_\chi^2) - 2E_\gamma)^{3/2}}{\sqrt{m_{P_q^0} - 2E_\gamma}}, \tag{4.146}$$

$$\begin{aligned}
\mathcal{B}(P_q^0 \to \chi_0\chi_0\gamma) &= \frac{f_P^2\alpha C_3 C_4 m_{P_q^0}^5}{6\Lambda^4\Gamma_{P_q^0}}\left(\frac{F_P}{4\pi}\right)^2 \left(\frac{1}{6}\sqrt{1 - 4x_\chi^2}(1 - 16x_\chi^2 - 12x_\chi^4)\right. \\
&\quad \left. - 12x_\chi^4 \log\frac{2x_\chi}{1 + \sqrt{1 - 4x_\chi^2}}\right).
\end{aligned} \tag{4.147}$$

We observe that Eqs. (4.146) and (4.147) are independent of $C_{1,2}$; this is due to the fact that $P^0 \to \gamma$ form factors of scalar and pseudoscalar currents vanish. In this manner, studies of $P_q^0 \to$ (missing energy) and $P^0 \to$ (γ+missing energy) processes probe complementary operators in the effective Hamiltonian of Eq. (4.143).

As the reader noted, we only considered the simplest transitions with the invisible final states. The effective Lagrangian of Eq. (4.143) can be applied to study other decays as well.

4.7 CP-violation and CP-violating observables

Quark systems allow for fruitful studies of patterns of CP-violation, which result in meaningful probes of New Physics. Since strong interactions make the predictions of transition rates difficult in such systems, it is the phenomenology of CP-violation that allows for such probes in both semileptonic and nonleptonic processes.

Let us get some bookkeeping out of the way by classifying studies of CP-violation with quark systems. It can be generally classified into three different categories,

(I) CP violation in the meson-antimeson mixing matrix (or "indirect" CP-violation). As we shall see below, introduction of $\Delta Q = 2$ transitions, either via SM or NP one-loop or tree-level NP amplitudes leads to non-diagonal entries in the meson-antimeson mass matrix,

$$\left[M - i\frac{\Gamma}{2}\right]_{ij} = \begin{pmatrix} A & p^2 \\ q^2 & A \end{pmatrix} \tag{4.148}$$

This type of CP violation is manifest when $R_m^2 = |p/q|^2 = (2M_{12} - i\Gamma_{12})/(2M_{12}^* - i\Gamma_{12}^*) \neq 1$.

(II) CP violation in the $\Delta Q = 1$ decay amplitudes (or "direct" CP-violation). This type of CP violation occurs when the absolute value of the decay amplitude for a meson or baryon state to decay to a final state f (A_f) is different from the one of the corresponding CP-conjugated amplitude. This can happen if the decay amplitude can be broken into at least two parts associated with different weak and strong phases,

$$A_f = |A_1|\, e^{i\delta_1} e^{i\phi_1} + |A_2|\, e^{i\delta_2} e^{i\phi_2}, \tag{4.149}$$

where ϕ_i represent weak phases ($\phi_i \to -\phi_i$ under CP-transformation), and δ_i represent strong phases ($\delta_i \to \delta_i$ under CP-transformation). This ensures that the CP-conjugated amplitude, $\overline{A}_{\overline{f}}$ would differ from A_f.

(III) CP violation in the interference of decays with and without mixing. This type of CP violation is possible for a subset of final states to which both neutral meson and antimeson can decay.

One of the most common observables for CP-violation is the asymmetry. For the B-meson states it can be defined as*

$$a_{CP}(f) = \frac{\Gamma(B \to f) - \Gamma(\overline{B} \to \overline{f})}{\Gamma(B \to f) + \Gamma(\overline{B} \to \overline{f})}. \tag{4.150}$$

This asymmetry is non-zero only if there are at least two different amplitudes with different weak and strong phases are present,

$$A(B \to f) \equiv A_f = |A_{f1}| e^{i\delta_1} e^{i\phi_1} + |A_{f2}| e^{i\delta_2} e^{i\phi_2} \tag{4.151}$$

We can now compute a_{CP},

$$a_{CP}(f) \sim \sin(\phi_1 - \phi_2) \sin(\delta_1 - \delta_2). \tag{4.152}$$

It is important to realize that the extraction of the fundamental parameters of CP-violation from Eq. (4.150) is quite complicated due to the fact that the final states are composed of hadrons. There are several methods that could be used to compute those non-leptonic decays amplitudes.

*Similar definitions for charmed or strange states can be obtained by trivial substitutions.

4.8 Non-leptonic decays: QCD challenges

Calculations of hadronic decay rates governed by four-fermion operators are quite complicated and often model-dependent. Simplified assumptions, such as factorization could be used to estimate the needed branching ratios. Dynamical approaches based on QCD factorization [18] or soft-collinear effective theory (SCET) [14] allow for systematic calculations of four-fermion matrix elements in $m_Q \to \infty$ limit. Such methods work well for the b-flavored mesons, as $m_b \gg \Lambda_{QCD}$. For the c-flavored states the agreement between calculations done with QCD factorization and experimental results are not that great. The main problem with reliable calculations of charmed meson decays is that they populate the energy range where non-perturbative quark dynamics is active. This leads to resonance effects that affect the phases of hadronic decay amplitudes, which makes predictions based on factorization quite unreliable.

Instead of predicting an absolute decay rate, it is often useful to obtain relations among several decay rates. These relations are helpful when some decay rates in a relation are measured, and some are unknown. This allows for a relation to be used to predict the unknown transition rate(s). The relations can be built based on some symmetries, such as standard flavor $SU(3)$ [168], or on an overcomplete set of universal quark-level amplitudes [99]. We shall discuss those methods below.

The partial width for a specific two-body decay of a charmed meson depends on both the invariant amplitude $\mathcal{A}$ and a phase space factor. For a specific two-body decay of a heavy state P_q into a PP final state,

$$\Gamma(P_q \to PP) = \frac{|\,\mathbf{p}\,|}{8\pi m_{H_q}^2} \left|\mathcal{A}(H_q \to PP)\right|^2, \tag{4.153}$$

where $|\mathbf{p}|$ is a center-of-mass 3-momentum of each final state particle. For a decay into a PV final state,

$$\Gamma(P_q \to PV) = \frac{|\,\mathbf{p}\,|^3}{8\pi m_{H_q}^2} \left|\mathcal{A}(H_q \to PV)\right|^2. \tag{4.154}$$

Note that in the case of PP final state the final state mesons are in the S-wave, while in the case of PV final state they are in a P-wave. This is why $|\mathcal{A}(P_q \to PP)|$ has dimension of energy, while $|\mathcal{A}(P_q \to PV)|$ is dimensionless.

4.8.1 QCD corrections in EFTs

In building effective theories of QCD we need to remember that the basis of operators of a given dimension will be enlarged due to color degrees of freedom possessed by each quark field. For example, operators of dimension 6 built out of four different quark fields q_i

$$\begin{aligned} Q_1 &= (\bar{q}_{1\alpha}\Gamma_\mu q_{2\beta})(\bar{q}_{3\beta}\Gamma_\mu q_{4\alpha}), \text{ and} \\ Q_2 &= (\bar{q}_{1\alpha}\Gamma_\mu q_{2\alpha})(\bar{q}_{3\beta}\Gamma_\mu q_{4\beta}), \end{aligned} \tag{4.155}$$

with $\alpha, \beta = 1, 2, 3$ being the color indices (summation is assumed), are not the same. Moreover, since gluons carry color quantum numbers, one would expect that perturbative QCD (pQCD) corrections will mix Q_1 into Q_2 and vise versa. Thus, below the scale associated with the W-boson mass we must consider additional operators describing electroweak transitions. In fact, the weak Hamiltonian describing transitions among q_i is

$$\mathcal{H}_{\text{eff}} = \frac{4G_F}{\sqrt{2}}\xi_{CKM}\left(C_1(\mu)Q_1(\mu) + C_2(\mu)Q_2(\mu)\right), \tag{4.156}$$

where ξ_{CKM} parameterizes relevant CKM parameters. We introduced an auxiliary scale, μ, that separates contributions from short distance physics, i.e. those in the Wilson coefficients C_i, from long distance physics, i.e. those in the operators Q_i.

Notice that QCD corrections to effective electroweak operators are oftentimes more important than the contributions of higher-dimensional operators, as these are further suppressed by powers of $1/M_W$). This is especially true for the description of low-energy processes, where $\alpha_s(\mu) \gg \mu^2/M_W^2$, and two-loop or even three-loop pQCD corrections could give more important contributions then the electroweak $1/M_W^4$ or possible New Physics $1/\Lambda^2$ effects. Thus, it is important to calculate pQCD corrections to the coefficients C_1 and C_2.

Matching at one loop in QCD

In order to derive a QCD-corrected effective Lagrangian, we need to calculate transition amplitudes (to be exact, amputated Green functions) in full and effective theories, $\mathcal{A}_{\text{full}}$ and $\mathcal{A}_{\text{eff}}$, to the desired order in α_s and match them at the energy scale and a kinematical point for external momenta p_i that we choose for matching full and effective theories to determine Wilson coefficients (for a review, see [135,149]),

$$\mathcal{A}_{\text{full}}(p_i, m_i, \mu) = \mathcal{A}_{\text{eff}}(p_i, m_i, \mu), \tag{4.157}$$

This matching procedure would need to be repeated every time we cross a threshold associated with various quarks' masses. At each such threshold, this would change the number of active quark flavors included in the running of α_s and might generate new operators.

Computing perturbative QCD corrections to four-fermion operators brings two potential concerns. First, we can expect the appearance of ultraviolet divergences. These divergences would appear from two sources: renormalization of external quark legs, and renormalization of composite operators Q_i. Those divergencies can be dealt with by employing methods of standard renormalization theory. Second, calculations of transition amplitudes in Eq. (4.157), in principle, include computations of nonperturbative hadronic matrix elements. However, since we are only interested in the Wilson coefficients C_i, we can choose *any* external states for the calculations of matrix elements of those operators. In particular, the simplest choice, perturbative quark states, will do the job.

In order to perform the matching, we need to compute one loop QCD corrections in the full electroweak theory, where quark currents $\bar{q}_{1\alpha}\Gamma_\mu q_{2\alpha}$ and $\bar{q}_{3\beta}\Gamma_\mu q_{4\beta}$ are connected by a propagator of W-boson. This involves calculations of six diagrams. In leading logarithmic approximation, assuming that all quarks have the same momentum p, and setting their masses to zero, we get a result for $\mathcal{A}_{\text{full}}$ (in dimensional regularization) [37]

$$\begin{aligned}
\mathcal{A}_{\text{full}} &= \frac{4G_F}{\sqrt{2}}\xi_{CKM}\left[\left(A_1 + \frac{B_1}{\epsilon}\right)\langle Q_2\rangle^T + A_2\langle Q_1\rangle^T\right], \\
A_1 &= 1 + 2C_F\frac{\alpha_s}{4\pi}\log\frac{\mu^2}{-p^2} + \frac{3}{N_c}\frac{\alpha_s}{4\pi}\log\frac{M_W^2}{-p^2}, \\
A_2 &= -\frac{3\alpha_s}{4\pi}\log\frac{M_W^2}{-p^2}, \quad B_1 = 2C_F\frac{\alpha_s}{4\pi},
\end{aligned} \tag{4.158}$$

where we used a well-known identity for the Gell-Mann matrices

$$2\delta_{\alpha\delta}\delta_{\gamma\beta} = \frac{2}{N_c}\delta_{\alpha\beta}\delta_{\gamma\delta} + \lambda^a_{\alpha\beta}\lambda^a_{\gamma\rho}, \tag{4.159}$$

where summation over a is implied. $\langle Q_i\rangle^T$ are the tree-level matrix elements of operators Eq. (4.155). Note divergent terms in Eq. (4.158) for $\epsilon \to 0$. Those are cancelled by QCD loop contributions on the external quark legs, which gives for the fully renormalized amplitude,

$$\mathcal{A}^r_{\text{full}} = \frac{4G_F}{\sqrt{2}}\xi_{CKM}\left[A_1\langle Q_2\rangle^T + A_2\langle Q_1\rangle^T\right]. \tag{4.160}$$

This is the amplitude that we shall match to the effective amplitude computed with the effective Hamiltonian of Eq. (4.156).

The effective theory contribution $\mathcal{A}_{\text{eff}}$ in the same approximation is given by [149]

$$\begin{aligned}\mathcal{A}_{\text{eff}} = \frac{4G_F}{\sqrt{2}}\xi_{CKM}\Big[C_1\Big(\Big(a_1+\frac{b_1}{\epsilon}\Big)\langle Q_1\rangle^T + \Big(a_2+\frac{b_2}{\epsilon}\Big)\langle Q_2\rangle^T\Big) \\ +C_2\Big(\Big(a_1+\frac{b_1}{\epsilon}\Big)\langle Q_2\rangle^T + \Big(a_2+\frac{b_2}{\epsilon}\Big)\langle Q_1\rangle^T\Big)\Big],\end{aligned} \tag{4.161}$$

where the coefficients are given by

$$\begin{aligned} a_1 &= 1 + 2C_F\frac{\alpha_s}{4\pi}\log\frac{\mu^2}{-p^2} + \frac{3}{N_c}\frac{\alpha_s}{4\pi}\log\frac{\mu^2}{-p^2}, \\ a_2 &= -\frac{3\alpha_s}{4\pi}\log\frac{\mu^2}{-p^2}. \\ b_1 &= \frac{\alpha_s}{4\pi}\left(2C_F + \frac{3}{N_c}\right), \quad b_2 = -\frac{3\alpha_s}{4\pi}. \end{aligned} \tag{4.162}$$

Notice that it is not possible to completely remove all of the divergences in the EFT amplitude Eq. (4.161) by wave function renormalization factors, as we are dealing with renormalization of composite operators. Additional renormalization factors are needed,

$$\langle Q_i\rangle^{(0)} = Z_\psi^{-2}Z_{ij}\langle Q_j\rangle. \tag{4.163}$$

Removing some of the UV divergencies with quark wave function renormalization factors Z_ψ and constructing Z_{ij} to remove the remaining divergences as

$$Z_{ij} = 1 + \frac{\alpha_s}{4\pi}\frac{1}{\epsilon}\begin{pmatrix} 3/N_c & -3 \\ -3 & 3/N_c \end{pmatrix} \tag{4.164}$$

gives us the renormalized amplitude,

$$\begin{aligned}\mathcal{A}^r_{\text{eff}} &= \frac{4G_F}{\sqrt{2}}\xi_{CKM}\Big[C_1\left(a_1\langle Q_1\rangle^T + a_2\langle Q_2\rangle^T\right) \\ &+ C_2\left(a_1\langle Q_2\rangle^T + a_2\langle Q_1\rangle^T\right)\Big],\end{aligned} \tag{4.165}$$

which is UV finite. Matching Eq. (4.160) and (4.165) we can now determine C_1 and C_2,

$$C_1(\mu) = -\frac{3\alpha_s}{4\pi}\log\frac{M_W^2}{\mu^2}, \quad C_2(\mu) = 1 + \frac{3}{N_c}\frac{\alpha_s}{4\pi}\log\frac{M_W^2}{\mu^2}. \tag{4.166}$$

It is interesting that perturbative QCD effects generated a new operator structure, Q_1! The numerical impact of QCD corrections is quite significant, as for a typical hadronic scale $\mu \sim 1$ GeV the logarithms in Eq. (4.166) are large, $\log M_W^2/\mu^2 \sim 8.8$. This might be a source of additional concerns, as α_s is not small at such low scales.

Renormalization group improvement

Calculation of QCD corrections to effective operators of Eq. (4.155) produced large logarithms of the form $\alpha_s \log(m_W^2/\mu^2)$ in Eq. (4.166). This makes the convergence of QCD perturbative series questionable, as for low scales the size of the log compensates the smallness of α_s. We might expect that at higher orders in α_s all corrections of the form $\alpha_s^n \log^n(m_W^2/\mu^2)$ would be large, so they must be resummed. This can be achieved with the help of renormalization group (RG) equations.

To do so, we first need to go to a basis of operators in which matrix Z_{ij} of Eq. (4.164) is diagonal. Diagonalization of this matrix will simplify solution of RG equations. It is not hard to convince yourself that in the basis given by

$$Q_\pm = \frac{1}{2}\left(Q_2 \pm Q_1\right), \quad C_\pm = C_2 \pm C_1 \tag{4.167}$$

the matrix Z_{ij} also becomes diagonal,

$$Z_\pm = \text{diag}\left(1 - \frac{2\alpha_s}{4\pi\epsilon}, 1 + \frac{4\alpha_s}{4\pi\epsilon}\right). \tag{4.168}$$

The large logarithms we had in Eq. (4.166) will also affect $C_\pm$, but resummation of these logs will be easier in the new basis. This resummation can be achieved with help of RG equation for $C_\pm$. Since C's play a role of the coupling constants in the effective Hamiltonian of Eq. (4.156), they should also be renormalized,

$$C_\pm = Z_\pm^C C_\pm^{(0)}, \tag{4.169}$$

where $C_\pm^{(0)}$ denote "bare" values for $C_\pm$. Let us take a logarithmic derivative of both sides of Eq. (4.169),

$$\mu\frac{dC_\pm}{d\mu} = \left(\mu\frac{dZ_\pm^C}{d\mu}\right)C_\pm^{(0)} + Z_\pm^C\left(\mu\frac{dC_\pm^{(0)}}{d\mu}\right). \tag{4.170}$$

Since "bare" couplings $C_\pm^{(0)}$ do not depend on renormalization scale μ, the second term in Eq. (4.170) is zero, so

$$\mu\frac{dC_\pm(\mu)}{d\mu} = \gamma_\pm C_\pm(\mu). \tag{4.171}$$

We introduced a matrix of *anomalous dimensions* for $C_\pm$ in Eq. (4.171),

$$\gamma_\pm(g_s) = \frac{\mu}{Z_\pm^C}\frac{dZ_\pm^C}{d\mu}. \tag{4.172}$$

It is interesting to note that the right-hand side of the above equation can be rewritten as a logarithmic derivative of $\log Z_\pm$! Expanding (diagonal) anomalous dimension matrix $\gamma_\pm$ as

$$\gamma_\pm(g_s) = \frac{d\log Z_\pm}{d\log\mu} = \gamma_\pm^{(0)}\frac{g_s^2}{16\pi^2} + \gamma_\pm^{(1)}\left(\frac{g_s^2}{16\pi^2}\right)^2 + \ldots \tag{4.173}$$

we obtain the needed anomalous dimensions for the Wilson coefficients as

$$\gamma_\pm^{(0)} = \text{diag}\left(4, -8\right). \tag{4.174}$$

Now, the solution of the RG equation of Eq. (4.171) can be obtained if we recall that $C_\pm(\mu)$ depend on μ both explicitly and via μ-dependence of the running coupling constant g_s,

$$\mu\frac{d}{d\mu} = \mu\frac{\partial}{\partial\mu} + \beta(g_s)\frac{\partial}{\partial g_s}, \tag{4.175}$$

where we introduced a standard QCD beta function, $\beta_s = \partial g_s/\partial \log\mu$. The solution of the RG equation is given by

$$C_\pm(\mu) = U_\pm(\mu, M_W) C_\pm(M_W) = \exp \int_{g(M_W)}^{g(\mu)} dg' \frac{\gamma_\pm(g')}{\beta(g')} C_\pm(M_W). \tag{4.176}$$

Expanding anomalous dimensions in $\alpha_s/4\pi$ we can find a solution in the leading logarithmic approximation (LLA),

$$C_\pm(\mu) = \left(\frac{\alpha_s(M_W)}{\alpha_s(\mu)}\right)^{\gamma_\pm/(2\beta_0)} C_\pm(M_W). \tag{4.177}$$

This formula achieves our goal: there are no longer any large logarithms in the expression for $C_\pm$. One can check that, if expanded in α_s, the expression derived in Eq. (4.166) is recovered.

Note that according to Eq. (4.167) $C_+(M_W) = C_-(M_W) = 1$. To complete the derivation of $C_\pm$ let us recall that β_0 depends on the number of flavors via $\beta_0 = 11 - 2n_f/3$, which ties together scale and the number of active flavors. For example, for $\mu = m_b$, where five quark flavors are active we get

$$C_+(m_b) = \left(\frac{\alpha_s(M_W)}{\alpha_s(m_b)}\right)^{6/23}, \quad C_-(m_b) = \left(\frac{\alpha_s(M_W)}{\alpha_s(m_b)}\right)^{-12/23}. \tag{4.178}$$

Numerically, $C_+(5 \text{ GeV}) \simeq 0.85$ is suppressed by QCD running, while $C_-(5 \text{ GeV}) \simeq 1.40$ is enhanced compared to $C_\pm(M_W) = 1$.

As we can see, the number of active flavors changes the values of Wilson coefficients, mainly due to the changing β_0. Thus, the evolution of $C_\pm$ depends on crossing quark thresholds. For example, evolving down to the scales near the charm quark mass $\mu = m_c$ should be done as

$$C_\pm(\mu) = U_\pm^{n_f=4}(m_c, m_b) U_\pm^{n_f=5}(m_b, M_W) C_\pm(M_W) \tag{4.179}$$

and similarly for the lower scales.

Complete basis. QCD and electroweak penguin operators

In order to write a complete set of operators that contribute to $b \to q_1\bar{q}_2 q_3$ process we have to follow a framework of effective field theory, where we have to write all possible dimension six operators to describe the non-leptonic low energy transitions of quarks. These new operators might mix among themselves under QCD renormalization. The reason why we did not see them in our analysis above is that it might happen that Q_1 and Q_2 mix into other operators, but not vise versa. We need to write out all possible operator structures.

This is exactly what has been done for the description of $\Delta Q = 1$ weak decays of $Q = b, c$, and s quarks. In this case, operator structures other than Q_1 and Q_2 can be generated in the Standard Model. One example is the so-called penguin operator, $(\bar{s}b)_{V-A} \sum_q (\bar{q}q)_{V-A}$, for $b \to s\bar{q}q$ decay. As it turns out, there are two types of such operators. Depending on the nature of the propagator connecting the $(\bar{s}b)_{V-A}$ current and $(\bar{q}q)_{V-A}$ current, one can classify those operators as *gluonic* or *electroweak* penguins.

Let us now write the full $\Delta Q = 1$ non-leptonic effective Hamiltonian. We shall only concentrate on the operators generated in the Standard Model. Any other operator structures that contribute to a given transition amplitude would indicate presence of BSM interactions. Concentrating on $\Delta b = 1$ transitions for simplicity

$$\mathcal{H}_{\text{eff}} = \frac{4G_F}{\sqrt{2}} \sum_{q=u,c} \lambda_q \left[C_1 Q_1^q + C_2 Q_2^q + \sum_{i=3,\dots,10} C_i Q_i + C_{7\gamma} Q_{7\gamma} + C_{8g} Q_{8g} \right] + h.c., \tag{4.180}$$

TABLE 4.2 Wilson coefficients for the $\Delta b = 1$ operator basis in Eq. (4.180) for $\mu = m_b(m_b) = 4.40$ GeV and $m_t = 170$ GeV (from [33]).

Wilson coefficient	LO	NLO (NDR) $\Lambda^{(5)} = 140$ MeV	LO	NLO (NDR) $\Lambda^{(5)} = 310$ MeV
C_1	−0.273	−0.165	−0.339	−0.203
C_2	1.125	1.072	1.161	1.092
C_3	0.013	0.013	0.016	0.016
C_4	−0.027	−0.031	−0.033	−0.039
C_5	0.008	0.008	0.009	0.009
C_6	−0.033	−0.036	−0.043	−0.046
C_7/α	0.042	−0.003	0.047	−0.001
C_8/α	0.041	0.047	0.054	0.061
C_9/α	−1.264	−1.279	−1.294	−1.303
C_{10}/α	0.291	0.234	0.360	0.288

where $\lambda_q = V_{qb}V_{qs}^*$ denotes the relevant CKM matrix elements. The current-current operators Q_1 and Q_2 for $\Delta b = 1$ transitions take the form

$$\begin{aligned} Q_1 &= (\bar{q}_\alpha b_\beta)_{V-A}(\bar{s}_\beta q_\alpha)_{V-A}, \text{ and} \\ Q_2 &= (\bar{q}_\alpha b_\alpha)_{V-A}(\bar{s}_\beta q_\beta)_{V-A}. \end{aligned} \tag{4.181}$$

Note that here we explicitly consider $b \to s$ decays and suppressed Dirac structures that are given by $(\bar{q}_1 q_2)_{V\pm A} = (1/2)(\bar{q}_1\gamma_\mu(1 \pm \gamma_5)q_2)$. Related transitions, such as $b \to d$ can be obtained by direct substitution of the appropriate quark fields and CKM matrix elements. We shall suppress color indices from now on for the structures like $(\bar{q}_\alpha q_\alpha)_{V\pm A} \to (\bar{q}q)_{V\pm A}$.

The gluonic penguin operators Q_{3-6} are given by

$$\begin{aligned} Q_3 &= (\bar{s}b)_{V-A}\sum_q(\bar{q}q)_{V-A}, \quad Q_4 = (\bar{s}_\alpha b_\beta)_{V-A}\sum_q(\bar{q}_\beta q_\alpha)_{V-A}, \\ Q_5 &= (\bar{s}b)_{V-A}\sum_q(\bar{q}q)_{V+A}, \quad Q_6 = (\bar{s}_\alpha b_\beta)_{V-A}\sum_q(\bar{q}_\beta q_\alpha)_{V+A}, \end{aligned} \tag{4.182}$$

while electroweak penguin operators Q_{7-10} are

$$\begin{aligned} Q_7 &= \frac{3e_q}{2}(\bar{s}b)_{V-A}\sum_q(\bar{q}q)_{V+A}, \; Q_8 = \frac{3e_q}{2}(\bar{s}_\alpha b_\beta)_{V-A}\sum_q(\bar{q}_\beta q_\alpha)_{V+A}, \\ Q_9 &= \frac{3e_q}{2}(\bar{s}b)_{V-A}\sum_q(\bar{q}q)_{V-A}, \; Q_{10} = \frac{3e_q}{2}(\bar{s}_\alpha b_\beta)_{V-A}\sum_q(\bar{q}_\beta q_\alpha)_{V-A}. \end{aligned} \tag{4.183}$$

Finally, there are two dipole operators,

$$\begin{aligned} Q_{7\gamma} &= -\frac{e}{8\pi^2}m_b\; \bar{s}\sigma_{\mu\nu}(1-\gamma_5)F^{\mu\nu}b, \text{ and} \\ Q_{8g} &= -\frac{g_s}{8\pi^2}m_b\; \bar{s}\sigma_{\mu\nu}(1-\gamma_5)G^{\mu\nu}b, \end{aligned} \tag{4.184}$$

In the Standard Model dipole operators with quarks of opposite helicity are suppressed by factors of the light quark mass, m_s. Therefore, their contribution is usually discarded as being small. There are models of New Physics where chirality flip occurs inside the loop diagram that generates the operators $Q_{7\gamma}$ and Q_{8g}. One way to search for those models would be to tag the helicity of the final state photon, which is a challenging measurement.

Even after resumming the logarithms, the scale dependence of Wilson coefficients remains strong (see Table 4.2). It is then advantageous to go beyond one loop matching. It would also be consistent to go beyond the leading order in the calculations of anomalous dimensions for the following renormalization group running. The next-to-leading order (NLO) calculation is interesting, as it brings new features in the discussion, such as the dependence of the final result on the renormalization scheme used in the calculation. The discussion of this, however, goes beyond the scope of this book, but can be found in a number of specialized reviews [33,37]. Here we present numerical results for the Wilson coefficients at LO and NLO (using Naive Dimensional Reduction (NDR) scheme) [33], which can be used for practical computations of decay rates.

The discussion of Wilson coefficients for photonics and gluonic dipole operators $Q_{7\gamma}$ and Q_{8g} is similar. Their values are $C_{7\gamma} = 0.299$ and $C_{8g} = 0.143$, which are quite large to affect some decays of B-mesons.

It should be noted that the techniques of calculating QCD corrections for four-fermion operators can be employed for other calculations. For example, analyses of meson mixing in the B-sector ($\Delta b = 2$ operators) can also be done in the same way, as can analyses of kaons and charmed meson decays.

4.8.2 $SU(3)_F$ flavor symmetry

One popular approach that was adopted for studies of hadronic decays involves the application of approximate symmetries, such as flavor $SU(3)_F$. This approach is based on the fact that the QCD Lagrangian acquires that symmetry in the limit where masses of all light quarks are the same. The $SU(3)_F$ analysis of decay amplitudes cannot predict their absolute values. However, at least in the symmetry limit, this approach can relate transition amplitudes for different decays, which could prove quite useful for experimental analysis. One potential difficulty with this approach, for charmed decays in particular, is related to the fact that available experimental data show that flavor $SU(3)_F$ symmetry is broken in charm transitions, so symmetry-breaking corrections should be taken into account [96,168,169].

In the flavor-symmetry approach, all particles are denoted by their $SU(3)_F$ representations. Beauty and charm quarks transform as singlets under flavor $SU(3)$. The fundamental representation of $SU(3)_F$ is a triplet, $\mathbf{3}$, so the light quarks u, d, and s belong to this representation with $(1,2,3) = (u,d,s)$. In this language, a B_c meson transforms as a singlet, while B_q and D mesons containing light quarks (u,d,s) form triplets and can be represented as row vectors,

$$B_i = \left(B^-, B_d^0, B_s^0\right), \quad D_i = \left(D^0, D^+, D_s\right). \tag{4.185}$$

The octet of pseudoscalar mesons M formed by the light quarks (u,d,s) can be represented by a $\mathbf{3} \times \mathbf{3}$ matrix M^i_j where the upper index represents the rows (quarks) and the lower index represents the columns (antiquarks). In this notation, the matrix M^i_j can be expressed as,

$$M = \begin{pmatrix} \frac{\pi^0}{\sqrt{2}} + \frac{\eta_8}{\sqrt{6}} & \pi^+ & K^+ \\ \pi^- & -\frac{\pi^0}{\sqrt{2}} + \frac{\eta_8}{\sqrt{6}} & K^0 \\ K^- & \overline{K}^0 & -\sqrt{\frac{2}{3}}\eta_8 \end{pmatrix}. \tag{4.186}$$

Similar matrix can be written for light vector mesons. In principle, physical states, such as η and η', are formed by mixing with an $SU(3)_F$ singlet, η_1 for the pseudoscalars and ω_1 for the vectors.

Flavor $SU(3)_F$ analysis of B-decays was done in [169], D-decays were considered in [168], while description of B_c decays can be found in [22]. As an example, let us consider applications of $SU(3)_F$ flavor symmetry to decays of charmed mesons.

The $\Delta C = -1$ part of the weak Hamiltonian has the flavor structure $(\bar{q}_i c)(\bar{q}_j q_k)$, so its matrix representation is written with a fundamental index and two antifundamentals, H^{ij}_k. This operator is a sum of irreducible representations contained in the product $\mathbf{3} \times \overline{\mathbf{3}} \times \overline{\mathbf{3}} = \overline{\mathbf{15}} + \mathbf{6} + \mathbf{3} + \overline{\mathbf{3}}$. In the limit in which the third generation is neglected, H^{ij}_k is traceless, so only the $\overline{\mathbf{15}}$ (symmetric

TABLE 4.3 Matrix representations of the operators $\overline{\mathbf{15}}$ and $\mathbf{6}$.

$H(\overline{\mathbf{15}})^{ij}_k$:	$H^{13}_2 = H^{31}_2 = 1$,	$H^{12}_2 = H^{21}_2 = s_1$,
	$H^{13}_3 = H^{31}_3 = -s_1$,	$H^{12}_3 = H^{21}_3 = -s_1^2$,
$H(\mathbf{6})^{ij}_k$:	$H^{13}_2 = -H^{31}_2 = 1$,	$H^{12}_2 = -H^{21}_2 = s_1$,
	$H^{13}_3 = -H^{31}_3 = -s_1$,	$H^{12}_3 = -H^{21}_3 = -s_1^2$.

on i and j) and $\mathbf{6}$ (antisymmetric on i and j) representations appear. That is, the $\Delta C = -1$ part of $\mathcal{H}_w$ may be decomposed as $\frac{1}{2}(\mathcal{O}_{\overline{\mathbf{15}}} + \mathcal{O}_{\mathbf{6}})$, where

$$\begin{aligned}
\mathcal{O}_{\overline{\mathbf{15}}} &= (\bar{s}c)(\bar{u}d) + (\bar{u}c)(\bar{s}d) + s_1(\bar{d}c)(\bar{u}d) \\
&\quad + s_1(\bar{u}c)(\bar{d}d) - s_1(\bar{s}c)(\bar{u}s) - s_1(\bar{u}c)(\bar{s}s) \\
&\quad - s_1^2(\bar{d}c)(\bar{u}s) - s_1^2(\bar{u}c)(\bar{d}s)\,, \\
\mathcal{O}_{\mathbf{6}} &= (\bar{s}c)(\bar{u}d) - (\bar{u}c)(\bar{s}d) + s_1(\bar{d}c)(\bar{u}d) \\
&\quad - s_1(\bar{u}c)(\bar{d}d) - s_1(\bar{s}c)(\bar{u}s) + s_1(\bar{u}c)(\bar{s}s) \\
&\quad - s_1^2(\bar{d}c)(\bar{u}s) + s_1^2(\bar{u}c)(\bar{d}s)\,,
\end{aligned} \tag{4.187}$$

and $s_1 = \sin\theta_C \approx 0.22$. The matrix representations $H(\overline{\mathbf{15}})^{ij}_k$ and $H(\mathbf{6})^{ij}_k$ have nonzero elements as shown in Table 4.3.

In the $SU(3)_F$ limit the effective Hamiltonian for the hadronic decays to two pseudoscalars $D \to PP$ can be written as

$$\begin{aligned}
\mathcal{H}^{SU(3)}_{\text{eff}} &= a_{\overline{\mathbf{15}}} D_i H(\overline{\mathbf{15}})^{ij}_k M^l_j M^k_l + b_{\overline{\mathbf{15}}} D_i M^i_l H(\overline{\mathbf{15}})^{lj}_k M^k_j \\
&+ c_{\mathbf{6}} D_i H(\mathbf{6})^{ij}_k M^l_j M^k_l + \delta\mathcal{H},
\end{aligned} \tag{4.188}$$

where $a_{\overline{\mathbf{15}}}$, $b_{\overline{\mathbf{15}}}$, and $c_{\mathbf{6}}$ are the reduced matrix elements that correspond to different $SU(3)$ representations and $\delta\mathcal{H}$ is an $SU(3)_F$-breaking term. Neglecting it for a moment, there is a number of amplitude relations that can be obtained from Eq. (4.188). In particular, it can be seen that it implies relations between $A_{D^0\to K^+K^-}$ and $A_{D^0\to\pi^+\pi^-}$ decay amplitudes. Reading off the amplitudes from Eq. (4.188),

$$\begin{aligned}
A_{D^0\to K^+K^-} &= s_1\left(-a_{\overline{\mathbf{15}}} - b_{\overline{\mathbf{15}}} + c_{\mathbf{6}}\right), \\
A_{D^0\to\pi^+\pi^-} &= s_1\left(a_{\overline{\mathbf{15}}} + b_{\overline{\mathbf{15}}} - c_{\mathbf{6}}\right),
\end{aligned} \tag{4.189}$$

so it appears that $|A_{D^0\to K^+K^-}| = |A_{D^0\to\pi^+\pi^-}|$. In practice, the corresponding branching fractions differ by a factor of three, which indicates that $SU(3)_F$ symmetry is broken in D-decays. Similar analysis can be performed for B decays with no such dramatic breaking of $SU(3)_F$.

It appears that a consistent approach should then include $SU(3)_F$-breaking corrections, which should be included in the analysis. For example, one could assume that $SU(3)_F$ breaking is proportional to light quark masses. In this case, it can be included in the analysis as a perturbation, suppressed by a typical hadronic scale $\Lambda_{\text{had}} \sim 1$ GeV, that transforms as $\mathbf{8} + \mathbf{1}$, as the quark mass operator belongs to the matrix representation $M^i_j = \text{diag}(m_u, m_d, m_s)$, which is an $\mathbf{8}$. Neglecting the masses of up and down quark, we can summarize the breaking in the matrix γ,

$$\gamma = \frac{m_s}{3\Lambda_{\text{had}}}\begin{pmatrix} 1 & 0 & 0 \\ 0 & 1 & 0 \\ 0 & 0 & 1 \end{pmatrix} + \frac{m_s}{3\Lambda_{\text{had}}}\begin{pmatrix} -1 & 0 & 0 \\ 0 & -1 & 0 \\ 0 & 0 & 2 \end{pmatrix} \tag{4.190}$$

where the first term transforms as a $\mathbf{1}$ and is absorbed into the reduced matrix elements of the $SU(3)_F$ conserving amplitudes. The second term gives rise to $SU(3)_F$ breaking. A complete analysis with broken $SU(3)_F$ is possible [168]. The $SU(3)_F$ breaking Hamiltonian can be obtained from the tensor product $\gamma \times (\mathcal{O}_{\overline{\mathbf{15}}} + \mathcal{O}_{\mathbf{6}})$, which can be decomposed into irreducible representations as

$$\begin{aligned} \mathbf{8} \times \overline{\mathbf{15}} &= \overline{\mathbf{42}} + \mathbf{24}^{(1)} + \overline{\mathbf{15}}^{(1)} + \overline{\mathbf{15}}^{(2)} + \overline{\mathbf{15}}^{(3)} + \mathbf{6}^{(1)} + \overline{\mathbf{3}}^{(1)}, \\ \mathbf{8} \times \mathbf{6} &= \mathbf{24}^{(2)} + \overline{\mathbf{15}}^{(4)} + \mathbf{6}^{(2)} + \overline{\mathbf{3}}^{(2)}, \end{aligned} \quad (4.191)$$

where, as before, the superscript distinguishes between representations of the same dimension, but different matrix elements. Different operators in Eq. (4.191) could be represented as tensors that induce different pattern of $SU(3)_F$ breaking. Interestingly, the two triplets would have one non-zero components each with

$$H_1(\overline{\mathbf{3}}^{(1)}) = -s_1, \quad H_1(\overline{\mathbf{3}}^{(2)}) = s_1, \quad (4.192)$$

inducing the breaking terms in the Hamiltonian of Eq. (4.188),

$$\delta\mathcal{H} = d_{\overline{\mathbf{3}}} D_i H(\overline{\mathbf{3}}^{(1)})^i M_j^k M_k^j + e_{\overline{\mathbf{3}}} D_i M_j^i M_k^j H(\overline{\mathbf{3}}^{(1)})^k, \quad (4.193)$$

so that the relations of Eq. (4.189) are modified as to

$$\begin{aligned} A_{D^0 \to K^+K^-} &= s_1 \left(-a_{\overline{\mathbf{15}}} - b_{\overline{\mathbf{15}}} + c_{\mathbf{6}} - 2\frac{m_s}{\Lambda_{\text{had}}} d_{\overline{\mathbf{3}}} - \frac{m_s}{\Lambda_{\text{had}}} e_{\overline{\mathbf{3}}} \right), \\ A_{D^0 \to \pi^+\pi^-} &= s_1 \left(a_{\overline{\mathbf{15}}} + b_{\overline{\mathbf{15}}} - c_{\mathbf{6}} - 2\frac{m_s}{\Lambda_{\text{had}}} d_{\overline{\mathbf{3}}} - \frac{m_s}{\Lambda_{\text{had}}} e_{\overline{\mathbf{3}}} \right). \end{aligned} \quad (4.194)$$

As can be seen from Eq. (4.194) the breaking term contributes equally to $A_{D^0 \to K^+K^-}$ and $A_{D^0 \to \pi^+\pi^-}$, but since the leading terms are of opposite sign, the amount of symmetry breaking needed to explain the factor of three difference between the experimental measurements of $\Gamma(D^0 \to K^+K^-)$ and $\Gamma(D^0 \to \pi^+\pi^-)$ does not have to be that large.

In some cases one does not need to employ the full formalism of $SU(3)_F$, but only rely on its subgroups. For example, the U-spin, a symmetry of the Lagrangian with respect to $s \to d$ quark interchange, can be employed to obtain several useful relations. For example, for the decays of D^0 meson into final states containing $M^0 = \pi^0$, η, and η', one can obtain

$$\frac{\mathcal{A}(D^0 \to K^0 M^0)}{\mathcal{A}(D^0 \to \overline{K}^0 M^0)} = -\tan^2\theta_C. \quad (4.195)$$

Equation (4.195) derives from the following argument. The initial state, D^0 contains c and $\bar{u}$ quarks, and so is a U-spin singlet. The CF transition $c \to su\bar{d}$ and DCS transition $c \to du\bar{s}$ produce $U = 1$ finals states with opposite $U_3 = 1$ in the decays of D^0 meson. The final state meson M^0 form a linear combination of U-spin singlet and triplet states, while neutral kaons are $U = 1$, $U_3 = \pm 1$ states. Thus, U-spin triplet part of M^0 cannot be produced, as it leads to the $U = 2$ final state. Thus, only the singlet part of M^0 can contribute to the transition, which leads to Eq. (4.195).

4.8.3 Flavor-flow (topological) diagram approach

Another useful approach to tackle hadronic decays of heavy mesons, which is equivalent to the $SU(3)_F$ amplitude method described above, is the flavor-flow (or topological $SU(3)$ approach). This approach involves a set of "quark diagrams", which shows the flow of flavor in the usual Feynman diagrams. It was originally developed for studies of B-meson decays [99] and proven to be equivalent to the approach described in the previous section in the exact $SU(3)_F$ symmetry limit [188]. The application of this method to D-decays [45] can even prove advantageous

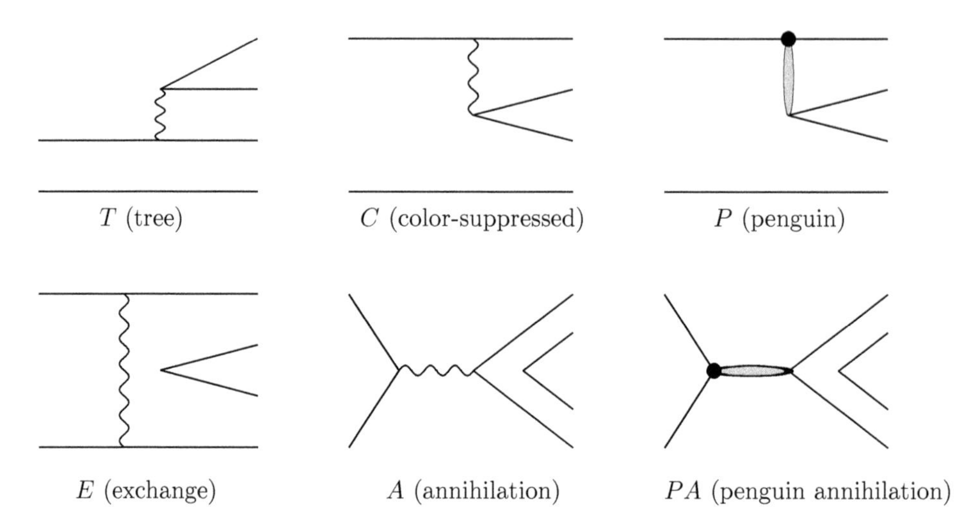

FIGURE 4.2 Flavor-flow diagrams representing the contributions of various decay topologies.

compared to flavor $SU(3)$ approach, as the number of unknown amplitudes grows rapidly if $SU(3)_F$-breaking is taken into account. Yet, at least in the case of linear $SU(3)_F$-breaking, those two approaches were proven to be equivalent [139].

In the topological flavor-flow approach each decay amplitude is parametrized according to the topology of Feynman diagrams (see Fig. 4.2): a color-favored tree amplitude (usually denoted by T), a color-suppressed tree amplitude (C), an exchange amplitude (E), a penguin amplitude (P), an annihilation amplitude (A), and a penguin annihilation amplitude (PA).

In order to describe charm meson decays in terms of these amplitudes, it is convenient to decompose initial and final states according to their $SU(3)_F$ structure. For instance, in the notation of [161], the following phase conventions are used:

1. Charmed mesons: $D^0 = -c\overline{u}$, $D^+ = c\overline{d}$, and $D_s = c\overline{s}$.
2. Pseudoscalar mesons: $\pi^+ = u\overline{d}$, $\pi^0 = \left(u\overline{u} - d\overline{d}\right)/\sqrt{2}$, $\pi^- = -d\overline{u}$, $K^+ = u\overline{s}$, $K^0 = d\overline{s}$, $\overline{K}^0 = s\overline{d}$, $K^- = -s\overline{u}$, $\eta = \left(s\overline{s} - u\overline{u} - d\overline{d}\right)/\sqrt{3}$, and $\eta' = \left(u\overline{u} + d\overline{d} - 2s\overline{s}\right)/\sqrt{6}$.
3. Vector mesons: $\rho^+ = u\overline{d}$, $\rho^0 = \left(u\overline{u} - d\overline{d}\right)/\sqrt{2}$, $\rho^- = -d\overline{u}$, $\omega^0 = \left(u\overline{u} + d\overline{d}\right)/\sqrt{2}$, $K^{*+} = u\overline{s}$, $K^{*0} = d\overline{s}$, $\overline{K}^{*0} = s\overline{d}$, $K^{*-} = -s\overline{u}$, and $\phi = s\overline{s}$.

As with the $SU(3)_F$ approach, this method does not provide absolute predictions for the branching fractions in D-meson decays. However, it provides relations among several decay amplitudes by matching the quark-level "flavor topology" graphs with the final states defined above. For example, a doubly-Cabibbo suppressed transition (DCS; or the one with the decay amplitude proportional to $\sin\theta_C^2 \sim \lambda^2$) $D^0 \to K^+\pi^-$ can proceed via a tree-level amplitude $T(c \to u\overline{s}d)$ and an exchange amplitude $E(c\overline{u} \to \overline{s}d)$. Matching those with the initial state meson $D^0 = -c\overline{u}$ and the final state mesons $K^+ = u\overline{s}$ and $\pi^- = -d\overline{u}$, one obtains the following amplitude relation,

$$A(D^0 \to K^+\pi^-) = T + E \equiv \frac{G_F}{\sqrt{2}} V_{us} V_{cd}^* \left(\mathcal{T}'' + \mathcal{E}''\right), \tag{4.196}$$

where we use calligraphic notation for the amplitudes with $G_F/\sqrt{2}$ and CKM-factors factored out. Similarly, for other transitions one obtains

$$\begin{aligned}
A(D^0 \to K^0\pi^0) &= \frac{1}{\sqrt{2}}(C - E) \\
&= \frac{1}{\sqrt{2}}\frac{G_F}{\sqrt{2}}V_{us}V_{cd}^* \; (\mathcal{C}'' - \mathcal{E}''), \\
A(D^0 \to \overline{K}^0\pi^0) &= \frac{1}{\sqrt{2}}(C - E) \\
&= \frac{1}{\sqrt{2}}\frac{G_F}{\sqrt{2}}V_{ud}V_{cs}^* \; (\mathcal{C} - \mathcal{E}), \\
A(D^+ \to K^0\pi^+) &= C + A = \frac{G_F}{\sqrt{2}}V_{us}V_{cd}^* \; (\mathcal{C}'' + \mathcal{A}''), \\
A(D^+ \to \overline{K}^0\pi^+) &= T + C = \frac{G_F}{\sqrt{2}}V_{ud}V_{cs}^* \; (\mathcal{T} + \mathcal{C}), \\
A(D^0 \to K^0\eta) &= \frac{1}{\sqrt{3}}C = \frac{1}{\sqrt{3}}\frac{G_F}{\sqrt{2}}V_{us}V_{cd}^* \; \mathcal{C}'',
\end{aligned} \tag{4.197}$$

and so on. Note that in Eq. (4.197) we denoted DCS amplitudes with double primes. Singly-Cabibbo-suppressed amplitudes (SCS; or the ones with the decay amplitude proportional to λ) are conventionally denoted by a single prime. Cabibbo-favored amplitudes (CF; or the ones with the decay amplitude proportional to $\lambda^0 \sim 1$) can be related to SCS and DCS amplitudes by proper scaling with $\tan\theta_C$. No penguin amplitudes contribute to CF or DCS decays, which is reflected in Eq. (4.197). The structure of SCS decay amplitudes is richer, involving P or PA amplitudes. Taking into account final state interaction, it might be difficult to introduce such amplitudes unambiguously.

One reason for the employed phase convention is a requirement that $SU(3)_F$ sum rules are satisfied. For example, for transitions $D^+ \to K^+\pi^0$, $D^+ \to K^+\eta$, and $D^+ \to K^+\eta'$, a sum rule

$$3\sqrt{2}A(K^+\pi^0) + 4\sqrt{3}A(K^+\eta) + \sqrt{6}A(K^+\eta') = 0 \tag{4.198}$$

can be written. With the flavor-flow parameterization,

$$\begin{aligned}
A(D^+ \to \overline{K}^+\pi^0) &= \frac{1}{\sqrt{2}}(T - A), \\
A(D^+ \to \overline{K}^+\eta) &= -\frac{1}{\sqrt{3}}\,T \\
A(D^+ \to \overline{K}^+\eta') &= \frac{1}{\sqrt{6}}(T + 3A)
\end{aligned} \tag{4.199}$$

the above sum rule gives $3(T - A) - 4T + (T + 3A) = 0$.

Thus, provided that a sufficient number of decay modes is measured, one can predict both branching fractions and amplitude phases for a number of transitions. Still, no prediction for absolute branching ratios are possible in this approach.

4.8.4 Factorization ansatz

The simplest way to estimate an absolute decay rate of a charmed meson is to employ a factorization ansatz. This ansatz implies that the amplitude for the hadronic transition can be written as a product of known form factors. Schematically, for a decay of a heavy meson H_q into two pseudoscalar states M_1 and M_2 (see Eq. (4.181)),

$$\begin{aligned}
\mathcal{A}(H_q \to M_1M_2) &= \langle M_1, M_2|\mathcal{Q}_2|H_q\rangle \\
&\sim \langle M_1|\left(\overline{q_1}_k\Gamma^\mu q_{2k}\right)|0\rangle \times \langle M_2|\left(\overline{q}_i\Gamma_\mu Q_i\right)|H_q\rangle,
\end{aligned} \tag{4.200}$$

which essentially means that the matrix element of a four-fermion operator is evaluated by inserting a complete set of states in between the currents with only a vacuum state retained*. This is a clear simplification, as we can now use parameterizations of quark currents developed in Sect. 4.5. In particular, if both M_1 and M_2 are pseudoscalar states,

$$\begin{aligned} \langle M_1|\bar{q}_1\gamma^\mu\gamma_5 q_2|0\rangle &= i f_{M_1} q^\mu, \quad \text{and} \\ \langle M_2|\bar{q}\Gamma^\mu Q|H_q\rangle &= f_+^{H_q\to M_2}(q^2)P^\mu + f_-^{H_q\to M_2}(q^2)q^\mu, \end{aligned} \tag{4.201}$$

where q^μ is the momentum of the M_1 meson and P^μ is the sum of H_q and M_2 momenta. This now means that

$$\mathcal{A}(H_q \to M_1 M_2) \sim f_{M_1}\left(m_{H_q}^2\right) f_+^{H_q\to M_2}(m_\pi^2), \tag{4.202}$$

provided the masses of the final state mesons $M_{1,2}$ are neglected.

Clearly, naive factorization of Eq. (4.200), while convenient, cannot be correct, as it assumes that renormalization scale and scheme dependence of a product of quark bilinears is the same as that of a four-fermion operator, which it is not. The situation can in principle be corrected, at least in the heavy-quark limit. In B-decays, a QCD factorization formula has been written that takes into account perturbative QCD corrections [18]. It is however not clear that this approach is applicable to charm decays, as charm quark might be too light for this approach to be applicable. Nevertheless, even naive factorization provides a convenient way to obtain an order-of-magnitude estimate for the D-meson decay rates.

4.9 Meson-antimeson oscillations

Studies of transitions that are dominated by loop amplitudes allow us to access the states that are kinematically unavailable at low energies. An example that we will pursue further in this section is meson-antimeson mixing. In the Standard Model, the $B^0-\overline{B}^0$ mixing observables probe CKM parameters associated with top quark without the need to produce it directly. Besides, if New Physics particles couple to the mixed meson system, they would affect the mixing observables as well. We can label mixing amplitudes by the change of the flavor quantum numbers, $\Delta Q = 2$ transitions, with $Q = s, c$, and b. We shall also develop mixing formalism for a generic neutral state H^0, with $H = B_q^0, D^0$, or K^0, specifying predictions of the mixing parameters separately for each state.

Let us note that studies of CP-violation greatly benefit from the inclusion of the $\Delta Q = 2$ transitions, as in the Standard Model both them and $\Delta Q = 1$ decays probe the same single CP-violating phase. This does not have to be the case in the general BSM scenario. Here we will develop the formalism for $H^0 - \overline{H}^0$ mixing.

Phenomenology of meson-antimeson oscillations.

The presence of a $\Delta Q = 2$ interaction couples the dynamics of H^0 and $\overline{H}^0$ states. As both H^0 and $\overline{H}^0$ states are not stable with respect to weak interactions, it makes sense to consider their time evolution simultaneously, similar to the discussion of coupled pendulums in classical mechanics. Let us then consider the state $|H(t)\rangle$,

$$|H(t)\rangle = \begin{bmatrix} a(t) \\ b(t) \end{bmatrix} = a(t)|H^0\rangle + b(t)|\overline{H}^0(t)\rangle. \tag{4.203}$$

*This is the reason why naive factorization ansatz is sometimes referred to as "vacuum saturation approximation."

Time evolution of the $|H(t)\rangle$ state is governed by a set of coupled Schrödinger equations,

$$i\frac{d}{dt}|H(t)\rangle = \left[M - i\frac{\Gamma}{2}\right]|H(t)\rangle \equiv \left[\begin{array}{cc} A & p^2 \\ q^2 & \bar{A} \end{array}\right]|H(t)\rangle. \qquad (4.204)$$

The Hamiltonian $\mathcal{H} = M - i\Gamma/2$ in Eq. (4.204) is clearly non-Hermitian, but this should not deter us from further discussion of the $H^0 - \overline{H}^0$ system. As non-Hermiticity implies non-conservation of probability, the H-states would decay, which is precisely what is expected. The Hermiticity of $\mathcal{H}$ can be restored by adding the states to which H decay, but that would make the discussion of $H^0 - \overline{H}^0$ unnecessarily complicated. Note that CPT-invariance requires the same matrix elements $A = \bar{A}$ on the main diagonal of the mass matrix, which implies that $M_{11} = M_{22}$, and $\Gamma_{11} = \Gamma_{22}$; violation of these conditions allows us to search for traces of CPT violation. The off-diagonal parameters p^2 and q^2 in Eq. (4.204) are, in general, complex.

An effective way to proceed is to diagonalize the Hamiltonian $\mathcal{H}$ in Eq. (4.204), which would also require us to switch from the flavor eigenstates H^0 and $\overline{H}^0$ to propagating or *mass* eigenstates, which form a basis in which $\mathcal{H}$ is diagonal,

$$|H_{L,H}\rangle = p|H^0\rangle \pm q|\overline{H^0}\rangle, \qquad (4.205)$$

where the indices $L(H)$ stand for "light" or "heavy". Neglecting CP violation leads to $p = q = 1/\sqrt{2}$. The mass eigenstates represent physical propagating states with their own masses and widths. This implies that the mass and width differences are observable,

$$\begin{aligned} \Delta M &= M_H - M_L, \\ \Delta\Gamma &= \Gamma_L - \Gamma_H. \end{aligned} \qquad (4.206)$$

Note that in these notations $m = (M_L + M_H)/2 = M_{11} = M_{22}$ and $\Gamma = (\Gamma_L + \Gamma_H)/2 = \Gamma_{11} = \Gamma_{22}$. Oftentimes, normalized mass and lifetime differences are employed,

$$x = \frac{\Delta M}{\Gamma}, \qquad y = \frac{\Delta\Gamma}{2\Gamma}, \qquad (4.207)$$

The mass matrix is diagonal in the mass basis, which can be found by transforming the flavor basis with help of a rotational matrix Q,

$$Q^{-1}\left[M - i\frac{\Gamma}{2}\right]Q = \left(\begin{array}{cc} M_L - i\Gamma_L/2 & 0 \\ 0 & M_H - i\Gamma_H/2 \end{array}\right). \qquad (4.208)$$

The transformation matrix Q can be found from Eqs. (4.204) and (4.205),

$$Q = \left(\begin{array}{cc} p & p \\ q & -q \end{array}\right), \quad \text{and} \quad Q^{-1} = \frac{1}{2pq}\left(\begin{array}{cc} q & p \\ q & -p \end{array}\right). \qquad (4.209)$$

In order to find the time evolution of the flavor eigenstates, we need to transform the evolution equation back to the flavor basis,

$$\left[\begin{array}{c} |B^0(t)\rangle \\ |\overline{B}^0(t)\rangle \end{array}\right] = Q\left(\begin{array}{cc} e^{-iM_L - \Gamma_L/2} & 0 \\ 0 & e^{-iM_H - \Gamma_H/2} \end{array}\right)Q^{-1}\left[\begin{array}{c} |B^0\rangle \\ |\overline{B}^0\rangle \end{array}\right] \qquad (4.210)$$

which leads to the following time dependence of the flavor eigenstates,

$$\begin{aligned} |H^0(t)\rangle &= g_+(t)|H^0\rangle + \frac{q}{p}g_-(t)|\overline{H}^0\rangle \\ |\overline{H}^0(t)\rangle &= \frac{p}{q}g_-(t)|H^0\rangle + g_+(t)|\overline{H}^0\rangle \end{aligned} \qquad (4.211)$$

where the time-dependence of the oscillation parameters is

$$\begin{aligned} g_+(t) &= e^{-imt}e^{-\Gamma t/2}\left[\cosh\frac{\Delta\Gamma t}{4}\cos\frac{\Delta M t}{2} - i\sinh\frac{\Delta\Gamma t}{4}\sin\frac{\Delta M t}{2}\right], \\ g_-(t) &= e^{-imt}e^{-\Gamma t/2}\left[-\sinh\frac{\Delta\Gamma t}{4}\cos\frac{\Delta M t}{2} + i\cosh\frac{\Delta\Gamma t}{4}\sin\frac{\Delta M t}{2}\right]. \end{aligned} \tag{4.212}$$

The formulas of Eq. (4.211) and (4.212) describe the oscillations of neutral flavor states for any values of the mixing parameters. In several cases, they could be simplified. For example, it is known experimentally that for the B^0 mesons $\Delta\Gamma/\Delta M \to 0$, so that formula (4.212) simplifies to

$$\begin{aligned} g_+(t) &= e^{-imt}e^{-\Gamma t/2}\cos\frac{\Delta M_B t}{2}, \\ g_-(t) &= ie^{-imt}e^{-\Gamma t/2}i\sin\frac{\Delta M_B t}{2}, \end{aligned} \tag{4.213}$$

while in case of D-mesons $\Delta\Gamma/\Delta M \to 1$ with both $\Delta M_D \ll \Gamma_D$ and $\Delta\Gamma_D \ll \Gamma_D$. It is then very convenient to rewrite the Eq. (4.212) in terms of the normalized mass and lifetime differences of Eq. (4.207) and expand in x_D and y_D, leading to the power-law dependence of $g\pm(t)$ for the D-mesons,

$$\begin{aligned} g_+(t) &= e^{-imt}e^{-\Gamma t/2}\left[1 + \frac{1}{8}(y - ix)^2(\Gamma t)^2\right], \\ g_-(t) &= -\frac{1}{2}e^{-imt}e^{-\Gamma t/2}(y - ix)(\Gamma t). \end{aligned} \tag{4.214}$$

Finally, we can relate the mass and width differences to the parameters of the original Hamiltonian. Since a matrix determinant is invariant under rotations, the secular equation reads

$$\left(\Delta M + \frac{i}{2}\Delta\Gamma\right)^2 = 4\left(M_{12} - i\frac{\Gamma_{12}}{2}\right)\left(M_{12}^* - i\frac{\Gamma_{12}^*}{2}\right). \tag{4.215}$$

Taking real and imaginary part of Eq. (4.215), we obtain two equations,

$$\begin{aligned} (\Delta M)^2 - \frac{1}{4}(\Delta\Gamma)^2 &= 4|M_{12}|^2 - |\Gamma_{12}|^2, \\ \Delta M \Delta\Gamma &= -4\mathrm{Re}\left(M_{12}\Gamma_{12}^*\right). \end{aligned} \tag{4.216}$$

It is easy to see that if $\Delta\Gamma/\Delta M \to 0$, as it is the case for B_d^0 mesons,

$$\Delta M_B = 2|M_{12}|. \tag{4.217}$$

Finally, we can also relate p and q to the mixing parameters,

$$\frac{q}{p} = -\frac{\Delta M + i\Delta\Gamma/2}{2M_{12} - i\Gamma_{12}} = -\frac{2M_{12}^* - i\Gamma_{12}^*}{\Delta M + i\Delta\Gamma/2} \tag{4.218}$$

This expression will be useful in determining CP-violating phases that appear in mixing amplitudes.

$\Delta B = 2$ effective weak Hamiltonians

The $\Delta B = 2$ effective weak Hamiltonian can be derived in the SM. The mixing amplitudes are dominated by the virtual top exchange, which is assured by large top quark mass and a combination of the relevant CKM parameters. In the SM, the operator basis includes one operator only [128],

$$\mathcal{H}_W^{\Delta B=2} = \frac{G_F^2}{4\pi^2} M_W^2 (\lambda_t^q)^2 \eta_2 S_0(x_t) Q_b^q \tag{4.219}$$

where $S_0(x_t)$ is the Inami-Lim function, η_2 are QCD corrections, and Q_b^q is the operator.

$$Q_b^q = \bar{b}_L \gamma_\mu q_L \, \bar{b}_L \gamma^\mu q_L \,, \qquad \text{for} \quad q = \{d, s\}\,, \tag{4.220}$$

We need to compute the matrix element of the operator Q_b^q. Let us, again, follow our recipe of vacuum saturation and compute a matrix element of such an operator. Consider first the current-current matrix element

$$\langle \bar{B}_q^0 | Q_b^q | B_q^0 \rangle = \langle \bar{B}_q^0 | \left(\bar{b}_L \gamma_\mu q_L\right)_1 \left(\bar{b}_L \gamma^\mu q_L\right)_2 | B_q^0 \rangle \tag{4.221}$$

where we again, following our recipe from Chapter 3, (temporarily) labeled the currents with subscripts 1 and 2 for the convenience of explanation. The vacuum saturation prescription tells us that the four-fermion matrix element can be computed by inserting a complete set of states $1 = \sum_I |I\rangle \langle I|$ in between the currents 1 and 2 and then discarding all contributions except for the vacuum one,

$$\begin{aligned} \langle \bar{B}_q^0 | \left(\bar{b}_L \gamma_\mu q_L\right)_1 \left(\bar{b}_L \gamma^\mu q_L\right)_2 | B_q^0 \rangle \;&\to\; \sum_I \langle \bar{B}_q^0 | \left(\bar{b}_L \gamma_\mu q_L\right)_1 |I\rangle \langle I| \left(\bar{b}_L \gamma^\mu q_L\right)_2 | B_q^0 \rangle \\ &\to\; \langle \bar{B}_q^0 | \left(\bar{b}_L \gamma_\mu q_L\right)_1 |0\rangle \langle 0| \left(\bar{b}_L \gamma^\mu q_L\right)_2 | B_q^0 \rangle \\ &=\; \frac{1}{4} \langle \bar{B}_q^0 | \bar{b} \gamma_\mu \gamma_5 q \,|0\rangle \langle 0| \,\bar{b} \gamma^\mu \gamma_5 q | B_q^0 \rangle. \end{aligned} \tag{4.222}$$

Just like in Chapter 3, it is clear that Eq. (4.222) is not a whole story. For example, since the b-field field operator b in the current 1 in Eq. (4.221) contains both creation and annihilation operators, it can not only act on the $\langle \bar{B}_q^0|$ state, but also on the $|B_q^0\rangle$ state, which is not possible after current separation in Eq. (4.222). The same is true for the light quark field q in the same current 1. Moreover, the b and q fields can, in principle, act on different states: b can act on $\langle \bar{B}_q^0|$, while q can act on $|B_q^0\rangle$, which is also not represented in Eq. (4.222)!

To account for this situation in the framework of vacuum saturation, it is convenient to again apply Fierz rearrangement formulas. This is where the similarities between the muonium and B-meson mixing matrix elements end. The main reason is the fact that quarks carry color. Indeed, for the operator Q_b^q under consideration,

$$\left(\bar{b}_{i,L} \gamma_\mu q_{i,L}\right) \left(\bar{b}'_{k,L} \gamma^\mu q'_{k,L}\right) = \left(\bar{b}_{i,L} \gamma_\mu q'_{k,L}\right) \left(\bar{b}'_{k,L} \gamma^\mu q_{i,L}\right), \tag{4.223}$$

where i and k are color indices and we again, temporarily, introduced primes to the second current. Notice that the operators on the left and the right of this equation have different color structures! We no longer can simply take a matrix element because the current carries the color and therefore is not defined in between color-less vacuum and meson states.

To correct the situation, we shall use the completeness identity for color matrices,

$$\delta_{i\ell} \delta_{kj} = \frac{1}{3} \delta_{ij} \delta_{k\ell} + \frac{1}{2} \lambda_{ij}^a \lambda_{k\ell}^a. \tag{4.224}$$

where λ^a are the Gell-Mann $SU(3)$ matrices. Now we can perform proper factorization of currents,

$$\begin{aligned} \langle \bar{B}_q^0 | Q_b^q | B_q^0 \rangle \;&=\; 2 \times \frac{1}{4} \langle \bar{B}_q^0 | \bar{b} \gamma_\mu \gamma_5 q \,|0\rangle \langle 0| \,\bar{b} \gamma^\mu \gamma_5 q | B_q^0 \rangle \\ &+\; 2 \times \frac{1}{4} \times \frac{1}{3} \langle \bar{B}_q^0 | \bar{b} \gamma_\mu \gamma_5 q \,|0\rangle \langle 0| \,\bar{b} \gamma^\mu \gamma_5 q | B_q^0 \rangle \\ &+\; 2 \times \frac{1}{4} \times \frac{1}{2} \langle \bar{B}_q^0 | \bar{b} \gamma_\mu \gamma_5 \lambda^a q \,|0\rangle \langle 0| \,\bar{b} \gamma^\mu \gamma_5 \lambda^a q | B_q^0 \rangle, \end{aligned} \tag{4.225}$$

where the last line's contribution is zero, because the matrix element of color-octet current $\bar{b}\gamma_\mu\gamma_5\lambda^a q$ in between two colorless states is zero.

$$\langle \bar{B}_q^0|Q_b^q|B_q^0\rangle = 2\times\left(\frac{1}{4}+\frac{1}{4}\times\frac{1}{3}\right)(-i)if_m^2 p_\mu p^\mu = \frac{2}{3}f_{B_q}^2 M_{B_q}^2 \hat{B}_{B_q}, \tag{4.226}$$

where we introduced a factor $\hat{B}_{B_q}$ to account for anything we did not take into account in this prescription. Generic New Physics contributions to $\Delta F = 2$ transitions, with $F = \{B, C\}$ generate several additional operators. This allows for model-independent studies of $\Delta F = 2$ processes where the Wilson coefficients at the matching scale are used as New Physics parameters.

Oscillation parameters: B-mesons

In a generic NP scenario there are eight $\Delta B = 2$ effective operators that can contribute to B_q-mixing. The operator basis that we will employ is [91, 129]

$$\begin{array}{ll} Q_1 = (\bar{b}_L\gamma_\mu q_L)\,(\bar{b}_L\gamma^\mu q_L)\,, & Q_5 = (\bar{b}_R\sigma_{\mu\nu}q_L)\,(\bar{b}_R\sigma^{\mu\nu}q_L)\,, \\ Q_2 = (\bar{b}_L\gamma_\mu q_L)\,(\bar{b}_R\gamma^\mu q_R)\,, & Q_6 = (\bar{b}_R\gamma_\mu q_R)\,(\bar{b}_R\gamma^\mu q_R)\,, \\ Q_3 = (\bar{b}_L q_R)\,(\bar{b}_R q_L)\,, & Q_7 = (\bar{b}_L q_R)\,(\bar{b}_L q_R)\,, \\ Q_4 = (\bar{b}_R q_L)\,(\bar{b}_R q_L)\,, & Q_8 = (\bar{b}_L\sigma_{\mu\nu}q_R)\,(\bar{b}_L\sigma^{\mu\nu}q_R)\,, \end{array} \tag{4.227}$$

where quantities enclosed in parentheses are color singlets, *e.g.* $(\bar{b}_L\gamma_\mu s_L) \equiv \bar{b}_{L,i}\gamma_\mu s_{L,i}$. These operators are generated at a scale Λ where the NP is integrated out. A non-trivial operator mixing then occurs via renormalization group running of these operators between the heavy scale M and the light scale μ at which hadronic matrix elements are computed.

We need to evaluate the B_q^0-to-$\overline{B}_q^0$ matrix elements of these eight dimension-six basis operators. This introduces eight *non-perturbative* B-parameters $\{B_i\}$ that require evaluation by means of QCD sum rules or QCD-lattice simulation. We express these in the form

$$\begin{array}{ll} \langle Q_1\rangle = \frac{2}{3}f_{B_q}^2 M_{B_q}^2 B_1\,, & \langle Q_5\rangle = f_{B_q}^2 M_{B_q}^2 B_5\,, \\ \langle Q_2\rangle = -\frac{5}{6}f_{B_q}^2 M_{B_q}^2 B_2\,, & \langle Q_6\rangle = \frac{2}{3}f_{B_q}^2 M_{B_q}^2 B_6\,, \\ \langle Q_3\rangle = \frac{7}{12}f_{B_q}^2 M_{B_q}^2 B_3\,, & \langle Q_7\rangle = -\frac{5}{12}f_{B_q}^2 M_{B_q}^2 B_7\,, \\ \langle Q_4\rangle = -\frac{5}{12}f_{B_q}^2 M_{B_q}^2 B_4\,, & \langle Q_8\rangle = f_{B_q}^2 M_{B_q}^2 B_8\,, \end{array} \tag{4.228}$$

where f_{B_q} is the B_q meson decay constant and $\langle Q_i\rangle \equiv \langle \bar{B}_q^0|Q_i|B_q^0\rangle$.

Oscillation parameters: D-mesons

For $D^0 - \overline{D^0}$ mixing calculations it is convenient to work with the normalized mass and lifetime differences defined in Eq. (4.207). They can be computed as absorptive and dispersive parts of a certain correlation function,

$$\begin{aligned} x_D &= \frac{1}{M_D\Gamma_D}\mathrm{Re}\Big[2\langle\overline{D}^0|\mathcal{H}_w^{|\Delta C|=2}|D^0\rangle \\ &+ \langle\overline{D}^0|i\int d^4x\,\mathrm{T}\left\{\mathcal{H}_w^{|\Delta C|=1}(x)\mathcal{H}_w^{|\Delta C|=1}(0)\right\}|D^0\rangle\Big], \end{aligned} \tag{4.229}$$

$$y_D = \frac{1}{M_D\Gamma_D}\mathrm{Im}\left[\langle\overline{D}^0|i\int d^4x\,\mathrm{T}\left\{\mathcal{H}_w^{|\Delta C|=1}(x)\mathcal{H}_w^{|\Delta C|=1}(0)\right\}|D^0\rangle\right]. \tag{4.230}$$

It is understood that only quarks whose masses are lighter than m_D can go on mass shell in Eq. (4.229) and provide nonzero value for the lifetime difference y_D.

The charm system is unique because x_D is *not* dominated by the contribution of the $\Delta C = 2$ operator that is local at the charm scale. It is very different from the case of B-mixing, where x is completely dominated by the top quark contribution. Since Glashow-Iliopoulos-Maiani guarantees that the mixing amplitude is proportional to the power of intrinsic quark mass running in the box diagram, suppression due to a combination of CKM greatly diminishes the contribution due to b-quark, the only heavy quark intermediate state possible in $D^0 - \overline{D^0}$ mixing. Thus, it is absolutely important to calculate the contribution due to the correlation functions in Eq. (4.229) with light intermediate s and d quarks.

The hardest problem in charm mixing is to properly evaluate the integrals in the above equations. This can be done in several ways, depending on whether one considers the decaying particle as heavy or light compared to the QCD's scale Λ_{QCD}. Since $m_c \simeq 1.3$ GeV, both approaches are possible for D-decays and mixing calculations.

If the decaying particle is heavy, it is possible to show [177] that the integrals in Eq. (4.229) are dominated by short distances, so a short-distance operator product expansion (OPE) can be used to evaluate the products of $|\Delta C| = 1$ Hamiltonians. Similar approaches worked very well for the calculations of lifetime differences of B_s mesons [17]. If the decaying particle is considered light, no short-distance expansion of operator products is possible, as the integrals are dominated by the long distances. However, only a few open channels are available for such light particles, so the calculations can be done by explicitly summing over the contributions from each of the channels. This approach worked well for kaon physics*. The number of available decay channels is quite large, but some predictions can nevertheless be made.

Before proceeding with the calculation of $D^0 - \overline{D^0}$ mixing amplitude, let us understand the underlying flavor symmetry structure. GIM mechanism implies that meson mixing amplitudes must be proportional to mass factors of quarks propagating in the loops providing $\Delta C = 2$ interactions. Neglecting for a moment the third generation, only s and d quarks give a contribution to x_D and y_D in the standard model. This means that GIM mechanism implies that $D^0 - \overline{D^0}$ mixing is an $SU(3)_F$-breaking effect, and predicting the Standard Model values of x_D and y_D depends crucially on estimating the size of $SU(3)_F$ breaking. The question is at what order in $SU(3)_F$-breaking parameter m_s does the effect become non-zero? To answer that, let us prove the following theorem [74],

THEOREM 4.1 *In the Standard Model, and in the limit $V_{ub} \to 0$, the $D^0 - \overline{D^0}$ mixing parameters x and y vanish in the flavor $SU(3)_F$ limit. The mixing parameters only appear at second order in the $SU(3)$ violating parameter m_s.*

PROOF 4.1 Let us look at the group-theoretical structure of mixing matrix elements $\langle \overline{D}^0 | \mathcal{H}_w \mathcal{H}_w | D^0 \rangle$ that define x_D and y_D. Here $\mathcal{H}_w$ denote the $\Delta C = -1$ part of the weak Hamiltonian. Let D be the field operator that creates a D^0 meson and annihilates a $\overline{D}^0$. Then the matrix element, whose $SU(3)$ flavor group theory properties we will study, may be written as

$$\langle 0 | D\, \mathcal{H}_w \mathcal{H}_w\, D | 0 \rangle \,. \tag{4.231}$$

Since the operator D is of the form $\bar{c}u$, it transforms in the fundamental representation of $SU(3)_F$, which we will represent with a lower index, D_i. We use a convention in which the correspondence between matrix indexes and quark flavors is $(1,2,3) = (u,d,s)$. The only nonzero element of D_i

*It is important to remember that this statement only refers to the bilocal part of the expressions for x and y. The mass difference in kaons is dominated by the contribution from heavy t and c quarks, i.e. by the $\mathcal{H}_w^{|\Delta C|=2}$.

is $D_1 = 1$. The $\Delta C = -1$ part of the weak Hamiltonian has the flavor structure $(\bar{q}_i c)(\bar{q}_j q_k)$, so its matrix representation is written with a fundamental index and two antifundamentals, H_k^{ij}. This operator is a sum of irreducible representations contained in the product $3 \times \overline{3} \times \overline{3} = \overline{15} + 6 + \overline{3} + \overline{3}$. In the limit in which the third generation is neglected, H_k^{ij} is traceless, so only the $\overline{15}$ and 6 representations appear. That is, the $\Delta C = -1$ part of $\mathcal{H}_w$ may be decomposed as $\frac{1}{2}(\mathcal{O}_{\overline{15}} + \mathcal{O}_6)$.

Since we are interested in $SU(3)_F$ breaking, let is introduce it through the quark mass operator $\mathcal{M}$, whose matrix representation is $M_j^i = \text{diag}(m_u, m_d, m_s)$. We set $m_u = m_d = 0$ and let $m_s \neq 0$ be the only $SU(3)$ violating parameter. All nonzero matrix elements built out of D_i, H_k^{ij} and M_j^i must be $SU(3)_F$ singlets.

We now prove that $D^0 - \overline{D^0}$ mixing arises only at second order in $SU(3)$ violation, by which we mean second order in m_s. First, we note that the pair of D operators is symmetric, and so the product $D_i D_j$ transforms as a 6 under $SU(3)$. Second, the pair of $\mathcal{H}_w$'s is also symmetric, and the product $H_k^{ij} H_n^{lm}$ is in one of the reps which appears in the product

$$\begin{aligned}\left[(\overline{15} + 6) \times (\overline{15} + 6)\right]_S &= (\overline{15} \times \overline{15})_S + (\overline{15} \times 6) + (6 \times 6)_S \\ &= (\overline{60} + \overline{24} + 15 + 15' + \overline{6}) + (42 + 24 + 15 + \overline{6} + 3) + (15' + \overline{6})\,.\end{aligned} \tag{4.232}$$

A direct computation shows that only three of these representations actually appear in the decomposition of $\mathcal{H}_w \mathcal{H}_w$. They are the $\overline{60}$, the 42, and the $15'$

$$DD = \mathcal{D}_6\,, \qquad \mathcal{H}_w \mathcal{H}_w = \mathcal{O}_{\overline{60}} + \mathcal{O}_{42} + \mathcal{O}_{15'}\,, \tag{4.233}$$

where subscripts denote the representation of $SU(3)_F$. Since there is no $\overline{6}$ in the decomposition of $\mathcal{H}_w \mathcal{H}_w$, there is no $SU(3)$ singlet which can be made with $\mathcal{D}_6$, and no $SU(3)$ invariant matrix element of the form (4.231) can be formed. This is the well-known result that $D^0 - \overline{D}^0$ mixing is *prohibited by* $SU(3)$ *symmetry.* Now consider a single insertion of the $SU(3)$ violating spurion $\mathcal{M}$. The combination $\mathcal{D}_6 \mathcal{M}$ transforms as $6 \times 8 = 24 + \overline{15} + 6 + \overline{3}$. There is still no invariant to be made with $\mathcal{H}_w \mathcal{H}_w$, thus $D^0 - \overline{D}^0$ mixing is *not induced at first order in* $SU(3)_F$ *breaking.* With two insertions of $\mathcal{M}$, it becomes possible to make an $SU(3)_F$ invariant. The decomposition of $\mathcal{D}\mathcal{M}\mathcal{M}$ is

$$\begin{aligned}6 \times (8 \times 8)_S &= 6 \times (27 + 8 + 1) \\ &= (60 + \overline{42} + 24 + \overline{15} + \overline{15}' + 6) + (24 + \overline{15} + 6 + \overline{3}) + 6\,.\end{aligned} \tag{4.234}$$

There are three elements of the 6×27 part which can give invariants with $\mathcal{H}_w \mathcal{H}_w$. Each invariant yields a contribution to $D^0 - \overline{D}^0$ mixing proportional to $s_1^2 m_s^2$. Thus, $D^0 - \overline{D^0}$ mixing arises only at *second order* in the $SU(3)$ violating parameter m_s [74], in the Standard Model x and y are generated only at second order in $SU(3)_F$ breaking,

$$x\,,\, y \sim \sin^2\theta_C \times [SU(3) \text{ breaking}]^2\,, \tag{4.235}$$

where θ_C is the Cabibbo angle. This result should be reproduced in all explicit calculations of $D^0 - \overline{D^0}$ mixing parameters.

The use of the OPE relies on local quark-hadron duality, and on the expansion parameter Λ/m_c being small enough to allow truncation of the series after the first few terms. Let us see what one can expect at leading order in $1/m_c$ expansion, i.e. assuming that the integrals in Eq. (4.229) are dominated by the short distances. The leading-order result is then generated by calculating the usual box diagram with intermediate s and d quarks.

Unitarity of the CKM matrix assures that the leading-order, mass-independent contribution due to s-quark is completely canceled by the corresponding contribution due to a d-quark. A non-zero contribution can be obtained if the mass insertions are added on each quark line in the box diagram. However, adding only one mass insertion flips the chirality of the propagating quarks, from being left-handed to right-handed. This does not give a contribution to the resulting amplitude, as right-handed quarks do not participate in the weak interaction. Thus, a second mass insertion is needed on each quark line. Neglecting m_d compared to m_s we see that the resulting contribution to x_D is $\mathcal{O}(m_s^2 \times m_s^2) \sim \mathcal{O}(m_s^4)$! It is easy to convince yourself that y_D has additional m_s^2 suppression due to on-shell propagation of left-handed quarks emitted from a spin-zero meson, which brings total suppression of y_D to $\mathcal{O}(m_s^6)$! An explicit calculation of the leading order mixing amplitude, as well as perturbative QCD corrections to it [92] agrees precisely with the hand-waving arguments above. Clearly, leading order contribution in $1/m_c$ gives "too much" of $SU(3)_F$ suppression compared to the theorem that was proven above [74].

Somewhat surprisingly, the resolution of this paradox follows from considerations of higher-order corrections in $1/m_c$. Among many higher-dimensional operators that encode $1/m_c$ corrections to the leading four-fermion operator contribution, there exists a class of operators that result from chirality-flipping interactions with background quark condensates. These interactions do not bring additional powers of light quark mass, but are suppressed by powers of Λ_{QCD}/m_c, which is not a very small number. The leading $\mathcal{O}(m_s^2)$ order of $SU(3)_F$ breaking is obtained from matrix elements of dimension twelve operators that are suppressed by $(\Lambda_{QCD}/m_c)^6$ compared to the parametrically-leading contribution in $1/m_c$ expansion! As usual in OPE calculation, the proliferation of the number of operators at higher orders (over 20) makes it difficult to pinpoint the precise value of the effect.

It is possible to calculate $D^0 - \overline{D^0}$ mixing rates by dealing explicitly with hadronic intermediate states which result from every common decay product of D^0 and $\overline{D}^0$. In the $SU(3)_F$ limit, these contributions cancel when one sums over complete $SU(3)$ multiplets in the final state. The cancellations depend on $SU(3)_F$ symmetry both in the decay matrix elements and in the final state phase space. While there are $SU(3)$ violating corrections to both of these, it is difficult to compute the $SU(3)_F$ violation in the matrix elements in a model-independent manner. As experimental data on nonleptonic decay rates become better and better, it is possible to use it to calculate y_D by directly inputting it into Eq. (4.229),

$$y_D = \frac{1}{\Gamma} \sum_n \int [\text{P.S.}]_n \, \langle \overline{D}^0 | \, \mathcal{H}_w \, | n \rangle \langle n | \, \mathcal{H}_w \, | D^0 \rangle \,, \tag{4.236}$$

where the sum is over distinct final states n and the integral is over the phase space for state n. Alternatively, with some mild assumptions about the momentum dependence of the matrix elements, the $SU(3)_F$ violation in the phase space depends only on the final particle masses and can be computed [74]. It was shown that this source of $SU(3)_F$ violation can generate y_D and x_D of the order of a few percent. The calculation of x_D relies on further model-dependent assumptions about off-shell behavior of decay form factors [74]. Restricting the sum over all final states to final states F which transform within a single $SU(3)_F$ multiplet R, the result is

$$y_D = \frac{1}{2\Gamma} \langle \overline{D}^0 | \, \mathcal{H}_w \left\{ \eta_{CP}(F_R) \sum_{n \in F_R} |n\rangle \rho_n \langle n| \right\} \mathcal{H}_w \, | D^0 \rangle \,, \tag{4.237}$$

where ρ_n is the phase space available to the state n, $\eta_{CP} = \pm 1$ [74]. In the $SU(3)_F$ limit, all the ρ_n are the same for $n \in F_R$, and the quantity in braces above is an $SU(3)_F$ singlet. Since the ρ_n depend only on the known masses of the particles in the state n, incorporating the true values of ρ_n in the sum is a calculable source of $SU(3)_F$ breaking.

This method does not lead directly to a calculable contribution to y, because the matrix elements $\langle n|\mathcal{H}_w|D^0\rangle$ and $\langle \overline{D}^0|\mathcal{H}_w|n\rangle$ are not known. However, CP symmetry, which in the Standard Model and almost all scenarios of New Physics is to an excellent approximation conserved in D decays, relates $\langle \overline{D}^0|\mathcal{H}_w|n\rangle$ to $\langle D^0|\mathcal{H}_w|\overline{n}\rangle$. Since $|n\rangle$ and $|\overline{n}\rangle$ are in a common $SU(3)_F$ multiplet, they are determined by a single effective Hamiltonian. Hence the ratio

$$\begin{aligned} y_{F,R} &= \frac{\sum_{n\in F_R}\langle \overline{D}^0|\,\mathcal{H}_w|n\rangle\rho_n\langle n|\mathcal{H}_w\,|D^0\rangle}{\sum_{n\in F_R}\langle D^0|\,\mathcal{H}_w|n\rangle\rho_n\langle n|\mathcal{H}_w\,|D^0\rangle} \\ &= \frac{\sum_{n\in F_R}\langle \overline{D}^0|\,\mathcal{H}_w|n\rangle\rho_n\langle n|\mathcal{H}_w\,|D^0\rangle}{\sum_{n\in F_R}\Gamma(D^0\to n)} \end{aligned} \tag{4.238}$$

is calculable, and represents the value which y_D would take if elements of F_R were the only channel open for D^0 decay. To get a true contribution to y_D, one must scale $y_{F,R}$ to the total branching ratio to all the states in F_R. This is not trivial, since a given physical final state typically decomposes into a sum over more than one multiplet F_R. The numerator of $y_{F,R}$ is of order s_1^2 while the denominator is of order 1, so with large $SU(3)_F$ breaking in the phase space the natural size of $y_{F,R}$ is 5%. Indeed, there are other $SU(3)_F$ violating effects, such as in matrix elements and final state interaction phases. Here we assume that there is no cancellation with other sources of $SU(3)_F$ breaking, or between the various multiplets which occur in D decay, that would reduce our result for y by an order of magnitude. This is equivalent to assuming that the D meson is not heavy enough for duality to enforce such cancellations. Performing the computations of $y_{F,R}$, we see [74] that effects at the level of a few percent are quite generic. Then, y_D can be formally constructed from the individual $y_{F,R}$ by weighting them by their D^0 branching ratios,

$$y_D = \frac{1}{\Gamma}\sum_{F,R} y_{F,R}\left[\sum_{n\in F_R}\Gamma(D^0\to n)\right]. \tag{4.239}$$

However, the data on D decays are neither abundant nor precise enough to disentangle the decays to the various $SU(3)_F$ multiplets, especially for the three- and four-body final states. Nor have we computed $y_{F,R}$ for all or even most of the available representations. Instead, we can only estimate individual contributions to y by assuming that the representations for which we know $y_{F,R}$ to be typical for final states with a given multiplicity, and then to scale to the total branching ratio to those final states. The total branching ratios of D^0 to two-, three- and four-body final states can be extracted from the Review of Particle Physics. Rounding to the nearest 5% to emphasize the uncertainties in these numbers, we conclude that the branching fractions for PP, $(VV)_{S\text{-wave}}$, $(VV)_{d\text{-wave}}$ and $3P$ approximately amount to 5%, while the branching ratios for PV and $4P$ are of the order of 10% [74].

One can see that y_D on the order of a few percent is completely natural, and that anything an order of magnitude smaller would require significant cancellations which do not appear naturally in this framework. The normalized mass difference, x_D, can then be calculated via a dispersion relation

$$x_D = -\frac{1}{\pi}\,\mathrm{P}\int_{2m_\pi}^{\infty} dE\,\frac{y_D(E)}{E-m_D} \tag{4.240}$$

that additionally contain guesses on the off-shell behavior of hadronic form-factors in $y_D(E)$ [74]. Here P denotes principal value. The result of the calculation yields $x_D \sim \mathcal{O}(1\%)$ [74].

4.9.1 Phenomenology of New Physics in nonleptonic B-decays

As we saw in the preceding sections, it is fairly difficult to keep soft QCD effects under control in a model-independent way. Yet, nonleptonic decays are ubiquitous, so there is a lot of experimental data that can be obtained by working with transitions with hadrons in the initial and final states. Is there any way to study New Physics with nonleptonic transitions?

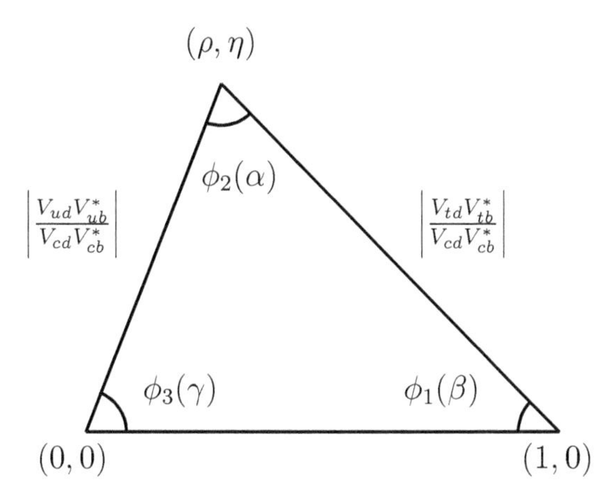

FIGURE 4.3 The Unitarity Triangle $V_{ud}V_{ub} + V_{cd}V_{cb}^* + V_{td}V_{tb}^* = 0$.

The answer is indeed positive. As was discussed in Sec. 1.2, one way to search for New Physics is by testing relations among the processes that are known to hold only in the Standard Model but might be violated in some BSM scenarios. In this case, the studies are related to the fact the only source of CP-violation in the Standard Model is the single phase of the CKM matrix (see discussion in Chapter 2). The strategy can be formulated as follows.

By construction, the CKM matrix is unitary, $V^\dagger V = 1$, which implies several relations among its matrix elements. Of special interest are the relations that involve products of the matrix elements that close onto non-diagonal matrix elements of the unit matrix and using an explicit form of the matrix in Eq. (2.50), they can be written as

$$\sum_{i,a\neq b} V_{ia}V_{ib}^* = 0, \tag{4.241}$$

where $(i,a,b) = \{u,d,c,s,t,b\}$. The summation is over i, which represents the up-type quarks $i = u,c,t$ if a and b are the down-type quarks. Likewise, i represents the down-type quarks $i = d,s,b$ if a and b are the up-type quarks. Since for three generations of quarks the sum involves three terms, such relations are often called *triangle relations*, as they define a triangle in complex space. For example, one of the most tested triangle relations is the one that can be obtained by multiplying the first and the third columns,

$$V_{ud}V_{ub} + V_{cd}V_{cb}^* + V_{td}V_{tb}^* = 0, \tag{4.242}$$

which is referred to as the Unitarity Triangle (UT). It is depicted in Fig. 4.3. The beauty of this particular UT is that it allows us to probe CKM matrix elements from all three generations of quarks. Also, all three sides of this triangle scale as λ^3, i.e., they are of similar length, so the triangle is not "squished". This is not so for other triangles obtained from the unitarity of the CKM matrix. It is conventional to normalize the sides of the triangle by dividing them by the length of the base, as depicted in Fig. 4.3. The angles of the UT can be defined as*

$$\begin{aligned} \phi_1(\text{or } \beta) &= \arg\left[-V_{cd}V_{cb}^*/V_{td}V_{tb}^*\right], \\ \phi_2(\text{or } \alpha) &= \arg\left[-V_{td}V_{tb}^*/V_{ud}V_{ub}^*\right], \\ \phi_3(\text{or } \gamma) &= \arg\left[-V_{ud}V_{ub}^*/V_{cd}V_{cb}^*\right]. \end{aligned} \tag{4.243}$$

While the triangle relation in Eq. (4.242) and the definitions of the angles above are independent of any particular parameterizations of the CKM matrix, the first and the third ratios in Eq. (4.243) define, respectively, the phases of V_{td} and V_{ub} in the Wolfenstein parameterization of the CKM matrix.

There are five other triangle relations that can be written. An important consequence of the fact that all of them are nontrivial is that all sources of CP-violation can be related to a single phase of the CKM matrix and thus all triangles should have the same area! This is an important test of the SM.

Another test of the SM follows from the measurements of the sides and angles of the UT of Eq. (4.242). As it is a simple triangle relation in the SM, all the triangle parameters can be fixed by measuring one angle and two sides, one side and two angles, or all three sides. Thus, overconstraining the triangle, i.e. measuring all its sides and angles simultaneously allows for tests of physics beyond the Standard Model.

It might appear that the measurements of the Unitarity triangle provide for a very limited test of New Physics: if various triangle measurements happen to be inconsistent, it might indicate that the "triangle" is not really a triangle, so it could be, for instance, a quadrilateral. This would only probe models with more than three generations of quarks that seem to be well constrained by other measurements.

This, however, is not a correct way to think about this problem. Indeed, any New Physics model would affect the transition amplitudes of the processes probing the triangle parameters. If we *insist* on interpreting all such measurements as the SM ones, the determinations of both absolute values of the CKM matrix elements and their phases would contain both the SM parts and the parts generated by New Physics. Any inconsistency in "closing" the triangle with experimentally-determined parameters would indicate the presence of BSM interactions. It is thus important to determine all angles and sides of the CKM Unitarity triangle. We will now discuss various methods of determination of the CKM angles.

CKM angle $\phi_1(\beta)$

The CKM angle $\phi_1(\beta)$ is sensitive to the phases associated with the top quark. Those can be accessed in $B^0 - \overline{B}^0$ oscillation measurements. Consider decays of neutral B mesons into a CP eigenstate, which can be generically called $f_{\rm CP}$. There are several possible final states that are CP-eigenstates, such as $J/\psi K_S$, $\pi^+\pi^-$, etc. As it turns out, the choice of the final state determines which CKM angle a given process can probe. We denote the amplitudes for such processes as [142]

$$A_f = \langle f_{\rm CP}|\mathcal{H}|B^0\rangle, \qquad \overline{A}_f = \langle f_{\rm CP}|\mathcal{H}|\overline{B}^0\rangle. \tag{4.244}$$

*It might seem strange that each angle of the Unitarity Triangle has two names. A story behind such a naming scheme echoes the story of a name for a J/ψ particle: two experimental collaborations (BaBar at SLAC and Belle at KEK) claimed different names for the same physical observable. It is rather unfortunate that they could not agree on a common naming scheme. It must be pointed out that a mnemonic way exists for remembering the angles of the "ϕ-convention": ϕ_i represents the angle opposite to the side of the UT defined by the generation index $|V_{id}V_{ib}^*|$. For example, the angle ϕ_1 is opposite to the side determined by $|V_{ud}V_{ub}^*|$ with the quark u belonging to the 1st generation, see Fig. 4.3.

Because of flavor oscillations, physical B^0 states $|B^0_{\rm phys}(t)\rangle$ would lead to the time-dependent decays amplitudes [23, 41],

$$\begin{aligned}
\langle f_{\rm CP}|\mathcal{H}|B^0_{\rm phys}(t)\rangle &= g_+(t)A_f + g_-(t)\frac{q}{p}\overline{A}_f = A_f\left[g_+(t) + \lambda_f g_-(t)\right], \\
\langle f_{\rm CP}|\mathcal{H}|\overline{B}^0_{\rm phys}(t)\rangle &= g_-(t)\frac{p}{q}A_f + g_+(t)\overline{A}_f = A_f\frac{p}{q}\left[g_-(t) + \lambda_f g_+(t)\right], \qquad (4.245)
\end{aligned}$$

with $g_\pm(t)$ defined in Eq. (4.213). The index "phys" in $B^0_{\rm phys}$ of Eq. (4.245) means that the state was tagged as B^0 at the time $t = 0$. The meaning of $\overline{B}^0_{\rm phys}$ is the same, except the state was tagged as $\overline{B}^0$. The (flavor) tagging can be done in several ways, depending on experimental conditions. e^+e^- flavor factories are usually run at the resonance energies, so a meson-antimeson pair is usually produced. For example, B-factories are usually tuned to run at energies that correspond to the $\Upsilon(4S)$ resonance, which lays just above $B_d\overline{B}_d$ threshold. This means that the produced state immediately decays $\Upsilon(4S) \to B\overline{B}$. Since there are two B-mesons present, the events could be selected such that one of them decays into a flavor-specific final state (say, semileptonically), so that the other state is immediately tagged as that of the opposite flavor. A similar technique is possible at the hadronic machines, with an added complication of several additional pions present in the event.

At hadronic machines, a technique known as *same-side tagging* is possible. Effective flavor tagging of the final state is possible if an excited B^* state is produced. In such a case, one can look for the sequence $B^{+(**)} \to B^0_d\pi^+$ with detection of the slow pion. The positive charge of the pion identifies the B-state as a B^0.

In writing Eq. (4.245) we also introduced λ_f for the final state f,

$$\lambda_f = \frac{q}{p}\frac{\overline{A}_f}{A_f}, \qquad (4.246)$$

which is a CKM convention-independent quantity. Squaring the amplitudes of Eq. (4.245), we can obtain time-dependent rates for initially pure B^0 or $\overline{B}^0$ states to decay into a CP eigenstate $f_{\rm CP}$,

$$\begin{aligned}
\Gamma(B^0_{\rm phys}(t) \to f_{\rm CP}) &= \frac{|A_f|^2 e^{-\Gamma t}}{2}\left[1 + |\lambda_f|^2 + (1 - |\lambda_f|^2)\cos(\Delta M t) - 2{\rm Im}\lambda_f \sin(\Delta M t)\right], \\
\Gamma(\overline{B}^0_{\rm phys}(t) \to f_{\rm CP}) &= \frac{|A_f|^2 e^{-\Gamma t}}{2}\left[1 + |\lambda_f|^2 + (1 - |\lambda|^2)\cos(\Delta M t) + 2{\rm Im}\lambda_f \sin(\Delta M t)\right] (4.247)
\end{aligned}$$

Such time-dependent rates can be used to form time-dependent CP-violating asymmetry, which is just a time-dependent version of Eq. (4.150)

$$a^f_{\rm CP}(t) = \frac{\Gamma(B^0_{\rm phys}(t) \to f_{\rm CP}) - \Gamma(\overline{B}^0_{\rm phys}(t) \to f_{\rm CP})}{\Gamma(B^0_{\rm phys}(t) \to f_{\rm CP}) + \Gamma(\overline{B}^0_{\rm phys}(t) \to f_{\rm CP})}, \qquad (4.248)$$

where we took into account the fact that $\overline{f}_{\rm CP} = f_{\rm CP}$. As stated before, we neglect $\Delta\Gamma$ for the B_d mesons, so that inserting Eqs. (4.247) into Eq. (4.248) we obtain

$$a^f_{\rm CP}(t) = \frac{1}{1 + |\lambda_f|^2}\left[(1 - |\lambda_f|^2)\cos(\Delta M t) - 2{\rm Im}\lambda_f \sin(\Delta M t)\right]. \qquad (4.249)$$

We are interested in extraction of the CKM angle $\phi_1(\beta)$ from the measurements in the B_d system. As can be seen from Eq. (4.218), in the limit $\Delta\Gamma \ll \Delta M$ or, equivalently, $\Gamma_{12} \ll M_{12}$, q/p is a pure phase and can be written as

$$\frac{q}{p} = -\sqrt{\frac{M^*_{12}}{M_{12}}} = e^{-2i\phi_M}, \qquad (4.250)$$

where the index M stands for "mixing". If we choose a decay channel that is completely dominated by a single amplitude, such as $B_d \to J/\psi K_S$ (so that $f = f_{\rm CP} = J/\psi K_S$), we will obtain

$$\overline{A}_f/A_f = \exp(-2i\phi_D), \tag{4.251}$$

or $|\overline{A}_f/A_f| = 1$. Then $|\lambda_{f_{\rm CP}}| = 1$! Note that in Eq. (4.251) the index D stands for "decay". This implies

$$a_{\rm CP}^{f_{\rm CP}}(t) = -{\rm Im}\lambda_{f_{\rm CP}} \sin(\Delta M t) = -\sin 2(\phi_M + \phi_D)\sin(\Delta M t), \tag{4.252}$$

where we used Eqs. (4.250) and (4.251) to obtain the final result. Note that separately ϕ_M and ϕ_D are convention-dependent, while their combination in Eq. (4.246) is convention-independent.

Why is the CP-violating asymmetry of Eq. (4.248) non-zero, even though the decay amplitude only has one component? This question is relevant, as we commented after Eq. (4.150) that two different amplitudes are required for the asymmetry to be non-zero. The answer is such that in this case the role of the second amplitude is played by the mixing. Since the final state of definite CP can be reached in both B^0 and $\overline{B}^0$ decays, it is the interference of $B \to f_{\rm CP}$ and $B \to \overline{B} \to f_{\rm CP}$ processes that produces a non-zero value of $a_{\rm CP}^{f_{\rm CP}}(t)$.

Let us now concentrate on a particular process, $B_d \to J/\psi K_S$, that satisfies all above requirements. In order to evaluate $a_{\rm CP}^{f_{\rm CP}}(t)$ we need a computation of $\lambda_{f_{\rm CP}}$. According to Eq. (4.246), there are several components to it.

The mixing in the B_d system is dominated by the top-quark intermediate states, as was discussed in Sec. 4.9. This means that the mixing phase that contributes to $\lambda_{f_{\rm CP}}$ is given by

$$\left.\frac{q}{p}\right|_{B_d} = \frac{V_{tb}^* V_{td}}{V_{tb} V_{td}^*}. \tag{4.253}$$

The decay $B_d \to J/\psi K_S$ is dominated by a tree-level amplitude with the underlying quark transition $b \to c\bar{c}s$. Discarding any other possible contributions, the decay amplitude for this process would be written as $A(B_d \to J/\psi K_S) = V_{cb}^* V_{cs} a(B_d \to J/\psi K_S)$, where $a(B_d \to J/\psi K_S)$ contains all hadronic uncertainties associated with a non-leptonic transition. An important observation is that this part of the amplitude is the same for B_d^0 and $\overline{B}_d^0$ transitions, $a(B_d \to J/\psi K_S) = a(\overline{B}_d \to J/\psi K_S)$, as we assumed that there is only one tree-level diagram that contributes to this decay! This means that

$$\frac{\overline{A}_{J/\psi K_S}}{A_{J/\psi K_S}} = \frac{V_{cb}^* V_{cs}}{V_{cb} V_{cs}^*}. \tag{4.254}$$

Since there is a quark-antiquark pair of the same flavor, a penguin amplitude also contributes. A careful analysis shows that this contribution is small. We could also observe that the penguin diagram's dominant contribution comes from the top quark, which follows from the GIM mechanism. This contribution would be proportional to the CKM combination $V_{tb} V_{ts}^*$, which, up to a sign, has the same phase as the tree diagram. This observation bolsters our assertion that led to the Eq. (4.254).

Our final point is that with a kaon in the final state one has to take into account the mixing phase in the K system. This phase is important, as K_S is an admixture of K^0 and $\overline{K}^0$. This is rather trivial, we simply need to multiply the definition of $\lambda_{f_{\rm CP}}$ in Eq. (4.246) by a $K^0\overline{K}^0$ mixing phase,

$$\left.\frac{q}{p}\right|_{K} = \frac{V_{cd}^* V_{cs}}{V_{cd} V_{cs}^*}. \tag{4.255}$$

Putting together all components from Eqs. (4.253), (4.254), and (4.255), we obtain

$$\lambda_{J/\psi K_S} = \frac{V_{tb}^* V_{td}}{V_{tb} V_{td}^*} \frac{V_{cb}^* V_{cs}}{V_{cb} V_{cs}^*} \frac{V_{cd}^* V_{cs}}{V_{cd} V_{cs}^*}. \tag{4.256}$$

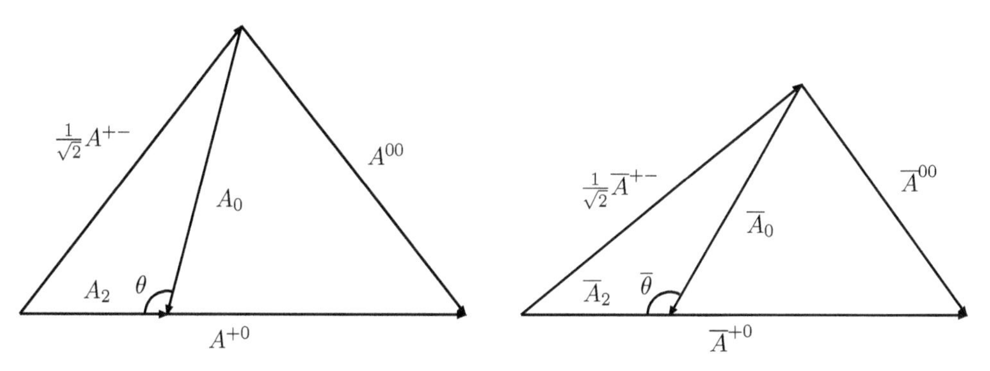

FIGURE 4.4 Graphical illustration of complex triangle relations of Eq. (4.262) leading to determination of the CKM angle $\phi_2(\alpha)$.

Taking its imaginary part, we see that it matches the definition of the angle $\phi_1(\beta)$ in Eq. (4.243),

$$\text{Im}\lambda_{J/\psi K_S} = -\sin(2\phi_1). \tag{4.257}$$

Thus, a measurement of the amplitude of time oscillations gives the angle $\phi_1(\beta)$ of the CKM matrix.

CKM angle $\phi_2(\alpha)$

As we could see, the choice of the decay mode in Eq. (4.252) can affect the phase $\phi_M + \phi_D$ in that equation. In fact, by choosing the oscillating system and the decay mode, we can probe all angles of the CKM matrix.

Let us now select the decay mode $B \to \pi^+\pi^-$. The final state is still a CP-eigenstate, so the formalism developed above still holds. Moreover, the mixing phase is still given by Eq. (4.253). What is different is the decay phase, which is now governed by the quark subprocess $b \to u\overline{u}d$. This process has both tree and penguin contributions. While this sounds like the previously discussed situation, it is really not. First, naive estimates show that the penguin contribution is not small, but rather of similar size to the tree contribution. Second, a major part of the penguin amplitude is governed by the CKM combination $V_{td}^*V_{tb}$, which has a different phase from the tree-level amplitude governed by $V_{ub}^*V_{ud}$.

Despite this, let us assume for a second that only the tree-level amplitude contributes to the $B \to \pi^+\pi^-$ decay. In such a case,

$$\frac{\overline{A}_{\pi^+\pi^-}}{A_{\pi^+\pi^-}} = \frac{V_{ud}^*V_{ub}}{V_{ud}V_{ub}^*}. \tag{4.258}$$

Again, putting together all components from Eqs. (4.253) and (4.258), we obtain

$$\lambda^T_{\pi^+\pi^-} = \frac{V_{tb}^*V_{td}}{V_{tb}V_{td}^*}\frac{V_{ud}^*V_{ub}}{V_{ud}V_{ub}^*}, \tag{4.259}$$

where the index T is there to remind us that we only looked at the ratio of the tree-level amplitudes. Taking its imaginary part, we see that it matches the definition of the angle $\phi_2(\alpha)$ in Eq. (4.243),

$$\text{Im}\lambda_{\pi^+\pi^-} = \sin(2\phi_2). \tag{4.260}$$

Thus, under the assumption of the single amplitude contributing to the decay $B_d \to \pi^+\pi^-$, a measurement of the amplitude of time oscillations gives the angle $\phi_2(\alpha)$ of the CKM matrix.

The real situation is, of course, more complicated. Yet, if the penguin amplitude is sizable, maybe we could eliminate it based on the additional arguments. As was shown in [100], it is possible to do so by using isospin symmetry relations, but additional decay rates, such as $B \to \pi^0\pi^0$ and $B^\pm \to \pi^0\pi^\pm$ would need to be measured as well.

To see how this and similar methods work, let's recall that the pions form an isospin triplet, but because of the Bose symmetry, only states of the total isospin $I = 0$ and $I = 2$ are possible for the $\pi\pi$ final state. Since the initial state has one light quark, its isospin is $I = 1/2$, which means that the effective Hamiltonian changes isospin as $\Delta I = 1/2$ or $\Delta I = 3/2$. Only the final state $\pi^+\pi^0$ can have $I = 2$, so the tree-level $B^+ \to \pi^+\pi^0$ only participates in $\Delta I = 3/2$ interaction. For other pion final states $I = 0$ and $I = 2$ are possible, from which we conclude that $\Delta I = 1/2$ operator gives rise to both tree and penguin contributions.

We can now expand the decay amplitudes

$$\begin{aligned} A^{+-} &= A(B_d \to \pi^+\pi^-), \\ A^{00} &= A(B_d \to \pi^0\pi^0), \\ A^{+0} &= A(B^+ \to \pi^+\pi^0) \end{aligned} \tag{4.261}$$

in terms of the amplitudes to decay into a pion state with the isospin state $I = 0$ (A_0) and $I = 2$ (A_2). Note that since A^{+0} contains only one amplitude, there is a simple relation of A^{+0} and its charge-conjugated version, $\overline{A}^{-0}$: $A^{+0} = \overline{A}^{-0}$. Using the appropriate Clebsh-Gordan coefficients and writing $\pi^+\pi^- = (\pi_1^+\pi_2^- + \pi_1^-\pi_2^+)/\sqrt{2}$, etc., we get

$$\begin{aligned} \frac{A^{+-}}{\sqrt{2}} &= A_2 - A_0, \\ A^{00} &= 2A_2 + A_0, \\ A^{+0} &= 3A_2. \end{aligned} \tag{4.262}$$

Inspection of Eq. (4.262) reveals the triangle relation among the complex amplitudes,

$$\begin{aligned} \frac{A^{+-}}{\sqrt{2}} + A^{00} - A^{+0} = 0, \\ \frac{\overline{A}^{+-}}{\sqrt{2}} + \overline{A}^{00} - \overline{A}^{-0} = 0, \end{aligned} \tag{4.263}$$

where the second line of Eq. (4.263) is just a charge-conjugated version of the first one. The triangle relations are shown in Fig. 4.4.

Now, to obtain $\phi_2(\alpha)$, we need to separate the penguin amplitudes that have different weak and strong phases. Since both tree and penguin amplitudes contribute to $B_d \to \pi^+\pi^-$ and $B_d \to \pi^0\pi^0$, we need to look more carefully at their time development. Since we made no assumptions about the structure of decay amplitudes in deriving Eq. (4.247), they are still valid. However, the structure of λ_f would be different. Using Eq. (4.246), (4.252), and (4.262) we get for $B_d \to \pi^+\pi^-$,

$$\lambda_{\pi^+\pi^-} = \frac{q}{p}\frac{\overline{A}_{\pi^+\pi^-}}{A_{\pi^+\pi^-}} = e^{2i(\phi_M+\phi_T)}\frac{\sqrt{2}|A_2|(1-\overline{A}_0/\overline{A}_2)}{\sqrt{2}|A_2|(1-A_0/A_2)} = e^{2i(\phi_M+\phi_T)}\frac{1-\overline{z}}{1-z}, \tag{4.264}$$

where ϕ_T is the weak phase of the A_2 amplitude, $A_2 = |A_2|\exp(i\delta_2)\exp(i\phi_T)$, $\overline{A}_2 = |A_2|\exp(i\delta_2)\exp(-i\phi_T)$, and we introduced complex ratios,

$$z = \frac{A_0}{A_2}, \quad \text{and} \quad \overline{z} = \frac{\overline{A}_0}{\overline{A}_2}. \tag{4.265}$$

Since the tree-level amplitude is generated by the quark process $b \to u\bar{u}d$, the weak phase of A_2 is the same as in Eq. (4.258), so $\phi_M + \phi_T = \phi_2(\alpha)$. This means that we can rewrite Eq. (4.264) as

$$\lambda_{\pi^+\pi^-} = \lambda^T_{\pi^+\pi^-} \frac{1-\bar{z}}{1-z} \tag{4.266}$$

Taking its imaginary part we can see that the coefficient of $\sin \Delta M t$ in Eq. (4.252) is

$$\mathrm{Im}\lambda_{\pi^+\pi^-} = \mathrm{Im}\left[e^{2\phi_2}\frac{1-|\bar{z}|e^{\pm i\bar{\theta}}}{1-|z|e^{\pm i\theta}}\right], \tag{4.267}$$

which reproduces Eq. (4.260) if penguin amplitude is neglected. It was pointed out [100] that those could be obtained from the geometrical triangle construction shown in Fig. 4.4. Determining $|A_0|$ from the triangle $A^{00} - A^{+0} - A_0$ we can employ the cosine theorem to obtain θ. Note that although $\cos\theta$ can be determined, $\sin\theta$ can not, which results in the sign ambiguity of θ. Geometrically, it is reflected in the fact that we do not know the orientation of the triangles in Fig. 4.4: they can be flipped along the line determined by the A^{+-} line. It follows that the phases of z and $\bar{z}$ are determined only up to a sign, leading to a four-fold ambiguity in determining $\phi_2(\alpha)$. Defining

$$m_{+-} \equiv \left|\frac{1-|\bar{z}|e^{\pm i\bar{\theta}}}{1-|z|e^{\pm i\theta}}\right|, \tag{4.268}$$

and its (four) possible phases as $\pm\epsilon_{+-}$ and $\pm\eta_{+-}$, the CKM angle $\phi_2(\alpha)$ can be determined as a solution of

$$\begin{aligned} \sin(2\phi_2 \pm \epsilon_{+-}) &= \frac{\mathrm{Im}\lambda_{\pi^+\pi^-}}{m_{+-}}, \\ \sin(2\phi_2 \pm \eta_{+-}) &= \frac{\mathrm{Im}\lambda_{\pi^+\pi^-}}{m_{+-}} \end{aligned} \tag{4.269}$$

with a four-fold ambiguity. To remove this ambiguity Ref. [100] proposes to study time-dependence of $B_d \to \pi^0\pi^0$ decay. While it is a challenging task experimentally, it leads to removal of the ambiguity. In this case, the coefficient of the $\sin(\Delta M t)$ term reads

$$\mathrm{Im}\lambda_{\pi^0\pi^0} = \mathrm{Im}\left[e^{2\phi_2}\frac{1-\frac{1}{2}|\bar{z}|e^{\pm i\bar{\theta}}}{1-\frac{1}{2}|z|e^{\pm i\theta}}\right], \tag{4.270}$$

which leads to another set of four equations,

$$\begin{aligned} \sin(2\phi_2 \pm \epsilon_{00}) &= \frac{\mathrm{Im}\lambda_{\pi^0\pi^0}}{m_{00}}, \\ \sin(2\phi_2 \pm \eta_{00}) &= \frac{\mathrm{Im}\lambda_{\pi^0\pi^0}}{m_{00}}. \end{aligned} \tag{4.271}$$

With exception of some special cases, Eqs. (4.269) and (4.271) determine $\sin 2\phi_2$ unambiguously.

There are other methods to determine $\phi_2(\alpha)$ with $\rho\rho$ or $\rho\pi$ final states. They might not provide a straightforward analysis discussed above, but their experimental studies are somewhat easier to perform. We encourage interested readers to pursue them in Refs. [103, 131, 173].

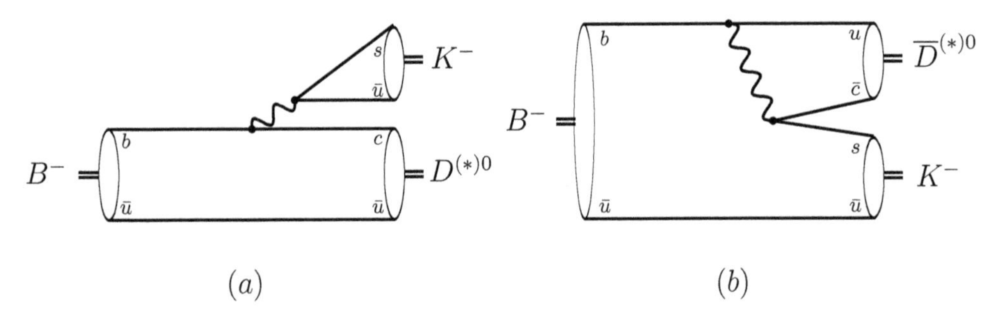

FIGURE 4.5 The flavor-flow diagrams illustrating the methods for extraction of the CKM angle $\phi_3(\gamma)$. In order for the diagrams to interfere $D^{(*)0}$ and $\overline{D}^{(*)0}$ must decay to the same final state.

CKM angle $\phi_3(\gamma)$

The methods described above involve processes that proceed via loop diagrams. While this might be a welcome development for the searches for New Physics particles that might give a similar contribution to the SM amplitudes, one might be interested in building the "reference" triangle that is obtained from the measurements that are dominated by the SM tree-level processes, at least in the b-quark sector. A method extracting $\phi_3(\gamma)$ that satisfies such requirements would be quite useful.

It turns out that there are several such methods. The most theoretically clean methods are based on the interference between $b \to \bar{u}cs$ and $b \to u\bar{c}s$ tree-level amplitudes. It might be rather surprising to look for such interference. It follows from Eq. (4.150) that in order to interfere two amplitudes must lead to the same final state, while $\bar{u}cs$ and $b \to u\bar{c}s$ are certainly different. Nevertheless, such common final state can be found at the hadronic level. The method of such type is exemplified by the processes $B^\pm \to D^{(*)}K^\pm$ followed by $D \to f$ decay, and $B^\pm \to \overline{D^{(*)}}K^\pm$ followed by $\overline{D} \to f$ (see Fig. 4.5). Here f is the final state that can be reached by the decays of both $D^{(*)}$ and $\overline{D^{(*)}}$. Interference between these two paths gives rise to CP-violating observables. Provided there is a sufficient number of observables, one can determine the decay rates and extract the unknown phase.

It is interesting that several such final states f exist and offer a variety of options. For example, both $D^0 \to K^+\pi^-$ and $\overline{D}^0 \to K^+\pi^-$ happen even without $D^0\bar{D}^0$ mixing: the first transition occurs rather often and is proportional to λ^0, while the second one is suppressed by λ^2. Here $\lambda \sim 0.2$ is the parameter of the Wolfenstein parameterization of the CKM matrix. Alternatively, both $D^0 \to \pi^+\pi^-$ and $\overline{D}^0 \to \pi^+\pi^-$ are possible and happen at a similar rate with the amplitudes proportional to λ^1. There are several methods that use this trick. We will describe them below.

Method of Gronau, London, and Wyler (GLW). The main idea behind the GLW method [101, 102] is to use the decay modes $B^\pm \to D^0K^\pm$ and $B^\pm \to \overline{D^0}K^\pm$, as depicted in Fig. 4.5. Since the D-meson is unstable, it can be reconstructed in two possible ways: as a flavor eigenstate D_{flav} or as a CP-eigenstate. Flavor eigenstate is reconstructed by either looking for a Cabibbo-favorite mode, such as $D^0 \to K^-\pi^+$ or employing semileptonic decay. A CP-eigenstate is defined such that the decay modes used to detect a D-meson are actually CP-eigenstates, such as $D \to \pi^+\pi^-$, to which both D^0 and $\overline{D}^0$ can decay. Reconstruction of a D-meson with such final states selects a linear superposition for the D-meson state,

$$|D_{CP^\pm}\rangle = \frac{1}{\sqrt{2}}\left(|D^0\rangle \pm |\overline{D}^0\rangle\right), \tag{4.272}$$

assuming that CP is conserved in D-decays, which is a good approximation. We can then construct four observables that encode the CP violation parameters,

$$\begin{aligned} R_{CP^\pm} &= 2\frac{\Gamma(B^- \to D_{CP^\pm}K^-) + \Gamma(B^+ \to D_{CP^\pm}K^+)}{\Gamma(B^- \to D_{\text{flav}}K^-) + \Gamma(B^+ \to D_{\text{flav}}K^+)} = 1 + r_B^2 \pm 2r_B \cos\delta_B \cos\phi_3, \\ A_{CP^\pm} &= \frac{\Gamma(B^- \to D_{CP^\pm}K^-) - \Gamma(B^+ \to D_{CP^\pm}K^+)}{\Gamma(B^- \to D_{CP^\pm}K^-) + \Gamma(B^+ \to D_{CP^\pm}K^+)} = \pm\frac{r_B}{R_{CP^\pm}} \sin\delta_B \sin\phi_3, \end{aligned} \tag{4.273}$$

where $r_B = \left|A(B^- \to \overline{D^0}K^-)/A(B^- \to D^0K^-)\right|$ is the absolute value of the ratio of decay amplitudes of $B^- \to \overline{D^0}K^-$ and $B^- \to D^0K^-$, and $\delta_B = \arg(A(B^- \to \overline{D^0}K^-)/A(B^- \to D^0K^-))$ is its strong phase. There are four observables defined by Eq. (4.273), which can be used to reconstruct three parameters: r_B, δ_B, and $\phi_3(\gamma)$. Note that there is a relation, which follows from the definitions of $R_{CP^\pm}$ and $A_{CP^\pm}$ in Eq. (4.273),

$$R_{CP^+}A_{CP^+} = -R_{CP^-}A_{CP^-}, \tag{4.274}$$

which implies that the system is not over-constrained, even though, naively, the number of observables exceeds the number of parameters.

Method of Atwood, Dunietz, and Soni (ADS). There is another way to construct the final state such that it would be common for both processes in Fig. 4.5. The ADS method [10] proposes to use D decays to non-CP eigenstate $K\pi$. In this method the final state can be reached in two ways: via CKM-favored $B^- \to D^0K^-$ followed by CKM-suppressed D-decay $D^0 \to K^+\pi^-$, or via CKM-suppressed $B^- \to \overline{D^0}K^-$ followed by Cabibbo-favored transition $\overline{D}^0 \to K^+\pi^-$. In this case two observables can be constructed,

$$\begin{aligned} R &= \frac{\Gamma(B^- \to D_{\to K^+\pi^-}K^-) + \Gamma(B^+ \to D_{\to K^-\pi^+}K^+)}{\Gamma(B^- \to D_{\to K^-\pi^+}K^-) + \Gamma(B^+ \to D_{\to K^+\pi^-}K^+)} \\ &= r_B^2 + r_D^2 + 2r_Br_D\cos(\delta_B + \delta_D)\cos\phi_3, \\ A &= \frac{\Gamma(B^- \to D_{\to K^+\pi^-}K^-) - \Gamma(B^+ \to D_{\to K^-\pi^+}K^+)}{\Gamma(B^- \to D_{\to K^+\pi^-}K^-) + \Gamma(B^+ \to D_{\to K^-\pi^+}K^+)} \\ &= 2\frac{r_Br_D}{R}\sin(\delta_B + \delta_D)\sin\phi_3, \end{aligned} \tag{4.275}$$

where r_B and δ_B are defined below Eq. (4.273). One issue in Eq. (4.275) that requires additional attention is the fact that D-meson decays to non-CP eigenstates, which means one needs to measure the ratio of the amplitudes of DCSD and CF decays, $r_D = |A(D^0 \to K^+\pi^-)/A(\overline{D}^0 \to K^+\pi^-)|$ and their relative phase δ_D. These hadronic parameters could either be obtained from the global fit of CP-violating B-decays or separately from D-mixing measurements. In the flavor SU(3) limit $r_D = \lambda^2$ and $\delta_D = 0$, however SU(3)-breaking corrections are quite significant to employ the results in the symmetry limit.

Method of Bondar, Giri, Grossman, Soffer, and Zupan (BGGSZ). One issue with both GLW and ADS methods described above lies with the fact that they both use two-body decays of the D-meson. BGGSZ method [86] makes use of the fact that many three-body, or *Dalitz*, decay modes of D mesons with final states accessible to both D^0 and $\overline{D}^0$ mesons have larger branching ratios, enhancing the experimental sample. As an example, let us consider the process $B^\pm \to D_{\to\pi^+\pi^-K_S}K^\pm$. We can write its amplitude as

$$\begin{aligned} A_{B^+}(s_+, s_-) &= \overline{A}_D(s_+, s_-) + r_Be^{i(\delta_B+\phi_3)}A_D(s_+, s_-), \\ A_{B^-}(s_+, s_-) &= A_D(s_+, s_-) + r_Be^{i(\delta_B+\phi_3)}\overline{A}_D(s_+, s_-) \end{aligned} \tag{4.276}$$

where we introduced the Dalitz variables $s_\pm = m^2_{K_S\pi^\pm}$. Also, $A_D(\overline{A}_D)$ is the amplitude of $D(\overline{D})$ decay into the final $K_S\pi^+\pi^-$ final state. The strong phase $\delta_B = \delta_B(s_+, s_-)$ varies across the Dalitz plot. The BGGSZ method has two variations, depending on the treatment of the D-meson

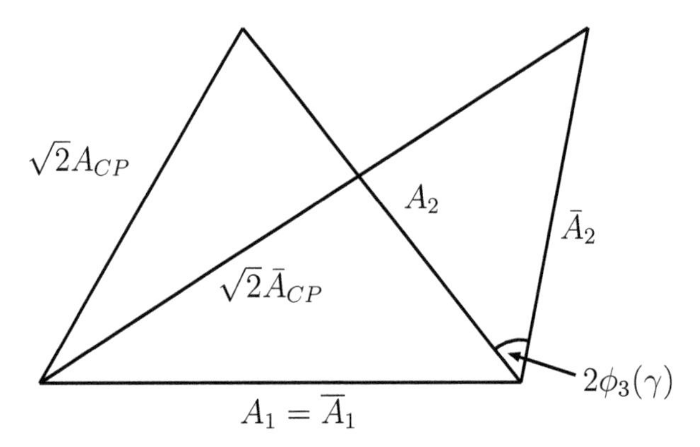

FIGURE 4.6 Graphical illustration of the $\phi_3(\gamma)$ extraction method with quantum-coherent initial state.

Dalitz plot. In one approach the Dalitz plot is divided ("binned") into sections with small strong phase variation, while in the second a model for the strong phase distribution is introduced. We invite the reader to consult Refs. [86] for details.

Coherent initial state method. A trick of using common decay modes of a meson to study $\phi_3(\gamma)$ can also be applied to the initial states if the initial state of neutral B_q mesons can be produced in a quantum-coherent state. This can be achieved if the pair comes from the decay of a $b\bar{b}$ meson such as the $\Upsilon(5S)$ (or $\Upsilon(4S)$). In this case, one has not only the option of tagging the flavor of the initial B_q, but also tagging it as an eigenstate of CP [75],

$$|B_q{}^{CP^\pm}\rangle = \frac{1}{\sqrt{2}}\left(|B_q^0\rangle \pm |\overline{B}_q^0\rangle\right). \tag{4.277}$$

This allows not only the decay amplitudes of flavor-specific initial states B_q^0 and $\overline{B}_q^0$ be measured,

$$\begin{aligned} A_1 &= A(B_q \to D_q^- P^+) = a_1 e^{i\delta_1}, \\ A_2 &= A(\overline{B}_q \to D_q^- P^+) = a_2 e^{-i\gamma} e^{i\delta_2}, \end{aligned} \tag{4.278}$$

but also the decay amplitudes of CP-tagged initial states

$$\begin{aligned} A_{\rm CP} &= A(B_q^{\rm CP} \to D_q^- P^+) = (A_1 + A_2)/\sqrt{2}, \\ \overline{A}_{\rm CP} &= A(B_q^{\rm CP} \to D_q^+ P^-) = (\overline{A}_1 + \overline{A}_2)/\sqrt{2} \end{aligned} \tag{4.279}$$

be extracted. Note that $P = K, \pi$ for $q = s, d$, respectively. The amplitude A_1 arises from the quark transition $\bar{b} \to \bar{c}u\bar{q}$, while the A_2 amplitude comes from $b \to u\bar{c}q$ and carries the relative weak phase $\phi_3(\gamma)$. The amplitudes A_i have strong phases δ_i. CP-conjugated amplitudes $\overline{A}_1$ and $\overline{A}_2$ for $\overline{B}_q$ decays can also be defined. It is clear that both CP-even and CP-odd amplitudes will yield triangle relations that contain the same information about $\phi_3(\gamma)$, so we do not need to differentiate between CP-even and CP-odd state of B_q.

The relations of Eq. (4.279) can be illustrated graphically in Fig. 4.6, where the amplitudes are pictured as vectors in the complex plane. As can be seen from the picture, the angle between A_2 and $\overline{A}_2$ is $2\phi_3(\gamma)$. The analytical solution can also be found [75]. Defining

$$\begin{aligned} \alpha &= \frac{2|A_{\rm CP}|^2 - |A_1|^2 - |A_2|^2}{2|A_1||A_2|}, \\ \overline{\alpha} &= \frac{2|\overline{A}_{\rm CP}|^2 - |\overline{A}_1|^2 - |\overline{A}_2|^2}{2|\overline{A}_1||\overline{A}_2|}, \end{aligned} \tag{4.280}$$

we can find the relation for $\phi_3(\gamma)$,

$$\sin 2\phi_3 = \pm\left(\alpha\sqrt{1-\overline{\alpha}^2} - \overline{\alpha}\sqrt{1-\alpha^2}\right). \quad (4.281)$$

Note that since $\Delta\Gamma_s/\Gamma_s$ is significant, the squared amplitudes $|A_i|^2$ and $|\overline{A}_i|^2$ in B_s system are proportional to partial rates, $|A_1|^2 \propto \Gamma(B_s \to D_s^- K^+)$ rather than to branching ratios.

4.9.2 Phenomenology of New Physics with nonleptonic D-decays

Phenomenological studies of New Physics with CP-violation in charm decays are rather different from the ones described above. This is mainly due to the fact that while overconstraining the CKM triangle is indeed possible in charm decays, it is rather challenging because charm CKM triangles are rather squished. Indeed, the "charm" CKM triangle is [24]

$$V_{us}^* V_{cs} + V_{ub}^* V_{cb} + V_{ud}^* V_{cd} = 0, \quad (4.282)$$

whose sides scale very differently with the Wolfenstein parameter $\lambda \sim 0.2$,

$$\begin{aligned} V_{us}^* V_{cs} &\sim \mathcal{O}(\lambda), \\ V_{ud}^* V_{cd} &\sim \mathcal{O}(\lambda), \quad (4.283) \\ V_{ub}^* V_{cb} &\sim \mathcal{O}(\lambda^5). \quad (4.284) \end{aligned}$$

This implies that while direct studies of the charm CKM triangle are possible, other methods might be preferable. Note that the "charmed" CKM triangle should have the same area as any other CKM triangle due to the unique source of CP-violation in the Standard Model.

D^0 decay processes that are easier to study experimentally contain all charged particles in the final state. Some of the most interesting ones include the doubly-Cabibbo-suppressed $D^0 \to K^+\pi^-$ decay, the singly-Cabibbo-suppressed $D^0 \to K^+K^-$ decay, the Cabibbo-favored $D^0 \to K^-\pi^+$ decay, and their three CP-conjugate decay processes. Let us now write down approximate expressions for the time-dependent decay rates that are valid for times $t \sim 1/\Gamma$. We take into account the experimental information that x, y and λ are small, and expand each of the rates only to the order that is relevant to experimental measurements:

$$\begin{aligned} \Gamma[D^0(t) \to K^+\pi^-] &= e^{-\Gamma t}|\bar{A}_{K^+\pi^-}|^2|q/p|^2 \\ &\times \left[|\lambda_{K^+\pi^-}^{-1}|^2 + [\mathrm{Re}(\lambda_{K^+\pi^-}^{-1})y + \mathrm{Im}(\lambda_{K^+\pi^-}^{-1})x]\Gamma t + \frac{1}{4}(y^2+x^2)(\Gamma t)^2\right], \\ \Gamma[\overline{D^0}(t) \to K^-\pi^+] &= e^{-\Gamma t}|A_{K^-\pi^+}|^2|p/q|^2 \\ &\times \left[|\lambda_{K^-\pi^+}|^2 + [\mathrm{Re}(\lambda_{K^-\pi^+})y + \mathrm{Im}(\lambda_{K^-\pi^+})x]\Gamma t + \frac{1}{4}(y^2+x^2)(\Gamma t)^2\right], \\ \Gamma[D^0(t) \to K^+K^-] &= e^{-\Gamma t}|A_{K^+K^-}|^2\left[1 + [\mathrm{Re}(\lambda_{K^+K^-})y - \mathrm{Im}(\lambda_{K^+K^-})x]\Gamma t\right], \quad (4.285) \\ \Gamma[\overline{D^0}(t) \to K^+K^-] &= e^{-\Gamma t}|\bar{A}_{K^+K^-}|^2\left[1 + [\mathrm{Re}(\lambda_{K^+K^-}^{-1})y - \mathrm{Im}(\lambda_{K^+K^-}^{-1})x]\Gamma t\right], \\ \Gamma[D^0(t) \to K^-\pi^+] &= e^{-\Gamma t}|A_{K^-\pi^+}|^2, \\ \Gamma[\overline{D^0}(t) \to K^+\pi^-] &= e^{-\Gamma t}|\bar{A}_{K^+\pi^-}|^2. \end{aligned}$$

where λ_f for $f = K\pi$ and KK was defined in Eq. (4.246). In particular, as $|q/p| = R_m$,

$$\lambda_{K^-\pi^+} = \sqrt{R}R_m e^{-i(\delta-\phi)}. \quad (4.286)$$

Studies of time-dependent rates of Eq. (4.285) allow to measure several interesting quantities. First, mixing parameters x and y can be determined,

$$\begin{aligned} \Gamma[D^0(t) \to K^+\pi^-] &= e^{-\Gamma t}|A_{K^-\pi^+}|^2 \\ &\times \left[R + \sqrt{R}R_m(y'\cos\phi - x'\sin\phi)\Gamma t + \frac{R_m^2}{4}(y^2+x^2)(\Gamma t)^2\right], \quad (4.287) \end{aligned}$$

where R is the ratio of DCS and Cabibbo favored (CF) decay rates. Since x and y are small, the best constraint comes from the linear terms in t that are also *linear* in x and y. A direct extraction of x and y from Eq. (4.287) is not possible due to unknown relative strong phase δ_D of DCS and CF amplitudes,

$$\begin{aligned} x' &= x\cos\delta_D + y\sin\delta_D, \\ y' &= y\cos\delta_D - x\sin\delta_D. \end{aligned} \tag{4.288}$$

While this strong phase can be measured independently, its value can also be determined as a parameter of a global fit, along with the CP-violating phase ϕ.

While CPT-symmetry requires the total widths of D and $\overline{D}$ to be the same, the partial decay widths $\Gamma(D \to f)$ and $\Gamma(\overline{D} \to \overline{f})$ could be different in the presence of CP-violation, which would be signaled by a non-zero value of the asymmetry

$$a_f = \frac{\Gamma(D \to f) - \Gamma(\overline{D} \to \overline{f})}{\Gamma(D \to f) + \Gamma(\overline{D} \to \overline{f})}, \tag{4.289}$$

which, depending on the final and initial state, can be generated by both $\Delta C = 1$ and $\Delta C = 2$ interactions. Contrary to the case of bottom quarks, standard model interactions do not produce a large CP-violating signal in the charmed meson system. This argument stems from the fact that all quarks that build up initial and final hadronic states in weak decays of charm mesons or baryons belong to the first two generations. This implies that, at the tree level, those transitions are governed by a 2×2 Cabibbo quark mixing matrix. This matrix is real, so no CP-violation is possible in the dominant tree-level diagrams which describe the decay amplitudes.

Asymmetries of Eq. (4.289) can be introduced for both charged and neutral D-mesons. In the latter case, a much richer structure becomes available due to the interplay of CP-violating contributions to decay and mixing amplitudes, which can play the role of a "second pathway" for the CP-asymmetry

$$\begin{aligned} a_f &= a_f^d + a_f^m + a_f^i, \\ a_f^m &= -R_f \frac{y'}{2}\left(R_m - R_m^{-1}\right)\cos\phi, \\ a_f^i &= R_f \frac{x'}{2}\left(R_m + R_m^{-1}\right)\sin\phi, \end{aligned}$$

where a_f^d, a_f^m, and a_f^i represent CP-violating contributions from decay, mixing, and interference between decay and mixing amplitudes, respectively. Note that for the final states that are also CP-eigenstates $f = \overline{f}$ and $y' = y$.

The combined asymmetry, formed as a difference of CP-asymmetries in $\pi^+\pi^-$ and K^+K^- channels,

$$\Delta A_{CP} = a_{KK} - a_{\pi\pi}, \tag{4.290}$$

is actually approximately double of the individual asymmetries due to $a_{KK} = -a_{\pi\pi}$ in the flavor $SU(3)_F$ limit. A proper interpretation of ΔA_{CP} in terms of fundamental CP-violating parameters is needed.

4.10 Kaon decays

Kaon decays serve as very sensitive tests of the SM and possible BSM effects. Out of many possible kaon decays, the transitions $K^+ \to \pi^+\nu\bar{\nu}$ and $K_L \to \pi^0\nu\bar{\nu}$ play a special role in NP searches with kaons. While challenging experimentally, they are relatively clean theoretically, and, being FCNC transitions, have a good chance of being affected by NP. We should expect that the matrix elements of these decays have a sensitivity to long-distance QCD effects and might be problematic to compute. As it turns out, flavor symmetries allow us to relate matrix elements of those processes to one another.

4.10.1 Rare decays and CP-violation. Grossman-Nir bound.

Let us discuss phenomenology of $K^+ \to \pi^+\nu\bar{\nu}$ and $K_L \to \pi^0\nu$ decays. As we shall see, a model-independent relation can be found that relates the two decay rates [105].

It is instructive to consider the CP-properties of the final state in these transitions. In general, in a three-body decay, the final state does not have a definite CP parity. Yet, the $\pi\bar{\nu}\nu$ is special. Since it is produced in the decay of a pseudoscalar kaon, the total angular momentum J of the final state must be zero. The left-handed nature of the neutrinos forces the $\bar{\nu}\nu$ state into a state with the same quantum numbers as Z^*, namely CP-even. Thus, a $\pi^0\bar{\nu}\nu$ system with a total $J = 0$ (and thus in a relative p-wave between a π and the neutrinos), a CP-odd pion, and a CP-even $\bar{\nu}\nu$ must be in a total CP-even configuration*. This fact leads to several interesting consequences. Since $K_{L,S}$ are defined as

$$|K_{L,S}\rangle = p|K^0\rangle \mp q|\overline{K}^0\rangle, \tag{4.291}$$

we can see that

$$\langle \nu\bar{\nu}\pi^0|H|K_{L,s}\rangle = pA \mp q\overline{A} = pA\left(1 \mp \lambda\right), \tag{4.292}$$

where λ is defined in Eq. (4.246), and the amplitudes A and $\overline{A}$ are defined as

$$\begin{aligned} A &= \langle \nu\bar{\nu}\pi^0|H|K^0\rangle, \\ \overline{A} &= \langle \nu\bar{\nu}\pi^0|H|\overline{K}^0\rangle. \end{aligned} \tag{4.293}$$

If we now consider the ratio of K_S and K_L decay rates into the $\pi^0\bar{\nu}\nu$ final state, we will find that

$$\frac{\Gamma(K_L \to \pi^0\bar{\nu}\nu)}{\Gamma(K_S \to \pi^0\bar{\nu}\nu)} = \frac{1+|\lambda|^2 - 2\text{Re}[\lambda]}{1+|\lambda|^2 + 2\text{Re}[\lambda]}. \tag{4.294}$$

As we argued above, $\pi^0\bar{\nu}\nu$ final state is a CP-eigenstate, so $\overline{A}/A = 1$ and thus λ is simply a phase. Moreover, if CP is conserved then $\Gamma(K_L \to \pi^0\bar{\nu}\nu) = 0$, as can be seen from Eq. (4.294) remembering that in this limit $|q/p| = 1$.

Introducing a CP-violating phase θ as from $\lambda = e^{2i\theta}$, we can see that Eq. (4.294) leads to

$$\frac{\Gamma(K_L \to \pi^0\nu\bar{\nu})}{\Gamma(K_S \to \pi^0\nu\bar{\nu})} = \frac{1-\cos 2\theta}{1+\cos 2\theta} = \tan^2\theta. \tag{4.295}$$

Note that θ does not have to be an SM phase. The logic that led us to Eq. (4.295) would not change in the more general case of New Physics affecting $K\overline{K}$ mixing.

It might be very challenging to measure $\Gamma(K_S \to \pi^0\nu\bar{\nu})$. Fortunately, we can relate its amplitude to the amplitude for another process that is easier to study experimentally, $K^+ \to \pi^+\bar{\nu}\nu$, which is related to $K_S \to \pi^0\nu\bar{\nu}$ via isospin transformation,

$$\sqrt{2}A(K_S \to \pi^0\nu\bar{\nu}) = A(K^+ \to \pi^+\nu\bar{\nu}). \tag{4.296}$$

*This statement is true up to small corrections. For instance, a chirality flip on the final state neutrino can produce a CP-even final state, but this contribution is suppressed by tiny neutrino masses. There are also higher-order effects, but they are suppressed by small quantities $\mathcal{O}(m_K^2/m_W^2)$

We can then form a CP-violating ratio,

$$a_{\rm CP}(K) = \frac{\Gamma(K_L \to \pi^0 \nu\bar{\nu})}{\Gamma(K^+ \to \pi^+ \nu\bar{\nu})} = \frac{1-\cos 2\theta}{2} = \sin^2\theta. \tag{4.297}$$

which is valid up to small isospin-breaking corrections. Note that if both kaon decay amplitudes are dominated by the SM top-quark contributions, then $\theta = \phi_1(\beta)$.

If we rewrite the ratio of Eq. (4.297) in terms of branching ratios, then the following bound would be true,

$$\mathcal{B}(K_L \to \pi^0\nu\bar{\nu}) \leq \frac{\tau_{K_L}}{\tau_{K^+} r_I}\mathcal{B}(K^+ \to \pi^+\nu\bar{\nu}) \leq 4.4\mathcal{B}(K^+ \to \pi^+\nu\bar{\nu}), \tag{4.298}$$

where τ_{K_i} are the lifetimes of kaon states, and $r_I \simeq 0.954$ is the isospin-breaking factor. The bound of Eq. (4.298) is referred to as a *Grossman-Nir* bound [105].

New Physics contributions can modify both the mixing and the decay amplitudes, so some models of NP would violate the bound of Eq. (4.298). This makes it a clean observable to study New Physics.

4.11 Flavor problem and minimal flavor violation

The apparent lack of the solution to the flavor problem prompted a rather interesting (but rather pessimistic) proposal: a Minimal Flavor Violation (MFV) hypothesis. This proposal assumes that the structure of flavor symmetry breaking in the Standard Model persists beyond the SM as well. In other words, BSM flavor-violating interactions are proposed to be linked with the Yukawa structure of the Standard Model. As a result, non-Standard Model contributions in FCNC transitions turn out to be suppressed to a level consistent with experiments even when the New Physics scale is only a few TeV [60].

The idea of MFV is formulated such that it is naturally implemented into the language of effective field theories. EFTs build with MFV restrictions establish unambiguous correlations among NP effects in various rare decays which can be experimentally tested.

The gist of the argument follows from an observation that, in the Standard Model, in the absence of fermion masses, its Lagrangian is invariant under a large global symmetry of flavor transformations, $\mathcal{G}_q \otimes \mathcal{G}_\ell \otimes U(1)^5$, where

$$\mathcal{G}_q = SU(3)_{Q_L} \otimes SU(3)_{U_R} \otimes SU(3)_{D_R}\,, \qquad \mathcal{G}_\ell = SU(3)_{L_L} \otimes SU(3)_{E_R}\,. \tag{4.299}$$

The $SU(3)$ groups refer to a rotation in flavor space among the three families of basic SM fields: the quark and lepton doublets, Q_L and L_L, and the three singlets U_R, D_R and E_R. Two of the five $U(1)$ groups can be identified with the total baryon and lepton number (not broken by the SM Yukawa interaction), while an independent $U(1)$ can be associated with the weak hypercharge. Under the MFV hypothesis, there is only one universal source of the breaking of flavor, and it is given by the SM Yukawa interactions,

$$\mathcal{L}_Y = \bar{Q}_L Y_D D_R H + \bar{Q}_L Y_U U_R H_c + \bar{L}_L Y_E E_R H \; + \; \text{h.c.} \tag{4.300}$$

Practically, MFV is implemented using a spurion analysis (see, e.g. [149]), where it is assumed that $\mathcal{G}_q$ is a good symmetry, but promoting $Y_{U,D}$ to be non-dynamical fields (spurions) with non-trivial transformation properties under this symmetry.

As a consequence, in the limit where we neglect light quark masses, the leading $\Delta Q = 2$ and $\Delta Q = 1$ FCNC amplitudes get exactly the same CKM suppression as in the SM [34, 60]

$$\mathcal{A}(q^i \to q^j)_{\rm MFV} = (V_{Ii}^* V_{Ij})\; \mathcal{A}_{\rm SM}^{(\Delta Q=1)} \left[1 + a_1 \frac{16\pi^2 M_W^2}{\Lambda^2}\right], \tag{4.301}$$

$$\mathcal{A}(M_{ij} - \bar{M}_{ij})_{\rm MFV} = (V_{Ii}^* V_{Ij})^2 \mathcal{A}_{\rm SM}^{(\Delta Q=2)} \left[1 + a_2 \frac{16\pi^2 M_W^2}{\Lambda^2}\right]. \tag{4.302}$$

where the $\mathcal{A}_{\text{SM}}^{(i)}$ are the SM amplitudes entering at one or more loops, I are the intermediate state quarks in the SM loop amplitudes, and the a_i are $\mathcal{O}(1)$ real parameters.

While MFV is a valid hypothesis with falsifiable predictions, it is not yet established as a guiding principle in building explicit NP models. Its validity could only be verified if we discover an NP model exhibiting such flavor breaking pattern as predicted by the MFV hypothesis.

4.12 Baryon number violating processes

As we pointed out in Chapter 2, one of the Sakharov conditions [165] explicitly requires baryon number violating processes to be present for the generation of the matter-antimatter asymmetry of the Universe. Such processes are possible in the Standard Model. We can define the baryon (B) and lepton (L) currents [35],

$$\begin{aligned} J_\mu^B &= \frac{1}{3}\sum_k \left[\overline{q}_L\gamma_\mu q_L + \overline{u}_R\gamma_\mu U_R + \overline{d}_R\gamma_\mu d_R\right], \\ J_\mu^L &= \sum_{k=1}^{N_f} \left[\overline{\ell}_L\gamma_\mu \ell_L + \overline{e}_R\gamma_\mu e_R\right], \end{aligned} \tag{4.303}$$

where k sums over the number of quark and lepton generations with $k_{\text{max}} = N_f = 3$. These currents represent anomalous symmetries: while conserved classically, due to the chiral nature of electroweak interactions, they are not conserved quantum-mechanically. In fact, their divergencies are given by

$$\partial^\mu J_\mu^B = \partial^\mu J_\mu^L = \frac{N_f}{32\pi^2}\left(-g^2 W_{\mu\nu}^I \widetilde{W}^{I\mu\nu} + (g')^2 B_{\mu\nu}\widetilde{B}^{\mu\nu}\right). \tag{4.304}$$

In any non-abelian gauge theory, including the electroweak $SU(2)_L$ the vacuum is not topologically trivial. In fact, there are infinitely many degenerate ground states, which differ in their value of the so-called Chern-Simons number N_{CS}, which in the Weinberg-Salam model are defined as

$$N_{\text{CS}}(t) = \frac{g^3}{92\pi^2}\int d^3r \epsilon_{ijk}\epsilon^{IJK} W^{Ii} W^{Jj} W^{Kk}, \tag{4.305}$$

where for vacuum-to-vacuum transitions W^{Aa} are pure gauge configurations. The Chern-Simons numbers N_{CS} are integers and so their difference ΔN_{CS} is expected to be an integer as well, $\Delta N_{\text{CS}} = \pm 1, \pm 2, \ldots$. The change in baryon and lepton number can be written as

$$\Delta B \equiv B(t_f) - B(t_i) = \int_{t_i}^{t_f} dt \int d^3r \partial^\mu J_\mu^B = N_f\left(N_{\text{CS}}(t_f) - N_{\text{CS}}(t_i)\right) \tag{4.306}$$

and represent points in field space are separated by a potential barrier whose height is given by the so-called *sphaleron* energy E_{sph}. In the Standard Model $N_f = 3$, so the smallest change of baryon (and lepton) number is $\Delta B = \Delta L = \pm N_f = \pm 3$.

While some proposals exist on observing sphalerons in high-energy collisions, the low energy effects of such transitions are unobservable. This should not deter us from looking for effects of baryon (and lepton) number violation at low energy: many models exist where such effects result from a decay of a "high-energy relic", a very heavy particle whose decay modes violate baryon or lepton number conservation. In fact, many Grand Unification models contain such particles in their spectra with masses around Λ_{GUT}. So, we will concern ourselves with searches for quantum effects of such particles.

4.12.1 Baryon number violating decays

Baryon number-violating transitions can occur naturally in Grand Unified Theories or some other models, such as R-parity-violating SUSY [140]. It is easy to identify experimental signatures for such processes: in order to write down decay channels that violate baryon number we simply need to make sure that the baryon number of the initial state does not match the baryon number of the final state. Since total angular momentum needs to be conserved, this implies that either a lepton or some new light (invisible) fermion [174] is present in the final state. For example, for a proton decay, processes such as $p \to e^+\pi^0$ or $p \to \bar{\nu}K^+$ prove to be common targets of experimental analyses.

Low energy effective Lagrangian for such processes were proposed in [178,179,181] in the late 20th century*. The basis for such operators has been formalized in SM EFT. In fact, baryon-number violating operators that exhibit $\Delta B = 1$ first appear at dimension six (see Table 2.6 in Chapter 2).

In principle, these operators lead to decays of any kind of baryons, including those with heavy quarks. Such baryons are only produced in accelerator experiments, so we do not expect that scales much higher than hundreds of TeV can be probed. Much better statistics can be obtained with the baryons available around us: protons and neutrons. This simplifies the flavor structure of those operators. In general, writing out all flavor and color indices such $\Delta B = \Delta L = 1$ operators can be written as

$$
\begin{aligned}
\mathcal{L}_{d=6} &= \frac{C^1_{abcd}}{\Lambda^2}\epsilon^{\alpha\beta\gamma}(\overline{d}^C_{a,\alpha}u_{b,\beta})(\overline{Q}^C_{i,c,\gamma}\epsilon_{ij}L_{j,d}) \\
&+ \frac{C^2_{abcd}}{\Lambda^2}\epsilon^{\alpha\beta\gamma}(\overline{Q}^C_{i,a,\alpha}\epsilon_{ij}Q_{j,b,\beta})(\overline{u}^C_{c,\gamma}\ell_d) \\
&+ \frac{C^3_{abcd}}{\Lambda^2}\epsilon^{\alpha\beta\gamma}\epsilon_{il}\epsilon_{jk}(\overline{Q}^C_{i,a,\alpha}Q_{j,b,\beta})(\overline{Q}^C_{k,c,\gamma}L_{l,d}) \\
&+ \frac{C^4_{abcd}}{\Lambda^2}\epsilon^{\alpha\beta\gamma}(\overline{d}^C_{a,\alpha}u_{b,\beta})(\overline{u}^C_{c,\gamma}\ell_d) + \text{h.c.}\,,
\end{aligned} \tag{4.307}
$$

where α, β, γ denote the color, i, j, k, l the $\mathrm{SU}(2)_L$, and a, b, c, d the family indices. In Eq. (4.307) u, d, and ℓ are the right-handed up-quark, down-quark, and lepton fields, while Q and L are the left-handed quark and lepton doublets. The Wilson coefficients C^x_{abcd} depend on a renormalization scale. This dependence could be important. In leading logarithmic approximation it has been computed in [4].

The explicit formulas for the decay rates depend on the chosen decay mode. They also depend on the values of $B \to M$ form factors that are not known very well, but were computed using a variety of methods, such as quark models. A set of relations between different baryon-number violating decay rates could be predicted,

$$
\begin{aligned}
\Gamma(p \to \ell^+_R\pi^0) &= \frac{1}{2}\Gamma(n \to \ell^+_R\pi^-) = \frac{1}{2}\Gamma(p \to \bar{\nu}\pi^+) = \Gamma(n \to \bar{\nu}\pi^0), \quad \text{and} \\
\Gamma(p \to \ell^+_L\pi^0) &= \frac{1}{2}\Gamma(n \to \ell^+_L\pi^-),
\end{aligned} \tag{4.308}
$$

where the index L or R on leptonic fields denotes polarization of the field. Generically, we expect the transition rate

$$
\Gamma \approx \alpha^2_{\text{NP}}\frac{m_p^5}{\Lambda^4}. \tag{4.309}
$$

*A nice theoretical summary with current experimental bounds can be found in [112,140]

It is important to have a good estimate of relevant hadronic matrix elements. But, naively, a conservative upper bound on the lifetime $\tau(p \to \pi^0 e^+) > 1.6 \times 10^{33}$ yields a naive bound on the scale of NP probed by proton decay, $\Lambda > 3 \times 10^{15}$ GeV for $\alpha_{\rm NP} \sim \alpha_{\rm GUT} \sim 1/30$! This is a mind-boggling scale, which is unlikely to be ever probed directly at particle accelerators. Three-body decay modes lead to weaker constraints.

Higher-dimensional baryon-number violating operators have also been written, see Table 2.7. They are suppressed by higher orders of Λ, so they don't probe such high NP scales. Yet, multi-body proton and neutron decays can provide interesting probes for such operators.

Naively, one would expect proton and neutron decays can not probe effective operators involving third-generation fermions because of kinematical constraints. This is not entirely true, as decays like $p \to \bar{\nu}_\tau K^+$ probe couplings of tau neutrinos. Another way to access baryon-number violating operators with heavy quarks or charged lepton is to consider neutron-antineutron oscillations, where those fermions contribute to the mass difference via off-mass shell diagrams. We shall consider $n - \bar{n}$ oscillations next.

4.12.2 Neutron-antineutron oscillations and their cousins

Neutron is not electrically charged, so a question might be posited if neutron-antineutron oscillations are possible. We already discussed muonium-anti-muonium oscillations in Chapter 3 and meson-anti-meson oscillations earlier in this chapter. The former phenomenon required violation of lepton flavor and is appreciably suppressed in the Standard Model, while the latter one is actually enhanced in the SM.

What about neutrons - or any other neutral baryons? The biggest difference with the systems discussed above lies in the fact that a neutron has a non-zero baryon number. Baryon number is conserved in the Standard Model, so neutron-antineutron transitions are forbidden. Yet, as we discussed in the previous section, baryon number symmetry is accidental and can be broken by some BSM interactions. This in turn could lead to $n - \bar{n}$ oscillations.

Baryon-antibaryon oscillations can then also happen with other neutral baryon states, such as Λ or Λ_b. The presence of an s or a b-quark in those states makes those systems potentially sensitive to other $\Delta B = 2$ interactions. Experimentally, those states are much more difficult to produce and, once produced, they have much shorter lifetimes than a neutron. Yet, searches for Λ or Λ_b oscillations can be done at colliders or in some fixed-target experiments. The lower numbers of produced states lead to much weaker constraints on the oscillation parameters. While we shall concentrate our discussion on the neutrons from now on, the developed formalism can be readily adapted to studies of oscillations of other neutral baryon states.

Phenomenology of $n - \bar{n}$ oscillations

Denoting the low energy effective Hamiltonian for $\Delta B = 2$ interaction as $\mathcal{H}_{\rm eff}^{\Delta B=2}$, we now can assume that its matrix element taken in between the neutron n and the anti-neutron $\bar{n}$ is non-zero [152], but is equal to some, in general complex, parameter δm.

$$\langle \bar{n} | \mathcal{H}_{\rm eff}^{\Delta B=2} | n \rangle = \langle n | \mathcal{H}_{\rm eff}^{\Delta B=2} | \bar{n} \rangle = \delta m, \tag{4.310}$$

where we, for simplicity, assumed CP conservation. The exact form of the Hamiltonian $\mathcal{H}_{\rm eff}^{\Delta B=2}$ is not important right now, we shall discuss it later in this chapter.

The complete low energy effective Hamiltonian $\mathcal{H}_{\rm eff}$ would also contain a baryon number-conserving piece $\mathcal{H}_{\rm eff}^{\Delta B=0}$,

$$\mathcal{H}_{\rm eff} = \mathcal{H}_{\rm eff}^{\Delta B=0} + \mathcal{H}_{\rm eff}^{\Delta B=2}, \tag{4.311}$$

which implies that the diagonal pieces of the mass matrix

$$\langle n | \mathcal{H}_{\rm eff}^{\Delta B=2} | n \rangle = M_{11}, \quad \langle \bar{n} | \mathcal{H}_{\rm eff}^{\Delta B=2} | \bar{n} \rangle = M_{22}. \tag{4.312}$$

While CPT invariance requires that $M_{11} = M_{22}$, it might not be the case if we also include a possibility of CPT-violating interactions. Note that since neutron is an unstable particle with a lifetime $\tau_n = 880$ s, the diagonal entries of the mass matrix are complex. Assuming CPT-invariance,

$$M_{11} = M_{22} = m_n - i\frac{\Gamma}{2}, \tag{4.313}$$

where m_n is neutron's mass, and $\Gamma = 1/\tau_n$* is its width. A compact mass term for the neutron Lagrangian can be written as

$$\mathcal{L} = m_n \bar{n} n + \frac{\delta m}{2} n^T C n, \tag{4.314}$$

where C is charge-conjugation matrix defined in Section 2.3.2.

Vacuum oscillations. We can combine everything in the 2×2 mass matrix, which in the basis $(n, \bar{n})$ has the form

$$\mathcal{M} = \begin{pmatrix} m_n - i\Gamma/2 & \delta m \\ \delta m & m_n - i\Gamma/2 \end{pmatrix}. \tag{4.315}$$

Just like in previously considered cases of meson-antimeson oscillations, the matrix in Eq. (4.315) is diagonalized by a rotation to the mass basis,

$$\begin{bmatrix} |n_1\rangle \\ |n_2\rangle \end{bmatrix} = \begin{pmatrix} \cos\theta & \sin\theta \\ -\sin\theta & \cos\theta \end{pmatrix} \begin{bmatrix} |n\rangle \\ |\bar{n}\rangle \end{bmatrix} \tag{4.316}$$

where for the vacuum oscillations the mixing angle is $\theta = \pi/4$. The (propagating) mass eigenstates can be written in terms of the particle/antiparticle ones as

$$|n_{1,2}\rangle = |n_\pm\rangle = \frac{1}{\sqrt{2}} \left(|n\rangle \pm |\bar{n}\rangle \right), \tag{4.317}$$

which also coincide with the CP eigenstates. If we assume that δm is real, that is, all New Physics particles that generate non-zero neutron-antineutron matrix elements of Eq. (4.310) are heavier than the neutron, the mass eigenstates would have the same width Γ and the masses

$$m_\pm = m_n \pm \delta m. \tag{4.318}$$

The notation of Eq. (4.310) is now clear, as those matrix elements are encoding the mass difference ΔM between the mass eigenstates in the system,

$$\Delta M = m_+ - m_- = \delta m, \tag{4.319}$$

which is a physical observable. Time development of the neutron-antineutron system is given by the solution of the Schrödinger equation,

$$i\frac{\partial}{\partial t} \begin{bmatrix} n \\ \bar{n} \end{bmatrix} = \mathcal{M} \begin{bmatrix} n \\ \bar{n} \end{bmatrix}, \tag{4.320}$$

which after diagonalization yields a usual exponential time evolution for the CP-eigenstates $|n_\pm(t)\rangle$, which leads to the oscillation probability for the n and $\bar{n}$ states,

$$P(n \to \bar{n})(t) = \sin^2\left[(\delta m)t\right] e^{-\Gamma t} \tag{4.321}$$

*Note that another notation $\lambda \equiv \Gamma$ is common in research literature [152].

for the state initially produced as a neutron and detected as an antineutron. Current experimental limits on δm suggest that there is a hierarchy,

$$\tau_{n\bar{n}} = \frac{1}{|\delta m|} \gg \tau_n. \tag{4.322}$$

This implies that the argument of the sine function is always small, which is expected as the mixing is driven by BSM physics. Expanding Eq. (4.321) we get

$$P(n \to \bar{n})(t) = [(\delta m)t]^2 e^{-\Gamma t} \tag{4.323}$$

In deriving the oscillation probabilities, we assumed that δm is real. One can show that, similar to meson-antimeson oscillations, one can expect the lifetime difference between mass eigenstates. Such lifetime difference would be generated by the on-shell intermediate states and generically be even more suppressed by powers of Λ.

Magnetic fields and $n\bar{n}$ oscillations. We discussed possible $n\bar{n}$ oscillations in a vacuum in the situation when no external fields are present. It is however important to realize that contrary to the case of mesons discussed earlier, neutrons are fermions with non-zero magnetic moment. This fact matters, as n and $\bar{n}$ interact differently with an external magnetic field $\vec{B}$ via their magnetic dipole moments $\vec{\mu}_{n,\bar{n}}$, as discussed in Eq. (3.2). As $\mu_n = -\mu_{\bar{n}} = -1.91\mu_N$, where $\mu_N = 2/(2m_N)$, magnetic field separates neutrons from antineutrons, leading to suppression of the oscillation probability. To see this, let us rewrite the $n\bar{n}$ Hamiltonian matrix in the $(n, \bar{n})$ basis, now taking magnetic field into account,

$$\mathcal{M}_B = \begin{pmatrix} m_n - \vec{\mu}_n \cdot \vec{B} - i\Gamma/2 & \delta m \\ \delta m & m_n + \vec{\mu}_n \cdot \vec{B} - i\Gamma/2 \end{pmatrix}. \tag{4.324}$$

Once again, diagonalizing it with the help of the rotation matrix from Eq. (4.316), we find that the mass eigenstates $|n_{1,2}\rangle$ are obtained if the rotation angle θ is given by

$$\tan(2\theta) = -\frac{\delta m}{\vec{\mu}_n \cdot \vec{B}}. \tag{4.325}$$

The mass eigenstates will be propagating with energies

$$E_{1,2} = m_n \pm \sqrt{\left(\vec{\mu}_n \cdot \vec{B}\right)^2 + (\delta m)^2} - i\Gamma/2. \tag{4.326}$$

Clearly, in the limit $\left|\vec{B}\right| \to 0$ we recover previous expression of Eq. (4.318) for the mass eigenstates.

The energy difference $\Delta E = E_1 - E_2$ is now

$$\Delta E = 2\sqrt{\left(\vec{\mu}_n \cdot \vec{B}\right)^2 + (\delta m)^2} \approx 2|\vec{\mu}_n \cdot \vec{B}|, \tag{4.327}$$

where the second approximate equality follows from the fact that even for a magnetic field of $|\vec{B}| \sim 10^{-10}$ T the combination $|\vec{\mu}_n \cdot \vec{B}| \sim 10^{-23}$ MeV, while experimental constraints on δm achieved in experiments are already such that $\delta m \leq 10^{-29}$ MeV. Thus, even a small ambient magnetic field can affect the time evolution.

At any rate, the oscillation probability in (any) external field, i.e. keeping both θ and ΔE, is

$$\begin{aligned} P(n \to \bar{n})(t) &= \sin^2(2\theta) \sin^2 [(\Delta E)t/2]^2 e^{-\Gamma t} \\ &= \frac{(\delta m)^2}{(\vec{\mu}_n \cdot \vec{B})^2 + (\delta m)^2} \sin^2 \left[\sqrt{\left(\vec{\mu}_n \cdot \vec{B}\right)^2 + (\delta m)^2 t}\right]^2 e^{-\Gamma t}, \end{aligned} \tag{4.328}$$

where in the second line we used a trigonometric identity connecting the sine and tan functions. It is interesting to note that if one sets the observation time t such that $|\vec{\mu}_n \cdot \vec{B}|t \ll 1$ and $t \ll \tau_n$, we would get for the Eq. (4.328),

$$P(n \to \bar{n})(t) \approx \frac{(\delta m)^2}{|\vec{\mu}_n \cdot \vec{B}|^2} \left(\vec{\mu}_n \cdot \vec{B}t\right)^2 = \left[(\delta m)t\right]^2, \tag{4.329}$$

which is equivalent to Eq. (4.323) for $t \ll \tau_n$. This is sometimes referred to as *quasi-free propagation.*

Matter effects in $n\bar{n}$ oscillations. The last topic in the phenomenology of $n\bar{n}$ oscillations that we are going to discuss is $n\bar{n}$ oscillations in ordinary matter, i.e. inside the atomic nuclei. Neutrons are stable inside the nuclei, as they are sitting in the collective potential provided by other nucleons. The diagonal entries of the potential are different for neutrons and antineutrons. Moreover, one of the diagonal entries of the potential is complex, as antineutrons annihilate, producing pions with a multiplicity $\langle n_\pi \rangle \approx 4-5$, so one can write

$$V_{\bar{n}} = V_{\bar{n}R} - iV_{\bar{n}R}, \tag{4.330}$$

where $V_{\bar{n}R}$ parameterizes annihilation of the antineutron. Denoting the neutron potential V_n, the mass matrix can be written as

$$\mathcal{M}_B = \begin{pmatrix} m_{n\text{eff}} & \delta m \\ \delta m & m_{\bar{n}\text{eff}} \end{pmatrix}, \tag{4.331}$$

where we introduced a short cut notation,

$$\begin{aligned} m_{n\text{eff}} &= m_n + V_n, \\ m_{\bar{n}\text{eff}} &= m_{\bar{n}} + V_{\bar{n}}. \end{aligned} \tag{4.332}$$

Again, the matrix in Eq. (4.331) can be diagonalized to obtain the mass eigenstates $|n_{1,2}\rangle$ by

$$\tan 2\theta = \frac{2\delta m}{\sqrt{(V_{nR} - V_{\bar{n}R})^2 + V_{\bar{n}I}^2}} \tag{4.333}$$

with the eigenvalues that describe energy eigenstates of mixed neutron-antineutron states,

$$E_{1,2} = m_n + \frac{1}{2}\left((V_{nR} + V_{\bar{n}R}) - iV_{\bar{n}I}\right) \pm \frac{1}{2}\sqrt{\left((V_{nR} - V_{\bar{n}R}) + iV_{\bar{n}I}\right)^2 - 4(\delta m)^2}. \tag{4.334}$$

As can be seen from Eq. (4.334), the energy eigenstates are complex, which leads to matter instability. Expanding in δm and isolating the imaginary part gives a width for the nuclei due to the oscillations taking place,

$$\Gamma_{\text{bound}} = \frac{2(\delta m)^2 V_{\bar{n}I}}{(V_{nR} - V_{\bar{n}R})^2 + V_{\bar{n}I}^2}, \tag{4.335}$$

which leads to the lifetime of a nucleus $\tau_{\text{bound}} \sim (\delta m)^{-2}$. Since δm is small compared to V, this lifetime is quite large.

SM EFT and $n - \bar{n}$ oscillations

A naive guess on the form of the quark-level effective Lagrangian that leads to neutron-antineutron oscillations can be made purely on dimensional grounds,

$$\mathcal{L}_{\text{eff}} \sim -\frac{1}{\Lambda^5}\left(dud\right)^2, \tag{4.336}$$

as, naively, the operator changing three quarks into three antiquarks must contain six quark operators (thus being of dimension 9) and accounting for the fact that overall dimension of a Lagrangian should be equal to four. This means that, naively, the oscillation probability that is proportional to δm^2,

$$\delta m = \langle n| \int d^3x \mathcal{H}_{\text{eff}} |\bar{n}\rangle \simeq \Lambda_{\text{QCD}}^6/\Lambda^5, \tag{4.337}$$

implies that one pays a steep price of Λ^{-10} suppression for probing high scales Λ! This is especially depressing when compared to the scales probed by baryon-number violating decays in Sec. 4.12.1. Note that, as usual, $\mathcal{H}_{\text{eff}} = -\mathcal{L}_{\text{eff}}$. The estimate of Eq. (4.337) was supported by an explicit computation in quark models, while lattice calculations show a somewhat larger number.

Still, there are models with light mediators where the naive scale Λ^5 is given by a combination of smaller scales. Alternatively, there are models where Eq. (4.336) is generated by heavy particles that do not mediate proton decay [9].

To be more precise, there are twelve dimension-9 operators that can be written in the so-called Wagman-Buchoff basis,

$$\mathcal{L}_{\text{eff}} = \frac{1}{\Lambda^5} \sum_{i=1}^{6} \left(c_i \mathcal{O}_i + \bar{c}_i \overline{\mathcal{O}}_i \right) + h.c., \tag{4.338}$$

where the operators are

$$\begin{aligned}
\mathcal{O}_1 &= \frac{1}{2}\epsilon_{ijk}\epsilon_{lmn} \left(\overline{u}_i^c P_R d_j\right)\left(\overline{u}_l^c P_R d_m\right)\left(\overline{u}_k^c P_R d_n\right), \\
\mathcal{O}_2 &= \epsilon_{ijk}\epsilon_{lmn} \left(\overline{u}_i^c P_L d_j\right)\left(\overline{u}_l^c P_R d_m\right)\left(\overline{u}_k^c P_R d_n\right), \\
\mathcal{O}_3 &= \frac{1}{2}\epsilon_{ijk}\epsilon_{lmn} \left(\overline{u}_i^c P_L d_j\right)\left(\overline{u}_l^c P_L d_m\right)\left(\overline{u}_k^c P_R d_n\right), \\
\mathcal{O}_4 &= \epsilon_{ijk}\epsilon_{lmn} \left(\overline{u}_i^c P_R d_j\right)\left(\overline{u}_l^c P_L d_m\right)\left(\overline{u}_k^c P_L d_n\right), \\
\mathcal{O}_5 &= \left(\epsilon_{ijk}\epsilon_{lmn} + \epsilon_{imn}\epsilon_{ljk}\right) \left(\overline{u}_i^c P_R d_j\right)\left(\overline{u}_l^c P_L d_m\right)\left(\overline{u}_k^c P_L d_n\right), \\
\mathcal{O}_6 &= \epsilon_{ijk}\epsilon_{lmn} \left(\overline{u}_i^c P_L d_j\right)\left(\overline{u}_l^c P_L d_m\right)\left(\overline{u}_k^c P_R d_n\right), \\
&+ \left(\epsilon_{ijk}\epsilon_{lmn} + \epsilon_{imn}\epsilon_{ljk}\right) \left(\overline{u}_i^c P_L d_j\right)\left(\overline{u}_l^c P_L d_m\right)\left(\overline{u}_k^c P_R d_n\right),
\end{aligned} \tag{4.339}$$

4.12.3 Experimental methods of detection

In reactor experiments with free neutrons one can estimate a number of antineutrons $N_{\bar{n}}$ produced after time t in a run of duration T_{run},

$$N_{\bar{n}} = P(n \to \bar{n}) N_n, \tag{4.340}$$

as follows from Eq. (4.323). Here N_n is the number of produced neutrons. Since the transition probability for a quasi-free propagation is proportional to t^2, as follows from Eq. (4.329), a "figure of merit" for observation of neutron oscillation is given by NT^2, where N is the number of neutrons produced and T is a quasi-free observation time.

4.13 Notes for further reading

Original idea on the conditions that any theory must satisfy in order to predict baryogenesis was put forward by A. Sakharov in [165]. Reviews of baryogenesis models can be found in [48]. The original proposal for baryogenesis via leptogenesis was given in [133]. A nice overview of the neutron EDM problem with the theoretical background can be found in [61]. A thorough discussion of discrete symmetries and their experimental studies can be found in [152].

Flavor physics is a huge subject, we provided most of the references in the text of this chapter. In addition, several textbooks provide more in-depth exposition of the problems discussed in this chapter [65, 135, 144, 149]. A good discussion of the convergence of perturbative expansion can be found in [16]. A good discussion of kaon physics can be found in [32], with minimal flavor violation hypothesis discussed in [60, 79].

A nice review of theoretical models and experimental implication of baryon number violating processes can be found in [82].

Problems for Chapter 4

1. **Problem 1.** Perform diagonalization of Eq. (4.315) to solve for the time development of $|n_\pm(t)\rangle$ states. Using the obtained result, prove the formula for the oscillation probability given in Eq. (4.321). Taking into account possible ambient fields present in an experiment, derive Eq. (4.328).
2. **Problem 2.** Derive time dependence of flavor eigenstates in $H^0\overline{H}^0$ mixing. To do so, use Eq. (4.209 to derive Eq. (4.210) employing relations between exponential and trigonometric functions and definitions in Eq. (4.206).

5 New Physics searches with neutrinos

5.1 Introduction

Neutrinos are some of the most mysterious particles in the Standard Model. Over 100 trillion neutrinos pass through our bodies every second without us noticing, implying minute interactions with matter. How well are those interactions described by the Standard Model? Most importantly, with such weak interactions between neutrinos and matter, are there any uses for neutrinos to study New Physics?

The question of how neutrinos acquire their masses is also very interesting. It's been known for a while now that neutrinos have tiny masses, much smaller than those of all known elementary particles. In principle, the non-zero values of the neutrino masses are easy to accept: there is no unbroken symmetry in the SM that protects neutrino from acquiring a mass*. The explanation of their *smallness*, if exists, is much more challenging.

As we discussed in Chapter 2.2.1, all electrically charged fermions in the Standard Model receive their masses after the spontaneous breaking of the electroweak symmetry. As the Higgs field acquires vacuum expectation value, Yukawa interaction terms of Eq. (2.25) that couple left- and right-handed fermions to the Higgs field, generate fermion mass matrices. Neutrinos are an exception. Since there are no right-handed neutrino fields included in the *minimal* Standard Model, no term like $Y^\nu \overline{L}_L H \nu_R$ could be written to give neutrino masses after the SSB.

In principle, there is no technical problem in extending the minimal SM by including the right-handed neutrinos and writing in additional Yukawa terms generating neutrino masses the same way all charged fermion masses are generated. Yet, this would introduce two (theoretical) issues that would need to be explained. First, it is hard to understand why neutrino masses, if coming from the same structure of the matrices (vY^ν), end up being so much smaller than all other fermion masses. Second, the introduction of the right-handed neutrinos means the

*An example of such symmetry in the Standard Model that remains unbroken is the $U(1)$ gauge symmetry that keeps photon massless.

fields that are SM gauge singlets are introduced. The particles associated with those fields, the *sterile* neutrinos, do not interact with other SM particles. It is then entirely possible to write a separate mass term for such sterile particles. The masses of the RH neutrino particles do not in general have to be small. In that case, the observable neutrino masses could be some functions of those masses, and apriori do not need to be small. Such RH mass terms could be forbidden by introducing an explicit new symmetry, such as the lepton number.

Alternatively, we can use the fact that neutrinos are the only neutral fermions in the Standard Model. As was shown by E. Majorana, one could write a mass term composed entirely out of the left-handed fields. The Majorana mass term would explicitly break the lepton number by two units, bringing in a host of new phenomena, not available in other sectors of the Standard Model. We shall address some of those phenomena here.

Most of the experimental insight into the properties of neutrinos comes from neutrino flavor oscillation experiments. We will discuss them in more detail below, but for now, it is important to discuss a generic setup for such experiments. It usually includes a source of neutrinos or antineutrinos, which could be man-made, such as a nuclear reactor or an accelerator, or natural, such as the Sun or a supernova. The source usually emits neutrinos (antineutrinos) with a continuous spectrum of energies produced in a weak interaction process, i.e. in a *flavor* eigenstate. A detector is usually placed at a distance L from the source and also detects neutrinos in flavor eigenstates. The neutrinos interact with the detector material either via charged-current interaction, where a neutrino with a well-defined flavor converts into a charged lepton of the same flavor or via neutral-current interaction (Z-boson exchange), which is detected by observing the system against which the neutrinos are recoiling. Since neutrinos are propagating as mass eigenstates, neutrino produced in one flavor eigenstate (say, a muon), can be detected as a neutrino of a different flavor (say, electron). A probability of such transition can be measured.

5.2 Neutrino masses and oscillations

5.2.1 Neutrino oscillations

As we know from our discussion of the quark sector of the Standard Model, quarks of the same electric charge but different flavors can mix. This phenomenon is encoded in the observable CKM matrix discussed in the previous chapter. Neutrinos do not carry an electric charge, so it should not be difficult to imagine that neutrinos would mix as well.

To simply introduce neutrino oscillations, let us first consider the case of two flavors*. As it will turn out, this model actually captures a lot of physics that happens in the realistic case of three neutrinos. Let us discuss oscillations of electron neutrino ν_e into another active neutrino state, say muon neutrino ν_μ. If the Standard Model Lagrangian or some BSM scenario contains terms that couple neutrinos of different flavors it should be seen in the time development of such states,

$$i\frac{d}{dt}\begin{pmatrix}|\nu_e(t)\rangle \\ |\nu_\mu(t)\rangle\end{pmatrix} = \hat{\mathcal{H}}\begin{pmatrix}|\nu_e(t)\rangle \\ |\nu_\mu(t)\rangle\end{pmatrix}. \tag{5.1}$$

In other words, the time evolution of the neutrino flavor eigenstate $|\nu_i(t)\rangle$ is governed by a Schrödinger equation of Eq. (5.1) with a non-diagonal 2×2 Hamiltonian $\hat{\mathcal{H}}$.

The presence of off-diagonal pieces in the effective Hamiltonian leads to a familiar

*We will partially follow the discussion in [130].

phenomenon of mixing. Neutrino flavor eigenstates mix into propagating mass eigenstates,

$$|\nu_i\rangle = \sum_\alpha U_{i\alpha} |\nu_\alpha\rangle, \tag{5.2}$$

where we used Latin (Greek) indices to indicate mass (flavor) eigenstates. The mixing matrix $U_{i\alpha}$ is the Pontecorvo-Maki-Nakagawa-Sakata (PMNS) mixing matrix. It is a unitary matrix, i.e.

$$|\nu_\alpha\rangle = \sum_i U^\dagger_{\alpha i} |\nu_i\rangle. \tag{5.3}$$

The mass eigenstates are orthogonal to each other. Since mass eigenstates diagonalize the Hamiltonian, they propagate according to a simple time evolution factor e^{-iE_it}, with $E_i = \sqrt{\mathbf{p}^2 + m_i^2}$. Flavor eigenstates would experience flavor oscillations, i.e., a state produced as one flavor eigenstate has a non-zero probability to be detected as some other flavor eigenstate.

For a two-flavor example, we can easily parameterize the 2×2 PMNS mixing matrix in terms of the sines and cosines of the neutrino mixing angle,

$$\begin{pmatrix} |\nu_e\rangle \\ |\nu_\mu\rangle \end{pmatrix} = \begin{pmatrix} \cos\theta & \sin\theta \\ -\sin\theta & \cos\theta \end{pmatrix} \begin{pmatrix} |\nu_1\rangle \\ |\nu_2\rangle \end{pmatrix}. \tag{5.4}$$

Let us assume that a neutrino state is produced in one of the flavor eigenstates at $t = 0$ with a definite 3-momentum $\mathbf{p}$. Since neutrino masses are small compared to their momenta in a typical neutrino oscillation experiment, we can expand

$$E_i = \sqrt{\mathbf{p}^2 + m_i^2} \simeq |\mathbf{p}| + \frac{m_i^2}{2|\mathbf{p}|} = |\mathbf{p}| + \frac{m_i^2}{2E}, \tag{5.5}$$

which is good at least up to the next order in this expansion. For a neutrino produced as an electron flavor eigenstate at $t = 0$ we can expect for $t > 0$,

$$|\nu(t)\rangle = U^*_{\alpha i} e^{-iE_it} |\nu_i\rangle = e^{-iE_1t} \cos\theta |\nu_1\rangle + e^{-iE_2t} \sin\theta |\nu_2\rangle. \tag{5.6}$$

We can ask for the probability of a neutrino that was originally produced as a ν_e to be detected as ν_μ. This probability, $P(\nu_e \to \nu_\mu; t) = |\langle\nu_\mu|\nu(t)\rangle|^2$, is

$$\begin{aligned} P(\nu_e \to \nu_\mu; t) &= \left|(-\sin\theta\langle\nu_1| + \cos\theta\langle\nu_2|)\left(e^{-iE_1t}\cos\theta|\nu_1\rangle + e^{-iE_2t}\sin\theta|\nu_2\rangle\right)\right|^2 \\ &= 2\cos^2\theta\sin^2\theta\left|e^{-iE_1t} + e^{-iE_2t}\right|^2 = \sin^2 2\theta \sin^2\left[\frac{\Delta m_{21}^2}{4E}t\right], \end{aligned} \tag{5.7}$$

where we introduced $\Delta m_{ik}^2 = m_i^2 - m_k^2$ and used Eq. (5.5) to arrive at the answer.

There are a couple of important points that we can make from looking at Eq. (5.7). First of all, the neutrino flavor oscillation probability is only non-zero if $\Delta m_{21}^2 \neq 0$. This means that experimental observation of flavor oscillations of the neutrinos implies that the neutrinos have masses. In all oscillation experiments, neutrinos are moving with a speed that is close to the speed of light $c = 1$, covering the distance $L \approx ct$. Converting distance to kilometers and leaving the energy E in GeV a convenient approximate formula can be derived,

$$P(\nu_e \to \nu_\mu; t) = \sin^2 2\theta \sin^2\left[1.27\Delta m_{21}^2 \frac{L}{E}\right]. \tag{5.8}$$

In case of full three-flavor oscillations we can derive a formula for the probability of oscillations similar to Eq. (5.7). The probability for a neutrino of flavor α to be detected as a neutrino of flavor β, $P(\nu_\alpha \to \nu_\beta; t) = \left|U_{\beta i} U^*_{\alpha i} e^{-iE_i t}\right|^2$, is

$$\begin{aligned} P(\nu_\alpha \to \nu_\beta; t) &= \sum_{k=1,3} |U_{\beta k}|^4 \, |U_{\alpha k}|^4 \\ &+ 2 \sum_{i<k} Re \left[U_{\beta i} U^*_{\beta k} U^*_{\alpha i} U_{\alpha k}\right] \cos\left(\frac{\Delta m^2_{ik} L}{2E}\right) \\ &+ 2 \sum_{i<k} Im \left[U_{\beta i} U^*_{\beta k} U^*_{\alpha i} U_{\alpha k}\right] \sin\left(\frac{\Delta m^2_{ik} L}{2E}\right) \end{aligned} \tag{5.9}$$

We want to emphasize again that an important lesson from Eq. (5.9) is that the observation of flavor oscillations indicates that neutrinos have masses! This is a clear departure from predictions of the minimal Standard Model! We will come back to the formula of Eq. (5.9) later to see if it can be simplified. We shall note, however, that studies of flavor oscillations do not reveal the absolute masses of neutrinos, only the differences of their squared values. We would need different types of experiments to reveal the values of absolute neutrino masses. Also, Eq. (5.9) would be a good place to discuss the observation of CP-violation in the neutrino sector.

5.2.2 Neutrino masses. Dirac and Majorana constructions.

Since flavor oscillations imply massive neutrinos, it might be interesting to discuss how exactly those neutrino masses are generated. It may well be so that the mechanism, if any, that keeps them small is related to the (inverse) large scale associated with New Physics.

Dirac mass

The phrase "Dirac neutrino mass" is usually referred to a mass parameter that mixes up left-handed and right-handed chiral fermion states.

$$\mathcal{L}_D = -\sum_{i=1}^{3} m_i^{(D)} \overline{\nu}_{iR} \nu_{iL} + h.c., \tag{5.10}$$

which can be equivalently rewritten as

$$\mathcal{L}_D = -\frac{1}{2} \sum_{i=1}^{3} m_i^{(D)} \left(\overline{\nu}_{iR} \nu_{iL} + \overline{\nu}_{iL} \nu_{iR}\right) + h.c., \tag{5.11}$$

where $m_i^{(D)}$ are the neutrino masses. This construction is exactly the same as any mass terms we have considered so far. Yet, it is important to emphasize that it involves a new field, the right-handed neutrino ν_R, that is *sterile* with respect to the rest of the Standard Model: besides the expression for neutrino masses, it does not enter any other parts of the SM Lagrangian and does not interact with any other field in the Standard Model. Dirac neutrinos (antineutrinos) have lepton number $L = +1(-1)$.

Dirac neutrino mass can be generated in a similar way to any other fermions in the Standard Mode, i.e. via a usual Yukawa interaction of the type we already discussed in Eq. (2.25),

$$\begin{aligned} \mathcal{L}_D &= -Y^\nu \overline{L}_L \widetilde{H} \nu_R + h.c. \\ &= -\frac{Y^\nu v}{\sqrt{2}} \overline{\nu}_L \nu_R + h.c. \end{aligned} \tag{5.12}$$

The Dirac neutrino mass would then be

$$m^{(D)} = -\frac{Y^\nu v}{\sqrt{2}}. \tag{5.13}$$

It follows that the Yukawa couplings associated with neutrinos are extremely small, $Y^\nu \sim \sqrt{2}m_\nu^{(D)}/v \sim 10^{-12}$! While technically there is nothing wrong with such small values of Yukawa couplings, it is rather unsettling that they span twelve orders of magnitude for various SM fermions!

Majorana mass

Dirac mass, defined by Eq. (5.10), requires both left- and right-handed neutrino fields to be present. In the minimal Standard Model, however, there are no right-handed neutrino fields. Naively, that means that no neutrino mass terms could be generated. However, one might ask a generic question: is it possible to construct a field "ψ_R" that, for all practical purposes, serves the role of the right-handed field, but is constructed from a left-handed field ψ_L? It turns out that it is indeed possible, as E. Majorana pointed out in 1937. Let us follow the argument by S. Boyd and construct such a field using the Dirac equation. Since any spinor field ψ can be written in terms of its left- and right-handed components, $\psi = \psi_L + \psi_R$, a Lagrangian for the field ψ with the mass m can be written in terms of the left- and right-handed fields,

$$\begin{aligned} \mathcal{L} &= \overline{\psi}\left(i\partial\!\!\!/ - m\right)\psi \\ &= \overline{\psi_R} i\partial\!\!\!/ \psi_R - m\overline{\psi_L}\psi_R + \overline{\psi_L} i\partial\!\!\!/ \psi_L - m\overline{\psi_R}\psi_L. \end{aligned} \tag{5.14}$$

This fact leads to two coupled Dirac equations, which follow from Euler-Lagrange procedure applied to the Lagrangian of Eq. (5.14),

$$\begin{aligned} i\partial\!\!\!/ \psi_L &= m\psi_R, \\ i\partial\!\!\!/ \psi_R &= m\psi_L. \end{aligned} \tag{5.15}$$

Taking Hermitian conjugation of the second equation in Eq. (5.15) and multiplying the result on the right by γ^0 we get

$$-i\partial_\mu \psi_R^\dagger \gamma^{\mu\dagger}\gamma^0 = m\psi_L^\dagger \gamma^0. \tag{5.16}$$

Using a property of gamma matrices $\gamma^{\mu\dagger}\gamma^0 = \gamma^0\gamma^\mu$ we obtain

$$-i\partial_\mu \overline{\psi_R}\gamma^\mu = m\overline{\psi_L}. \tag{5.17}$$

Transposing both sides of this equation and using the property of the charge conjugation operator $\mathcal{C}\gamma^{\mu T} = -\gamma^\mu \mathcal{C}$ we obtain

$$i\partial\!\!\!/ \, \mathcal{C}\,\overline{\psi_R}^T = m\,\mathcal{C}\,\overline{\psi_L}^T. \tag{5.18}$$

Now, if we compare Eq. (5.18) to the first equation in Eq. (5.15) we see that in order to have the same structure we have to identify*

$$\psi_R = \mathcal{C}\overline{\psi_L}^T \equiv \psi_L^c. \tag{5.19}$$

This is amazing! Indeed, one can show that for the left-handed projector P_L the field indeed satisfies the projection equation,

$$P_L\psi_R = \left[\mathcal{C}\overline{\psi_L}^T\right] = 0. \tag{5.20}$$

Note that the notation in terms of the charge-conjugated field ψ_L^c is very common. Purists of field theory would claim at this point that the most natural description of the Majorana field

*This statement is clearly only possible for an electrically neutral field.

can be obtained using a two-component spinor formalism. They would, indeed, be correct! Yet, we will be using more familiar four-spinor formalism.

Now that we have written a right-handed field in terms of its left-handed projection, we can write the whole Majorana field ψ as

$$\psi = \psi_L + \psi_L^c. \tag{5.21}$$

Taking charge conjugation of ψ

$$\psi^c = (\psi_L + \psi_L^c)^c = \psi_L^c + \psi_L = \psi, \tag{5.22}$$

we note that the field ψ is *self-conjugated.* In other words, Majorana neutrino is its own antiparticle! Since the SM neutrinos and antineutrinos have opposite lepton quantum numbers, it immediately implies that the interactions of Majorana neutrinos violate lepton number conservation by two units.

The discussion above implies that the mass term in the neutrino Lagrangian can also be written as

$$\mathcal{L}_M = -\frac{1}{2}\sum_{i=1}^{3} m_i^{(M)}\ \overline{\nu_{iL}^c}\ \nu_{iL} + h.c., \tag{5.23}$$

The factor of one-half is there to account for double counting since the displayed mass term is identical to its Hermitian conjugate term.

A comment is in order. All particles in the Standard Model receive their masses via interactions with the Higgs field. However, since the SM left-handed neutrino fields are assigned weak isospin $I = 1/2$ and weak hypercharge $Y = -1$, a Yukawa-like term like $\overline{\nu_L^C} H \nu_L$ is not a singlet under $SU(2)_L \times U(1)$ gauge symmetry. That is unless the H field has weak isospin $I = -1$ and weak hypercharge $Y = -2$. But then H is not the SM Higgs field! Thus, experimental confirmation of the Majorana nature of the neutrino mass would clearly signal the presence of physics beyond the Standard Model.

It is possible to generate Majorana neutrino mass at higher orders in SM EFT by adding more Higgs doublets. It already works by adding only one Higgs field, i.e. at the level of dimension-5 operators. We discussed this fact in Chapter 2.1. There is a single operator of dimension 5, the Weinberg operator, which can generate Majorana mass for the neutrinos. This operator defines the SM EFT operator class $\psi^2 H^2$, as in Eq. (2.76). It is built out of two fermion and two Higgs fields, both of which transform as doublets under the electroweak $SU(2)$ group. For the flavor-diagonal term $p = q$ we get after spontaneous symmetry breaking

$$\mathcal{L}^{(5)} = \frac{C^{(5)}}{\Lambda}\epsilon_{jk}\epsilon_{mn} H^j H^m \left(L^k\right)^T C L^n = -\frac{C^{(5)} v^2}{2\Lambda}\nu_L \nu_L^c. \tag{5.24}$$

It is interesting to note that $\Lambda > 10^{14}$ GeV if neutrino masses $m_\nu^{(M)} < 1$ eV and setting $C^{(5)} \sim 1$. This scale is unlikely to be probed by particle accelerators any time soon.

To conclude this section we point out that similar construction is possible for the right-handed neutrinos as well. That is to say, if right-handed neutrinos do exist, we can write out a Majorana mass term for them in total analogy with the construction described above,

$$\mathcal{L}'_M = -\frac{1}{2}\sum_{i=1}^{3} m_{R,i}^{(M)}\ \overline{\nu_{iR}^c}\ \nu_{iR} + h.c. \tag{5.25}$$

The idea for such construction seems weird though, as the whole idea behind constructing Majorana mass term was to get away from the introduction of right-handed neutrinos! But right-handed neutrinos are the SM singlets, which means that they do not receive their mass from the Higgs mechanism. What this means is that their mass can, in principle, take any value. It can even be very large, say, $\sim 10^9$ GeV! While this fact might seem to only exacerbate the problem of small neutrino masses, it actually offers a nice way to solve it!

5.2.3 Neutrino mass puzzle: see-saw mechanism

While technically it does not present a problem, one might be curious as to why neutrino masses are so much smaller than the masses of all other fermions in the Standard Model. An elegant solution for this is provided by a see-saw mechanism.

Let us collect the information presented in the previous two sections. Neutrinos can have a Dirac mass, but in order for them to acquire it, right-handed neutrinos have to be introduced. To ensure that the Dirac neutrino masses come out small compared to all other fermion masses, ultra-small Yukawa couplings need to be introduced. We can also introduce Majorana masses for both left-handed and right-handed neutrinos. The LH neutrino Majorana mass is automatically small, as it is generated only by a dimension-five operator in SM EFT. Yet, the RH Majorana neutrino mass can be very large. Assuming that neutrinos have both Dirac and Majorana masses, the total mass term can be summarized in the following matrix Lagrangian,

$$\mathcal{L}_M = -\frac{1}{2}\left(\overline{\nu_L^c},\ \overline{\nu_R}\right)\begin{pmatrix} m_L^{(M)} & m^{(D)} \\ m^{(D)} & m_R^{(M)} \end{pmatrix}\begin{pmatrix} \nu_L \\ \nu_R^c \end{pmatrix}, \tag{5.26}$$

where we combined Eqs. (5.11) and (5.23) into one matrix equation. The mass matrix is not diagonal in Eq. (5.26). To obtain physical propagating neutrino mass eigenstates we must diagonalize it. The diagonal values are given by

$$m_{1,2} = \frac{m_L^{(M)} + m_R^{(M)}}{2} \pm \frac{1}{2}\sqrt{\left(m_L^{(M)} - m_R^{(M)}\right)^2 + 4m_D^2}\ . \tag{5.27}$$

If Dirac masses generated by the Higgs mechanism yield values comparable to quark or lepton masses in the MeV range, LH Majorana masses are tiny (so they can be neglected, $m_R^{(M)} = 0$), and RH Majorana masses are very large, then the following hierarchy of physical neutrino masses can be observed:

$$\begin{aligned} m_1 &= \frac{m_D^2}{m_R^{(M)}}, \\ m_2 &= m_R^{(M)}\left(1 + \frac{m_1}{m_R^{(M)}}\right) \approx m_R^{(M)}. \end{aligned} \tag{5.28}$$

which means that if $m_D \sim 10^3$ eV, and $m_R^{(M)} \sim 10^{15}$ eV, then small value of the neutrino mass $m_1 \sim 10^{-3}$ eV is automatically generated. The described construction goes under the name the *see-saw mechanism.*

5.2.4 Experimental methods of detection: neutrinoless double-β decay

One of the most interesting questions in neutrino physics is the nature of neutrino mass. Being the only neutral fermion in the Standard Model, is it of Majorana type? As we can see no neutrino oscillation experiment can answer this question at the moment because the lepton number is conserved in the oscillation experiments.

In order to probe the Majorana nature of neutrino mass, experiments that probe violation of lepton number must be pursued. Such experiments are those looking for neutrinoless double-β decay [25], or the transition

$$\beta\beta_{0\nu}: \qquad {}^A_Z X \rightarrow {}^A_{Z+2} X + 2e^-, \tag{5.29}$$

where ${}^A_Z X$ stands for a nucleus with the atomic number Z and atomic mass A. A candidate for such nuclei is ${}^{76}_{32}Ge$. A quark-level Feynman diagram is shown in Fig. 5.1. It must be proportional

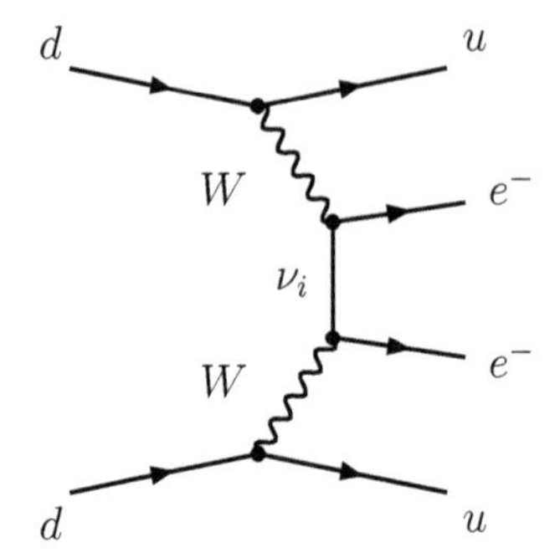

FIGURE 5.1 Quark-level transition that induces neutrinoless double beta decay.

to $m_{\nu_i} G_F^2$, an extremely rare process involving several scales, from the nuclear scale of nuclei transition to extremely high scale of neutrino mass formation. This make the computation of the rate non-trivial [46, 64].

5.3 Neutrinos and leptonic CP violation

Neutrino oscillation experiments study the time evolution of neutrinos. In a sense, this time evolution is quite similar to the time evolution of neutral mesons considered in the previous chapter. It would then be interesting to know if there are observables or combinations of observables that signal CP-violation in the leptonic system. Comparing the probability of the oscillation process with its CP-conjugate would do just that, so

$$P\left(\nu_\alpha \to \nu_\beta\right) \neq P\left(\overline{\nu}_\alpha \to \overline{\nu}_\beta\right). \tag{5.30}$$

To derive the probability of antineutrino oscillations, let us take another look at Eq. (5.9). As we discussed in Sec. 2.3, CP-transformation of the oscillation probability is equivalent to replacement of the mixing matrix U with its complex conjugate matrix U^* in the expression of that probability. As can be seen, this results in a very similar expression, but with one exception: the coefficient in front of the term proportional to $\sin\left(\Delta m_{ik}^2 L/(2E)\right)$ changes sign! That means that it is possible to see CP-violation in neutrino oscillations, provided the matrix U is complex.

Since the CPT theorem requires that the process and its CPT-conjugate one have the same probabilities to occur,

$$P\left(\nu_\alpha \to \nu_\beta\right) = P\left(\overline{\nu}_\beta \to \overline{\nu}_\alpha\right), \tag{5.31}$$

we can also formulate a condition at which T-symmetry would also be broken,

$$P\left(\nu_\alpha \to \nu_\beta\right) \neq P\left(\nu_\beta \to \nu_\alpha\right). \tag{5.32}$$

Again, looking at Eq. (5.9), we note that the time-reversed expression for the oscillation probability can be obtained by interchanging $\alpha \leftrightarrow \beta$. Examining Eq. (5.9) we conclude that the resulting expression is equivalent to $P\left(\nu_\alpha \to \nu_\beta\right)$ with one exception: the coefficient in front of $\sin\left(\Delta m_{ik}^2 L/(2E)\right)$ changes sign. In other words, time reversal leads to the same observable effects in the oscillation probability as CP-transformation, which is in full accord with the CPT theorem.

Let us now look at the most general parameterization of the PMNS matrix to see how many complex phases it contains.

5.3.1 CP-violation: Phases of the PMNS mixing matrix

Let's recall our discussion from Sec. 2.4. There, we concluded that the only source of CP-violation in the quark sector of the Standard Model is the (sole) phase of the CKM quark mixing matrix. A similar discussion can be held in the case of neutrino mixing matrix, so in order to study CP-violation in neutrino system we need to see how many CP-violating phases one can have in the Pontecorvo-Maki-Nakagawa-Sakata neutrino mixing matrix. While the answer to this question might at first seem trivial: same as in the case of the CKM matrix, we should check what happens if neutrinos have Majorana masses. We shall address this below, but before doing so let us state the answer: in the most general case of N neutrino flavors, one can have $N(N-1)/2$ mixing angles and $(N-1)(N-2)/2$ "Dirac phases". If neutrino masses are of the Majorana type, there are $(N-1)$ additional "Majorana" phases.

Indeed, we can follow the discussion of Sec. 2.4 to state that a complex $N \times N$ matrix contains $2N^2$ real parameters. As the neutrino mixing matrix is unitary, the unitarity condition $U_{\alpha k} U^\dagger_{k\beta} = \delta_{\alpha\beta}$ puts N^2 conditions on the elements of U. As explained in Sec. 2.4, $N(N+1)/2$ of those parameters are classified as angles and the rest $N^2 - N(N+1)/2 = N(N-1)/2$ as phases. Not all of those phases are physical, so some can be removed by rephasing the neutrino flavor or mass eigenstates. This precisely where the difference between the number of phases of Dirac and Majorana neutrinos shows up.

Let's consider $N \times N$ generalization of Eq. (5.4),

$$\begin{pmatrix} |\nu_e\rangle \\ |\nu_\mu\rangle \\ |\nu_\tau\rangle \\ \cdots \end{pmatrix} = \begin{pmatrix} U^*_{e1} & U^*_{e2} & U^*_{e3} & \cdots \\ U^*_{\mu 1} & U^*_{\mu 2} & U^*_{\mu 3} & \cdots \\ U^*_{\tau 1} & U^*_{\tau 2} & U^*_{\tau 3} & \cdots \\ \cdots & \cdots & \cdots & \cdots \end{pmatrix} \begin{pmatrix} |\nu_1\rangle \\ |\nu_2\rangle \\ |\nu_3\rangle \\ \cdots \end{pmatrix}. \tag{5.33}$$

We can always redefine the unobservable phases of the neutrino matrix elements by rephasing each flavor eigenstate. In the case of $N \times N$ neutrino mixing matrix we can charge phases of each flavor eigenstate, $|\nu_\alpha\rangle \to e^{i\phi_\alpha} |\nu_\alpha\rangle$. This would change phases of the neutrino production (or detection) matrix elements, as neutrinos are produced as flavor eigenstates. This removes N phases.

In the case of the Dirac neutrino masses, we can rephase the mass eigenstates, which can make one row of the matrix U in Eq. (5.33) real. This removes additional $N-1$ phases. Note that there are only $N-1$ conditions, not N, as the overall phase of that row is already fixed by rephasing of the corresponding flavor eigenstate. This leads to $N(N+1)/2 - N - (N-1) = (N-1)(N-2)/2$ physical phases, in complete agreement with our parameter counting in Sect. 2.4!

Parameter counting is slightly different in the case of the Majorana neutrinos. While the rephasing of the flavor eigenstates is unchanged and also leads to the elimination of N parameters, the rephasing of the mass eigenstates is not possible. This is so because the Majorana mass term of Eq. (5.23) contains two left-handed fields, so rephasing $\nu_{L,i}$ in $M(\nu^c_{L,i}\nu_{L,i} + h.c.)$ does not make the mass real! Thus, the number of phases remaining in the Majorana case is $N(N+1)/2 - N = N(N-1)/2$! Another way to say it is that there are additional $N-1$ "Majorana phases".

In the physically interesting case of three generations one has one "Dirac" and two "Majorana" phases present. In the case of two generations, there are no "Dirac" phases present, but there is one "Majorana" phase. It is important to note that the Majorana phases cancel in flavor evolution, as they affect an entire column of the PMNS matrix. A three-generation PNMS matrix can be written in a way similar to the "standard" parameterization of the CKM matrix of Eq. (2.52), with the exception of the Majorana phases,

$$U = \begin{pmatrix} 1 & 0 & 0 \\ 0 & c_{23} & s_{23} \\ 0 & -s_{23} & c_{23} \end{pmatrix} \begin{pmatrix} c_{13} & 0 & s_{13}e^{-i\delta} \\ 0 & 1 & 0 \\ -s_{13}e^{i\delta} & 0 & c_{213} \end{pmatrix} \begin{pmatrix} c_{12} & s_{12} & 0 \\ -s_{12} & c_{12} & 0 \\ 0 & 0 & 1 \end{pmatrix} \begin{pmatrix} e^{i\alpha_1/2} & 0 & 0 \\ 0 & e^{i\alpha_2/2} & 0 \\ 0 & 0 & 1 \end{pmatrix}$$
$$= \begin{pmatrix} c_{12}c_{13} & s_{12}c_{13} & s_{13}e^{-i\delta} \\ -s_{12}c_{23} - c_{12}s_{13}s_{23}e^{i\delta} & c_{12}c_{23} - s_{12}s_{13}s_{23}e^{i\delta} & c_{13}s_{23} \\ s_{12}s_{23} - c_{12}s_{13}c_{23}e^{i\delta} & -c_{12}s_{23} - s_{12}s_{13}c_{23}e^{i\delta} & c_{13}c_{23} \end{pmatrix} \begin{pmatrix} e^{i\alpha_1/2} & 0 & 0 \\ 0 & e^{i\alpha_2/2} & 0 \\ 0 & 0 & 1 \end{pmatrix}, \quad (5.34)$$

where $c_{ij} = \cos\theta_{ij}$, $s_{ij} = \sin\theta_{ij}$, δ is the Dirac phase, and α_i are the Majorana phases. Because of the experiments that are probing different angles, the first matrix in Eq. (5.34) is sometimes called the "atmospheric" mixing matrix, the second one – the "reactor" mixing matrix, and the third one – the "solar" mixing matrix. It must be pointed out that, contrary to the quark CKM mixing matrix, the two angles, θ_{12} and θ_{23}, are quite large.

One can explicitly see it in our example of two-generation mixing: while one Majorana phase could, in fact, be present, we did not explicitly include any phases in the 2×2 PMNS matrix of Eq. (5.4). A single possible Majorana phase would appear twice, multiplying both elements either in the first or in the second column. In calculating the oscillation probability of Eq. (5.7) each of those elements is multiplied by its complex-conjugated, resulting in no observable physical effects in the flavor oscillation experiments.

5.3.2 Neutrino oscillations, again

Let us come back to the oscillation probability of Eq. (5.9). It clearly has a structure: a constant term and oscillating sine and cosine contributions. Using the unitarity of the PMNS matrix it is easy to show that the constant term would be equal to one for the flavor-diagonal probabilities $\alpha = \beta$ and zero for $\alpha \neq \beta$. Let us introduce the constants,

$$\begin{aligned} J_{ik}^{\alpha\beta} &= -Im\left[U_{\alpha i}U_{\alpha k}^* U_{\beta i}^* U_{\beta k}\right], \\ A_{\alpha\beta}^{ik} &= -4Re\left[U_{\alpha i}U_{\beta i}^* U_{\alpha k}^* U_{\beta k}\right]. \end{aligned} \quad (5.35)$$

As it turns out (see problem 2 at the end of this chapter), anti-symmetric properties of the matrix $J_{ik}^{\alpha\beta}$ in both upper and lower indices allows relating all matrix elements of $J_{ik}^{\alpha\beta}$ to a single one,

$$\begin{aligned} J &= J_{12}^{e\mu} = -Im\left[U_{e1}U_{\mu1}^* U_{e2}^* U_{\mu2}\right] \\ &= c_{13}^2 2 s_{13}s_{12}c_{12}s_{23}c_{23}\sin\delta. \end{aligned} \quad (5.36)$$

called the Jarlskog invariant. CP-violation is manifested by the non-zero phases of the PMNS matrix. The Jarlskog invariant is zero if CP is conserved in the neutrino sector.

Then the most general form for the probability of detecting neutrino of flavor β if neutrino of flavor α has been emitted,

$$\begin{aligned} P(\nu_\alpha \to \nu_\beta) &= A_{\alpha\beta}^{12}\sin^2\left(\frac{\Delta_{12}^2 L}{4E}\right) + A_{\alpha\beta}^{23}\sin^2\left(\frac{\Delta_{23}^2 L}{4E}\right) + A_{\alpha\beta}^{13}\sin^2\left(\frac{\Delta_{13}^2 L}{4E}\right) \\ &\pm 8\,J\sin\left(\frac{\Delta_{12}^2 L}{4E}\right)\sin\left(\frac{\Delta_{23}^2 L}{4E}\right)\sin\left(\frac{\Delta_{13}^2 L}{4E}\right). \end{aligned} \quad (5.37)$$

As follows from the Eq. (5.37), the observable CP-violating effects are possible if all angles θ_{ik} and a Dirac phase δ are non-trivial, i.e. not equal to 0 or π. In addition, the oscillation parameters $\left(\Delta_{ij}^2 L/(4E)\right)$ should also be sufficiently different from zero, such that experimental observation is possible.

5.3.3 Experimental methods of detection: neutrino oscillations

Experimental techniques for studies of neutrino properties are well developed. Methods of detections vary depending on the source of neutrinos: solar, atmospheric, reactor, and accelerator experiments vary in size and complexity. A multitude of techniques developed for neutrino physics is well described in a variety of texts and reviews. We will quote only a few of them, e.g. [88,130].

5.4 Electromagnetic properties of neutrinos

How dark are neutrinos? Even though they are electrically neutral and, at least at the scales smaller than a TeV, point-like, they can still participate in electromagnetic interactions. Similarly to electrically neutral atoms, they in fact can interact with the SM gauge fields. The effective interactions of neutrinos with gauge fields are fairly restricted. Yet, quantum loop effects containing charged particles can induce such electromagnetic characteristics as electrical or magnetic dipole moments. These properties would be especially interesting if those charged particles did not exist in the Standard Model spectrum. Thus, studies of the electromagnetic properties of neutrinos provide us with a powerful tool in searches for physics beyond the Standard Model. While it would be admittedly difficult to study electromagnetic interactions of neutrinos produced in a lab, astrophysical neutrinos provide plenty of such opportunities, as they propagate large distances in electromagnetic fields of astrophysical origins, such as galactic magnetic fields. It would be also interesting to see if the electromagnetic properties of Dirac neutrinos were different from those of their Majorana counterparts.

5.4.1 Interactions with a single photon

To determine how neutrinos interact electromagnetically, we shall write the most general expression for the electromagnetic current operator in a way similar to Eq. (3.3). If we, for now, concentrate on the case of neutrino interactions with a single photon, we can write

$$\langle \nu(p_f)|j^\mu|\nu(p_i)\rangle = \overline{u}(p_f)\Gamma^\mu(P,q)u(p_i), \tag{5.38}$$

where, as before, $q = p_f - p_i$, and $\Gamma^\mu_\nu(P,q)$ is a vertex function. We can write the most general decomposition of the vertex function as

$$\begin{aligned} \Gamma^\mu(P,q) \equiv \Gamma^\mu(q) &= F_1(q^2)q^\mu + F_2(q^2)q^\mu\gamma_5 + F_3(q^2)\gamma^\mu \\ &+ F_4(q^2)\gamma^\mu\gamma_5 + F_5(q^2)\sigma^{\mu\nu}q_\nu + F_6(q^2)\epsilon^{\mu\nu\alpha\beta}q_\nu\sigma_{\alpha\beta}, \end{aligned} \tag{5.39}$$

where the choice of the Dirac structure for the $F_6(q^2)$ form factor was made for convenience of the arguments that will follow momentarily.

Not all of the form factors in Eq. (5.39) are independent. As follows from the requirement of gauge invariance, the electromagnetic current must be a conserved quantity, $\partial_\mu j^\mu = 0$, which implies for the matrix element in Eq. (5.38) that

$$q_\mu \, \overline{u}(p_f)\Gamma^\mu(P,q)u(p_i) = 0. \tag{5.40}$$

We can see from Eq. (5.39) that

$$F_1(q^2)q^2 + F_1(q^2)q^2\gamma_5 + 2m_\nu F_4(q^2)\gamma_5 = 0. \tag{5.41}$$

While for the on-shell photon this immediately implies that $F_4(q^2) = 0$, we can actually derive a slightly more useful relation if we recall that the first term is multiplied by a 4×4 unit matrix $\mathbb{1}$ in the Dirac space, and $\mathbb{1}$ and γ_5 are linearly independent, so

$$F_1(q^2) = 0, \quad F_4(q^2) = -\frac{q^2}{2m}F_2(q^2). \tag{5.42}$$

Thus, the vertex function Γ^μ of Eq. (5.39) can be defined in terms of four form factors, which makes it consistent with the requirements of Lorentz and gauge invariance.

Another requirement that the neutrino EM current must satisfy is hermiticity. This follows from the fact that the Lagrangian and the electromagnetic field to which the current couples are Hermitian, so the matrix elements of Eq. (5.38) are such that

$$\langle \nu(p_f)|j^\mu|\nu(p_i)\rangle = \langle \nu(p_i)|j^\mu|\nu(p_f)\rangle^*. \tag{5.43}$$

This implies that the function $\Gamma_\mu(q)$ satisfies the following condition [89],

$$\Gamma_\mu(q) = \gamma^0 \Gamma_\mu(-q)^\dagger \gamma^0. \tag{5.44}$$

The condition of Eq. (5.44), applied to Eq. (5.39) allows us to conclude that the form factor F_2 is real, while the form factors F_3, F_5, and F_6 are purely imaginary. Keeping this in mind, it would be more convenient to redefine the form factors as

$$\begin{aligned} \Gamma^\mu(q) &= F_Q(q^2)\,\gamma^\mu - F_M(q^2)\,i\sigma^{\mu\nu}q_\nu \\ &+ F_E(q^2)\,\sigma^{\mu\nu}q_\nu\gamma_5 + F_A(q^2)\left(q^2\gamma^\mu - q^\mu \not{q}\right)\gamma_5, \end{aligned} \tag{5.45}$$

where we defined $F_Q = F_3$ (charge form factor), $F_M = iF_5$ (magnetic dipole form factor), $F_E = -2iF_6$ (electric dipole form factor), and $F_A = -F_2/(2m)$ (anapole form factor).

We can also determine the physical properties of neutrino that those form factors define at $q^2 = 0$. For the real photon, $F_Q(0) = 0$, as neutrino is not electrically charged. $F_M(0) = \mu^{(\nu)}$ represents neutrino magnetic dipole moment, while $F_E(0) = d^{(\nu)}$ is neutrino electric dipole moment. $F_A(0)$ gives neutrino's anapole moment.

It is also interesting to see how the form factors behave under discrete symmetry transformations. As shown in Table 2.2, a fermionic vector current is odd under the combined CP transformation. Also, as it follows from Eq. (2.41), the electromagnetic field A_μ is odd under CP transformation. The combination of those two facts, together with the simplicity of the QED vacuum, implies that CP is conserved in electromagnetic interactions.

If we are to require that CP is conserved in one-photon electromagnetic interactions of neutrinos, we must also require that the current in Eq. (5.38) would also be odd under CP transformation. This implies for the vertex function Γ_μ,

$$\Gamma_\mu(q) \xrightarrow{CP} -\Gamma_\mu(q_P). \tag{5.46}$$

Alternatively, we could see what would happen if we explicitly apply the constraints Eq. (2.42) to $j^\mu(x)$,

$$\Gamma_\mu(q) \xrightarrow{CP} \gamma^0 C \Gamma_\mu^T(q_P) C^\dagger \gamma^0, \tag{5.47}$$

where we defined $q_P^\mu = q_\mu$. Application of transformation described in Eq. (5.47) to Eq. (5.45) leads to

$$\begin{aligned} \Gamma^\mu(q) \xrightarrow{CP} & -\left[F_Q(q^2)\,\gamma^\mu - F_M(q^2)\,i\sigma^{\mu\nu}q_\nu\right. \\ & \left. - F_E(q^2)\,\sigma^{\mu\nu}q_\nu\gamma_5 + F_A(q^2)\left(q^2\gamma^\mu - q^\mu\not{q}\right)\gamma_5\right], \end{aligned} \tag{5.48}$$

Comparing Eq. (5.48) with Eq. (5.46) we must conclude that, as in previous chapters, $F_E(q^2) = 0$, provided that CP is required to be conserved. The contrary is also true: if CP is broken then $F_E(q^2)$ does not vanish! This would imply that neutrino's EDM would also be non-zero.

5.4.2 Higher order interactions: Rayleigh operators

At higher orders, neutrinos can have interactions that lead to processes similar to light scattering off the neutral atoms, the so-called Rayleigh process. Such scattering can be induced by the dimension-7 "Rayleigh operators",

$$\mathcal{L} \supset \frac{C^{(7)}_{1,\alpha\beta}}{\Lambda^3}\frac{\alpha}{12\pi}(\bar{\nu}_\alpha P_L \nu_\beta)F_{\mu\nu}F^{\mu\nu} + \frac{C^{(7)}_{2,\alpha\beta}}{\Lambda^3}\frac{\alpha}{8\pi}(\bar{\nu}_\alpha P_L \nu_\beta)F_{\mu\nu}\tilde{F}^{\mu\nu}. \tag{5.49}$$

The most important difference is that these operators are composed of two Lorentz scalars, $\bar{\nu}\nu$ and FF (or $F\widetilde{F}$). It is thus possible that the NP can contribute very differently to the Rayleigh operators than it does to neutrino magnetic moments and to neutrino mass matrix.

5.5 Standard neutrino interactions

Neutrinos that are observed in the oscillation experiments are not produced in the vacuum. For example, neutrinos produced in the Sun have to traverse inside until they reach the vacuum of space. The presence of matter in which neutrinos (or antineutrinos) travel, leads to modification of the oscillation formulas derived earlier in this chapter. In this section we concentrate on such effects, limiting our attention to the Standard Model interactions.

5.5.1 Neutrinos in matter: propagation and MSW effect

Neutrinos propagating in the matter can experience charged-current (CC) interactions, as well as neutral-current (NC) interactions. We will not consider NC interactions in this section because they lead to terms that are irrelevant for the oscillation probabilities. As for the CC interactions, it is important to consider two different classes of those. First, *incoherent* scatterings on atoms simply lead to a decrease in neutrino beam's intensity. The amplitude of each incoherent scattering is suppressed by the Fermi constant G_F and, in general, can be neglected.

The logic changes when *coherent* scattering is present. While the amplitude for coherent forward scattering is also suppressed by G_F , it can be enhanced by a large number of scattering centers. Thus, it can potentially give observable contributions to the oscillation probabilities. Thus, matter effects are important if the matter densities are large, like in the Sun, where electron neutrinos are produced. The constant density of matter gives a decent approximation, so this is the case we will be considering. If the density of matter changes along the path of neutrinos, the effective mixing of neutrinos grows up to some maximum value at some value of the density matter, which can lead to a resonant conversion of one type of neutrinos to another one.

In order to see the effect of neutrino-matter scattering in the SM, let us recall that ordinary matter contains atomic electrons. Charged current scattering off those electrons is described in SM EFT by a Fermi model Lagrangian,

$$\begin{aligned}\mathcal{L}_{CC} &= -\frac{4G_F}{\sqrt{2}}\left(\bar{e}_L\gamma^\mu \nu_{eL}\right)\left(\overline{\nu_e}_L\gamma_\mu e_L\right) \\ &= -\frac{G_F}{\sqrt{2}}\left(\bar{e}\gamma^\mu\left(1-\gamma_5\right)e\right)\left(\overline{\nu_e}\gamma^\mu\left(1-\gamma_5\right)\nu_e\right),\end{aligned} \tag{5.50}$$

where we used Fierz identity to group neutrinos to one current. Neutrinos propagating through matter interact with electrons, whose collective effect (mean-field potential) can be parameterized by computing the expectation value of an electron current in a "background matter field." Since the ordinary matter inside the Sun and inside the Earth is unpolarized, we can integrate over all directions. This implies that only one component of the electron current survives,

$$\langle \bar{e}\gamma^0 e\rangle = \langle e^\dagger e\rangle = N_e, \tag{5.51}$$

where the second equality follows from the fact that $e^\dagger e$ is just an electron number operator counting the number of electrons.

It then follows that the motion of electron neutrinos (antineutrinos) through matter would be equivalent to the motion in mean-field potential. This leads to the contribution to the energy of a single neutrino (antineutrino) that is given by

$$V_{\nu_e/\overline{\nu}_e} = \pm\sqrt{2}G_F N_e. \tag{5.52}$$

Note that the antineutrino's potential acquires an additional minus sign, which is related to the fact that the weak isospin of an antineutrino has an opposite sign to that of a neutrino.

Now let us see the effect of matter interactions on a simpler case of two-flavor neutrino mixing. First, let us rewrite the Schrödinger equation of Eq. (5.1). Since neutrinos move with the speed close to the speed of light, $x = ct \equiv t$ in natural units, so

$$i\frac{d}{dx}\begin{pmatrix}|\nu_e\rangle \\ |\nu_\mu\rangle\end{pmatrix} = \mathcal{H}\begin{pmatrix}|\nu_e\rangle \\ |\nu_\mu\rangle\end{pmatrix}. \tag{5.53}$$

where the Hamiltonian $\mathcal{H}$ now acquires a part that corresponds to the matter potential V, $\mathcal{H} = \mathcal{H}_{\text{vac}} + V$, which is different for different neutrino flavors. As we have seen from Eq. (5.1), the vacuum Hamiltonian is diagonal in the mass basis, so in the flavor basis it is

$$\mathcal{H}_{\text{vac}} = U\begin{pmatrix}E_1 & 0 \\ 0 & E_2\end{pmatrix}U^\dagger \tag{5.54}$$

where the energies E_i are given in Eq. (5.5). Just for the sake of illustration, let us show that the Hamiltonian can be manipulated if we remember that one can always redefine a matrix Hamiltonian by subtracting a diagonal matrix. That diagonal matrix plays no part in oscillations, so can be dropped! In other words,

$$\begin{aligned}
\mathcal{H}_{\text{vac}} &= \frac{1}{2E}U\begin{pmatrix}m_1^2 & 0 \\ 0 & m_2^2\end{pmatrix}U^\dagger + \lambda I \\
&= \frac{\Delta m_{21}^2}{2E}\begin{pmatrix}\cos\theta & \sin\theta \\ -\sin\theta & \cos\theta\end{pmatrix}\begin{pmatrix}-1 & 0 \\ 0 & +1\end{pmatrix}\begin{pmatrix}\cos\theta & -\sin\theta \\ \sin\theta & \cos\theta\end{pmatrix} + \lambda' I \\
&= \begin{pmatrix}-\cos 2\theta & \sin 2\theta \\ \sin 2\theta & \cos 2\theta\end{pmatrix} + \lambda' I,
\end{aligned} \tag{5.55}$$

where I is a 2×2 unit matrix and $\lambda^{(\prime)}$ are some functions of angles and other quantities. As we pointed out above, any terms proportional to the unit matrix can be dropped, so the exact form of $\lambda^{(\prime)}$ are irrelevant. Now, the matter potential term can go through the same set of transformations,

$$\begin{aligned}
\mathcal{H} &= \begin{pmatrix}-\cos 2\theta & \sin 2\theta \\ \sin 2\theta & \cos 2\theta\end{pmatrix} + \begin{pmatrix}V_{\nu_e} & 0 \\ 0 & 0\end{pmatrix} \\
&= \frac{V}{\beta}\begin{pmatrix}-\cos 2\theta + \beta & \sin 2\theta \\ \sin 2\theta & \cos 2\theta\end{pmatrix},
\end{aligned} \tag{5.56}$$

where $V \equiv V_{\nu_e}$, $\beta = 2EV/\Delta m_{21}^2$, and we, once again, dropped all terms that are proportional to the unit matrix. In Eq. (5.56), i.e. neglected flavor-independent contributions that affect the phase factors to the S-matrix. Once again, the Hamiltonian is no longer diagonal, so in order to define propagating eigenstates we need to diagonalize it. It is rather easy to do so for the 2×2 case, so we write the Hamiltonian as

$$\mathcal{H} = U_m\begin{pmatrix}E_{1M} & 0 \\ 0 & E_{2M}\end{pmatrix}U_m^\dagger, \tag{5.57}$$

where U_M is the PMNS mixing matrix in matter, where the vacuum mixing angle is replaced by θ_M,

$$\begin{aligned}\sin 2\theta_M &= \frac{\sin 2\theta}{\sqrt{\sin^2 2\theta + (\cos 2\theta - \beta)^2}},\\ \cos 2\theta_M &= \frac{\cos 2\theta - \beta}{\sqrt{\sin^2 2\theta + (\cos 2\theta - \beta)^2}}.\end{aligned} \tag{5.58}$$

The eigenvalues of the Hamiltonian in Eq. (5.56) are

$$E_{(1,2)M} = \frac{V_{\nu_e}}{2} + \frac{V}{\beta}\left(1 \mp \sqrt{\sin^2 2\theta + (\cos 2\theta - \beta)^2}\right). \tag{5.59}$$

It is easy to see that in the limit $V \to 0$ ($\beta \to 0$) we recover the vacuum oscillation solution.

The oscillation probability in the matter would become,

$$P(\nu_e \to \nu_\mu) = \sin^2 2\theta_M \sin^2\left(\frac{\Delta m_{21}^2 L}{4E}\sqrt{\sin^2 2\theta + (\cos 2\theta - \beta)^2}.\right) \tag{5.60}$$

It is interesting to note that such matter effects can lead to the observable effect, first pointed out by Mikheyev and Smirnov, and independently, by L. Wolfenstein. It is now known as the MSW effect.

As can be seen from Eq. (5.60), the effective mixing angle $\sin 2\theta_M$ becomes unity as $\beta = \cos 2\theta$, or when $2VE = \Delta m_{21}^2 \cos 2\theta$. It is possible to realize this condition for $\beta > 0$ if $\theta < \pi/4$. The conditions are reversed for antineutrinos because $V < 0$.

If we plot $\sin^2 2\theta_M$ as a function of β we would see that it exhibits a resonance-like curve with a maximum equal to one. This is the so-called "MSW resonance" which has a half-width at half maximum equal to $\sin 2\theta$, so it becomes narrower with smaller $\sin 2\theta$. This phenomenon is quite important in understanding solar neutrino oscillations.

5.6 Non-standard neutrino interactions

The previous section described neutrino oscillations and matter effects assuming that only SM interactions among propagating neutrinos and surrounding matter. This does not have to be the case. Just like in the case of multiple measurements of the CKM matrix elements in the previous chapter, studies of neutrino propagation can shed some light on possible BSM interactions of neutrinos. Indeed, the fact that neutrinos are massive particles already indicates the presence of physics beyond the Standard Model. Such non-standard neutrino interactions, or NSIs, represent New Physics effects beyond the mass generation.

5.6.1 NSIs: matter effects

We have already discussed possible new operators that affect charged-current interactions. Here we shall concentrate on neutral currents,

$$\mathcal{L}_{\text{NSI}} = -\frac{2G_F}{\sqrt{2}}\sum_{f,\alpha,\beta}\left(\overline{\nu}_\alpha \gamma_\mu P_L \nu_\beta\right)\left(\epsilon_{\alpha\beta}^{fV}\overline{f}\gamma^\mu f + \epsilon_{\alpha\beta}^{fA}\overline{f}\gamma^\mu\gamma_5 f\right) \tag{5.61}$$

where we only wrote out effective operators that correspond to the left-handed neutrino propagation.

Such a new interaction leads to a rich phenomenology in both scattering experiments and neutrino oscillation experiments. Since oscillation phenomenology is generally quite distinct from scattering phenomenology, the NSI framework provides a convenient way to relate New Physics models in both cases.

The vector component of NSIs affects oscillations by providing a new flavor-dependent matter effect. The Hamiltonian for this is

$$H = \frac{1}{2E} U \begin{pmatrix} 0 & 0 & 0 \\ 0 & \Delta m^2_{21} & 0 \\ 0 & 0 & \Delta m^2_{21} \end{pmatrix} U^\dagger + \sqrt{2} G_F N_e \begin{pmatrix} 1 - \epsilon_{ee} & \epsilon_{e\mu} & \epsilon_{e\tau} \\ \epsilon^*_{e\mu} & \epsilon_{\mu\mu} & \epsilon_{\mu\tau} \\ \epsilon^*_{e\tau} & \epsilon^*_{\mu\tau} & \epsilon_{\tau\tau} \end{pmatrix} \tag{5.62}$$

where U is the standard PMNS lepton mixing matrix, and $\sqrt{2} G_F N_e$ is the Wolfenstein matter potential discussed above. Note that 1 in the $1 - \epsilon_{ee}$ term is due to the standard charged current matter potential. Since the flavor-changing NSI terms can be complex, at the Hamiltonian level there are 9 new parameters, of which 8 are testable by oscillations [104].

5.6.2 NSIs and neutrino magnetic moments

It is interesting to note that NSIs can give contributions to other areas of studies of neutrino interactions. As an example, we will consider studies of electromagnetic properties of neutrinos described in Sec. 5.4.1.

Recall that the neutrino magnetic moment (NMM) $\mu_{\alpha\beta}$ can be defined by the Hermitian form factor $f^M_{\alpha\beta}(0) \equiv \mu_{\alpha\beta}$ of the term

$$-f^M_{\alpha\beta}(q^2)\, \bar{\nu}_\beta(p_2)\, i\sigma_{\mu\nu} q^\nu \nu_\alpha(p_1) \tag{5.63}$$

in the effective electromagnetic current of the neutrino. Here $\alpha, \beta = e, \mu, \tau$ are flavor indices and the momentum transfer is $q = p_2 - p_1$. The relation between NMMs in the flavor basis and the mass basis can be written as

$$\mu^2_{\alpha\beta} = \sum_{i,j,k} U^*_{\alpha j} U_{\beta k} e^{-i\Delta m^2_{jk} L/2E} \mu_{ij} \mu_{ik}, \tag{5.64}$$

where $i, j, k = 1, 2, 3$, $\Delta m^2_{jk} = m^2_j - m^2_k$ are the neutrino squared-mass differences, $U_{\ell i}$ is the leptonic mixing matrix, E is the neutrino energy, L is the baseline, and for simplicity we omitted the electric dipole moment contribution.

In the Standard Model, minimally extended to include Dirac neutrino masses, NMM is suppressed by small masses of observable neutrinos due to the left-handed nature of weak interaction. The diagonal and transition magnetic moments are calculated in the SM to be

$$\mu^{\text{SM}}_{ii} \approx 3.2 \times 10^{-20} \left(\frac{m_i}{0.1 \text{ eV}}\right) \mu_B \tag{5.65}$$

and

$$\mu^{\text{SM}}_{ij} \approx -4 \times 10^{-24} \left(\frac{m_i + m_j}{0.1 \text{ eV}}\right) \sum_{\ell = e,\mu,\tau} \left(\frac{m_\ell}{m_\tau}\right)^2 U^*_{\ell i} U_{\ell j}\, \mu_B, \tag{5.66}$$

respectively, where $\mu_B = e/(2m_e) = 5.788 \times 10^{-5}$ eV T^{-1} is the Bohr magneton.

Currently, the strongest experimental bound on NMM is far from the SM value

$$\mu_\nu < 3 \times 10^{-12}\, \mu_B. \tag{5.67}$$

It has been obtained from the constraint on energy loss from globular cluster red giants, which can be cooled faster by the plasmon decays due to NMM that delays the helium ignition. This bound can be applied to all diagonal and transition NMMs.

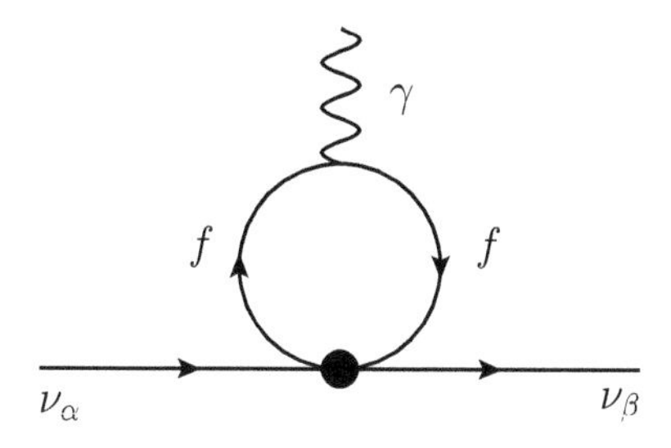

FIGURE 5.2 Effective diagram for magnetic moment of neutrino induced by tensorial NSI, indicated by the large dot.

It can be seen that among all possible $\nu_\alpha \nu_\beta f f$ interactions, the lowest order contribution to NMM can be generated through the one-loop diagram, shown in Fig. 5.2, with the dimension-six tensor operator,

$$\frac{\epsilon^{fT}_{\alpha\beta}}{\Lambda^2}(\bar{\nu}_\beta \sigma_{\mu\nu} \nu_\alpha)(\bar{f}\sigma^{\mu\nu} f), \tag{5.68}$$

where in the case of Majorana neutrinos $\bar{\nu}_\beta = \bar{\nu}^c_\beta$. In particular, interactions of neutrinos with quarks q via the operator

$$\frac{\epsilon^{q}_{\alpha\beta}}{\Lambda^2}(\bar{\nu}_\beta \sigma_{\mu\nu} \nu_\alpha)(\bar{q}\sigma^{\mu\nu} q), \tag{5.69}$$

where $\epsilon^q_{\alpha\beta} \equiv \epsilon^{qT}_{\alpha\beta}$ is real, can generate neutrino's magnetic moments

$$\mu_{\alpha\beta} = \mu^0_{\alpha\beta} - \sum_q \epsilon^q_{\alpha\beta} \frac{N_c Q_q}{\pi^2} \frac{m_e m_q}{M^2} \log\left(\frac{M^2}{m_q^2}\right) \mu_B. \tag{5.70}$$

Here $N_c = 3$ is the number of colors, Q_q and m_q are electric charge and mass of the quark, respectively. Here and later $\mu^0_{\alpha\beta}$ denotes the subleading part of the NMM that is not enhanced by the large logarithm. We note that this formula reproduces the leading order in the exact result, which can be derived in the model with scalar leptoquarks!

Similarly, for the interactions of neutrinos with charged leptons ℓ,

$$\frac{\epsilon^{\ell}_{\alpha\beta}}{M^2}(\bar{\nu}_\beta \sigma_{\mu\nu} \nu_\alpha)(\bar{\ell}\sigma^{\mu\nu} \ell) \tag{5.71}$$

with $\epsilon^\ell_{\alpha\beta} \equiv \epsilon^{\ell T}_{\alpha\beta}$, we have

$$\mu_{\alpha\beta} = \mu^0_{\alpha\beta} + \sum_\ell \frac{\epsilon^\ell_{\alpha\beta}}{\pi^2} \frac{m_e m_\ell}{M^2} \log\left(\frac{M^2}{m_\ell^2}\right) \mu_B. \tag{5.72}$$

We notice that the dominant logarithmic terms, such as in Eqs. (5.70) and (5.72), may not contribute to NMM in certain models, e.g., in the SM, due to a mutual compensation between the relevant diagrams.

5.7 Notes for further reading

An excellent review of neutrino properties can be found in the Guinti and Kim's book [88]. Neutrinoless double beta decay is discussed at length in [46,64]. There is also an excellent review discussing theory and experiment associated with this process [25]. A thorough discussion of NSI of neutrino could be found in the excellent review [2].

Problems for Chapter 5

1. **Problem 1.** Using the properties of the charge conjugation operator, $P_L C = C P_L^T$, which you can derive using your favorite representation of the Dirac matrices, prove Eq. (5.20),

$$P_L \psi_R = \left[C \overline{\psi_L}^T \right] = 0. \tag{5.73}$$

This shows that the right-handed field can indeed be defined as Eq. (5.19).

2. **Problem 2.** Show that the coefficients of the $\sin\left(\frac{\Delta m_{ik}^2 L}{2E}\right)$ in Eq. (5.9) can be related to a single combination of PMNS matrix elements, a convenient choice for which is the Jarlskog invariant J,

$$J = c_{13}^2 2 s_{13} s_{12} c_{12} s_{23} c_{23} \sin\delta. \tag{5.74}$$

To do so, consider the coefficient

$$J_{ik}^{\alpha\beta} = -Im \left[U_{\alpha i} U_{\alpha k}^* U_{\beta i}^* U_{\beta k} \right], \tag{5.75}$$

and prove that it is antisymmetric in both upper (flavor) and lower (mass) indices, which implies relations among its coefficients. Then use the unitarity of the PMNS matrix to show that the remaining matrix elements are the same.

6

New Physics searches with Higgs and gauge bosons

6.1 Introduction

Availability of high-energy and high-luminosity accelerators allowed for the creation of large datasets of decays of gauge $W^{\pm}$ and Z^0 bosons. Access to large sets of data containing Higgs decays also became possible. Since there appears to be a gap between the New Physics scale Λ and a variety of SM-related scales, it might be advantageous to closely look at various processes involving electroweak gauge and Higgs bosons as initial states: NP corrections to the SM processes are more pronounced at high energy processes. Analyses of decays of top quarks might provide a better probe for NP particles whose couplings to the SM particles depend on their masses.

Most importantly, extensive studies of Higgs decay patterns are imperative. Being the only fundamental scalar in the SM, it provides us with a number of puzzles, including the one at the center of the hierarchy problem: why is its mass so small? Thorough studies of its couplings to other SM particles might shed light on the solution of this puzzle.

In this section we look at some of the processes involving on-shell gauge and Higgs bosons, just touching upon the extensive body of work done by many physicists over the years.

6.2 Precision electroweak physics. S, T, and U parameters

In order to cast a wide net in identifying possible NP contributions, we usually consider situations where some observables would receive larger contributions than others. There could be models where BSM states' couplings to final state fermions are suppressed. Examples of such models could be those with Higgs-like scalar states or strongly coupled theories with pseudo-Goldstone bosons that have gauge quantum numbers, but no flavor quantum numbers. In principle, any NP particles that are charged under the $SU(2) \times U(1)$ gauge groups can contribute to such corrections. In those cases NP models would mainly produce vacuum-polarization-type, or *oblique*, corrections to experimental observables*. To define such observables, let us first recall that a two-point function for a gauge boson can be defined as

$$\Pi^{\mu\nu}_{AB}(q^2) = \Pi_{AB} g^{\mu\nu} + \Delta_{AB} q^\mu q^\nu, \tag{6.1}$$

where $AB = \{\gamma\gamma, \gamma Z, ZZ, W^+W^-\}$. The Δ_{AB} part is irrelevant for further discussion, as it is usually coupled to a conserved current or suppressed by masses of external fermions, which can be seen by applying Dirac equation.

To discuss the way New Physics might affect the electroweak oblique corrections, let us define observables, which we denote by a hat on top of quantities. For example, the EM fine structure constant $\hat{\alpha}$ can be obtained from $e^+e^- \to \gamma^* \to e^+e^-$ scattering, $\hat{G}_F$ can be extracted from muon decay, $\hat{m}_Z$ and $\hat{m}_W$ (W and Z boson masses) and $\hat{\Gamma}_{\ell^+\ell^-}$ (leptonic width of the Z-boson), and $\hat{s}^2_{\text{eff}}$ (effective value of $\sin^2\theta_{\text{W}}$) can be derived from the collider measurements. Can these observables be predicted from the parameters of the Standard Model Lagrangian with given precision or other, New Physics parameters are required to explain the experimental data? We will address this question below, mainly following [180].

At tree level all six observables can be expressed in terms of only three SM parameters: g_1 (the hypercharge coupling), g_2 (the electroweak coupling) of Eq. (2.4), and v, the Higgs vacuum expectation value of Eq.(2.17). As we know from [65,147,186], the first two can be related to the electric charge and various trigonometric functions of the Weinberg mixing angle $c_W \equiv \cos\theta_{\text{W}}$ and $s_W \equiv \sin\theta_{\text{W}}$ of Eq. (2.10), $g_1 = e/c_{\text{W}}$ and $g_2 = e/s_{\text{W}}$. Alternatively, we can just use e, s_{W}, and v.

In order to illustrate the method, we can compute all six observables above in terms of the three constants of our set at tree level,

$$\begin{aligned}
\hat{\alpha} &= \frac{e^2}{4\pi}, & \hat{G}_F &= \frac{1}{\sqrt{2}v^2}, \\
\hat{m}^2_Z &= \frac{e^2v^2}{4s^2_{\text{W}}c^2_{\text{W}}}, & \hat{m}^2_W &= \frac{e^2v^2}{4s^2_{\text{W}}}, \\
\hat{s}^2_{\text{eff}} &= s^2_{\text{W}}, & \hat{\Gamma}_{\ell^+\ell^-} &= \frac{v}{96\pi}\frac{e^2}{s^3_{\text{W}}c^3_{\text{W}}}\left[\left(-\frac{1}{2}+2s^2_{\text{W}}\right)^2+\frac{1}{4}\right],
\end{aligned} \tag{6.2}$$

and then see if the relations are consistent. In fact, we can do the following: we can express e, s_{W}, and v in terms of, say, $\hat{\alpha}$, $\hat{G}_F$, and $\hat{m}^2_Z$ and then substitute them into the remaining observables.

The first three equations then give us the following expressions,

$$e^2 = 4\pi\hat{\alpha}, \quad v^2 = \frac{1}{\sqrt{2}\hat{G}_F}, \quad \text{and} \quad s^2_{\text{W}}c^2_{\text{W}} = s^2_{\text{W}}(1-s^2_{\text{W}}) = \frac{\pi\hat{\alpha}}{\sqrt{2}\hat{G}_F\hat{m}^2_Z}, \tag{6.3}$$

where the last equation also allows us to obtain an expression for s^2_{W}. Now using Eqs. (6.2) and

*Contrary, the vertex and box-type corrections, which depend on the identity of the initial and final state fermions, are referred to as *nonoblique* corrections.

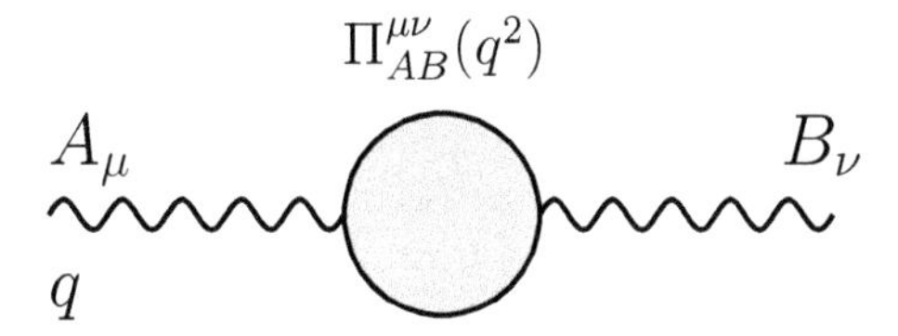

FIGURE 6.1 Vacuum polarization graphs for various gauge bosons in the Standard Model. Here $AB = \gamma\gamma$, WW, ZZ, and γZ.

(6.3), we obtain the predictions for the rest of the observables [180],

$$
\begin{aligned}
\hat{m}_W^2 &= \frac{\pi\sqrt{2}\hat{\alpha}}{\hat{G}_F}\left(1-\sqrt{1-\frac{4\pi\hat{\alpha}}{\sqrt{2}\hat{G}_F\hat{m}_Z^2}}\right)^{-1}, \\
\hat{s}^2_{\text{eff}} &= \frac{1}{2}\left(1-\sqrt{1-\frac{4\pi\hat{\alpha}}{\sqrt{2}\hat{G}_F\hat{m}_Z^2}}\right), \\
\hat{\Gamma}_{\ell^+\ell^-} &= \frac{\sqrt{2}\hat{G}_F\hat{m}_Z^3}{12\pi}\left[\left(\frac{1}{2}-\sqrt{1-\frac{4\pi\hat{\alpha}}{\sqrt{2}\hat{G}_F\hat{m}_Z^2}}\right)^2+\frac{1}{4}\right].
\end{aligned}
\tag{6.4}
$$

What have we achieved so far? We obtained relations among experimental observables that is true in the Standard Model – exactly what we need to check for possible New Physics, which we described as "method III" in the Introduction (see Chapter 1).

What's left to do – and we leave this as an exercise for the reader – is to plug in the latest experimental data from [1] and see if the observables on the left-hand side of Eqs. (6.4) are statistically consistent with the predictions obtained by plugging in the experimental values of $\hat{\alpha}$, $\hat{G}_F$, and $\hat{m}_Z$, again, from the Ref. [1], to the right-hand side of Eq. (6.4).

Having done this, we discover something rather remarkable: the predictions on the right-hand side of Eq. (6.4) differ from the data on the left-hand by tens of sigmas! Following our logic, we must conclude that the SM is incompatible with the experiment, and there must be New Physics affecting some of our observables, so the relations of Eq. (6.4) must be modified.

The bitter truth is that indeed, the relations of Eq. (6.4) must be modified, but not necessarily by New Physics contributions. It appears that the experimental data have better precision than theoretical expressions, as the obtained relations were derived at tree-level in the perturbative expansion of the SM. Higher-order perturbative corrections must be included to truly judge if NP affects our observables in a significant way. Let us sketch a discussion of how this is done.

If we *assume* that the major higher-order contributions come in the form of the oblique corrections (see Fig. 6.1), we can see how they affect the relations between the observables.

The self-energies of gauge bosons shift the position of the propagator pole defining their masses,

$$
m_V^2 \to m_V^2 + \Pi_{VV}(q^2 = m_V^2). \tag{6.5}
$$

Since photon is massless, $\Pi_{\gamma\gamma}(0) = 0$. For the Z and W bosons at one loop,

$$
\begin{aligned}
\hat{m}_Z &= \frac{e^2v^2}{4s^2c^2} + \Pi_{ZZ}(m_Z^2), \\
\hat{m}_W &= \frac{e^2v^2}{4s^2} + \Pi_{WW}(m_W^2).
\end{aligned}
\tag{6.6}
$$

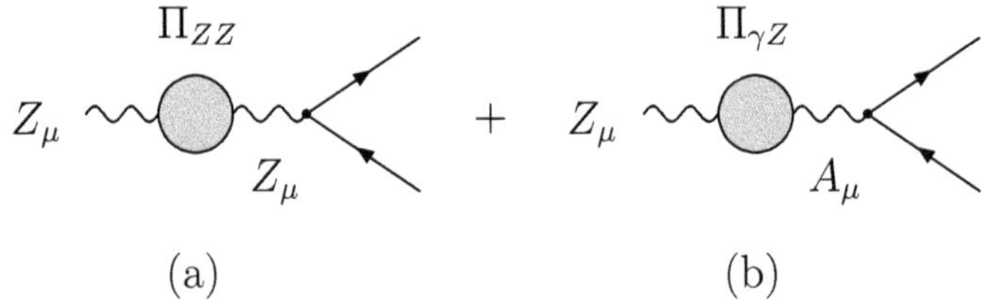

FIGURE 6.2 Vacuum polarization corrections to A_{LR} and $\Gamma_{\ell^+\ell^-}$. Note that only (b) is needed for $\hat{s}^2_{\text{eff}}$ computations (see text).

Further, we can compute corrections to α. An appropriate experimental observable could be read off the expression for the e^+e^- scattering in the limit of the small transferred momentum squared, $q^2 \to 0$ that probes Coulomb potential

$$-i \left. \frac{4\pi\hat{\alpha}}{q^2} \right|_{q^2\to 0} = \frac{-ie^2}{q^2} \left[1 + \frac{\Pi_{\gamma\gamma}(q^2)}{q^2} \right]_{q^2\to 0}. \tag{6.7}$$

If we can expand $\Pi_{\gamma\gamma}(q^2)$ around $q^2 = 0$, we get

$$\Pi_{\gamma\gamma}(q^2) = \Pi_{\gamma\gamma}(0) + \Pi'_{\gamma\gamma}(0) q^2 + \ldots \tag{6.8}$$

Now, since $\Pi_{\gamma\gamma}(0) = 0$, we can define

$$\Pi'_{\gamma\gamma}(0) \equiv \lim_{q^2\to 0} \frac{\Pi_{\gamma\gamma}(q^2)}{q^2}, \tag{6.9}$$

which results in an expression for the α that can be read off as a coefficient of $(-i4\pi)/q^2$,

$$\hat{\alpha} = \frac{e^2}{4\pi} \left(1 + \Pi'_{\gamma\gamma}(0) \right). \tag{6.10}$$

We would like to remind the reader that our goal is not to compute a complete set of one-loop corrections to the cross-sections or decay widths. We are trying to extract the observables that are most sensitive to the oblique corrections. With that in mind, prediction for $\hat{G}_F$ is

$$\hat{G}_F = \frac{g_1^2}{4\sqrt{2} m_W^2} \left[1 + \left. \frac{\Pi_{WW}(q^2)}{q^2 - m_W^2} \right|_{q^2\to 0} \right] = \frac{1}{\sqrt{2} v^2} \left(1 - \frac{\Pi_{WW}(0)}{m_W^2} \right), \tag{6.11}$$

which can be read off the muon lifetime. Note that the only correction that we included is the vacuum polarization of the W-boson.

Similarly, $\hat{s}^2_{\text{eff}}$ can be obtained from the asymmetry of polarized leptons produced at the Z-pole A_{LR}, which is defined as a difference of cross-sections of the left-handed leptons and the right-handed ones divided by their sum [180],

$$\hat{s}^2_{\text{eff}} = s^2_{\text{W}} - s_{\text{W}} c_{\text{W}} \frac{\Pi_{\gamma Z}(m_Z^2)}{m_Z^2}, \tag{6.12}$$

which is extracted experimentally from

$$\hat{A}_{LR} \equiv \frac{\sigma_L - \sigma_R}{\sigma_L + \sigma_R} = \frac{(1/2 - \hat{s}^2_{\text{eff}})^2 - (\hat{s}^2_{\text{eff}})^2}{(1/2 - \hat{s}^2_{\text{eff}})^2 + (\hat{s}^2_{\text{eff}})^2}. \tag{6.13}$$

The oblique corrections to A_{LR} come from the diagrams in Fig. 6.2. Finally, the last of the parameters from Eq. (6.2) to be corrected by one-loop oblique corrections is $\hat{\Gamma}_{\ell^+\ell^-}$, which also comes from the diagrams Fig. 6.2

$$\hat{\Gamma}_{\ell^+\ell^-} = \frac{\sqrt{2}\hat{G}_F\hat{m}_Z^3}{12\pi}\left[\left(\frac{1}{2} - \sqrt{1 - \frac{4\pi\hat{\alpha}}{\sqrt{2}\hat{G}_F\hat{m}_Z^2}}\right)^2 + \frac{1}{4}\right]. \tag{6.14}$$

The last step to take is to express observables in terms of observables (see [180] for the details),

$$\begin{aligned}
\hat{m}_W &= \frac{\pi\hat{\alpha}(\hat{m}_Z^2)}{\sqrt{2}\hat{G}_F\hat{s}_0^2}\left[1 - \frac{\Pi_{\gamma\gamma}(m_Z^2)}{m_Z^2} - \frac{c_0^2}{c_0^2 - s_0^2}\delta_S - \frac{\Pi_{WW}(0)}{m_W^2} + \frac{\Pi_{WW}(m_W^2)}{m_W^2}\right] \\
\hat{s}^2_{\text{eff}} &= \hat{s}_0^2 + \frac{s_0^2c_0^2}{c_0^2 - s_0^2}\left[\frac{\Pi_{ZZ}(m_Z^2)}{m_Z^2} - \frac{\Pi_{WW}(0)}{m_W^2} - \frac{(c_0^2 - s_0^2)}{s_0c_0}\frac{\Pi_{\gamma Z}(m_Z^2)}{m_Z^2} - \frac{\Pi_{\gamma\gamma}(m_Z^2)}{m_Z^2}\right] \\
\hat{\Gamma}_{\ell^+\ell^-} &= \hat{\Gamma}^0_{\ell^+\ell^-}\left[1 - \frac{as_0^2c_0^2}{c_0^2 - s_0^2}\frac{\Pi_{ZZ}(m_Z^2)}{m_Z^2} + \left(1 + \frac{as_0^2c_0^2}{c_0^2 - s_0^2}\right)\frac{\Pi_{WW}(0)}{m_W^2}\right. \\
&+ \left. as_0c_0\frac{\Pi_{\gamma Z}(m_Z^2)}{m_Z^2} - \frac{\Pi_{ZZ}(0)}{m_Z^2} + a\frac{s_0^2c_0^2}{c_0^2 - s_0^2}\frac{\Pi_{\gamma\gamma}(m_Z^2)}{m_Z^2}\right]
\end{aligned} \tag{6.15}$$

where the short-hand notation is used [180],

$$\hat{s}_0^2\hat{c}_0^2 = \frac{\pi\hat{\alpha}(\hat{m}_Z^2)}{\sqrt{2}\hat{G}_F\hat{m}_Z^2}, \tag{6.16}$$

and the constant a is defined as

$$a = \frac{-8(-1 + 4s_0^2)}{(-1 + 4s_0^2)^2 + 1} \simeq 0.636. \tag{6.17}$$

A convenient parametrization of one-loop oblique corrections is given by the Peskin-Takeuchi *STU* formalism [148]. It is a rather economic framework: for heavy NP such that $m_Z/\Lambda \ll 1$ all oblique corrections to all Z-pole observables can be written in terms of only three universal parameters: S, T, and U.

In terms of the self-energy Π's, the S, T and U parameters are

$$\begin{aligned}
S &= \frac{\alpha}{4s_W^2}\left[c_W^2\left(\frac{\Pi_{ZZ}(m_Z^2)}{m_Z^2} - \frac{\Pi_{ZZ}(0)}{m_Z^2} - \frac{\Pi_{\gamma\gamma}(m_Z^2)}{m_Z^2}\right)\right. \\
&\quad \left. - \frac{c_W}{s_W}(c_W^2 - s_W^2)\left(\frac{\Pi_{\gamma Z}(m_Z^2)}{m_Z^2} - \frac{\Pi_{\gamma Z}(0)}{m_Z^2}\right)\right],
\end{aligned} \tag{6.18}$$

$$T = \frac{1}{\alpha}\left[\frac{\Pi_{WW}(0)}{m_W^2} - \frac{\Pi_{ZZ}(0)}{m_Z^2} - 2\frac{s_W}{c_W}\frac{\Pi_{\gamma Z}(m_Z^2)}{m_Z^2}\right], \tag{6.19}$$

$$\begin{aligned}
U &= \frac{\alpha}{4s_W^2}\left[\frac{\Pi_{WW}(m_W^2)}{m_W^2} - \frac{\Pi_{WW}(0)}{m_W^2} - c_W^2\left(\frac{\Pi_{ZZ}(m_Z^2)}{m_Z^2} - \frac{\Pi_{ZZ}(0)}{m_Z^2}\right)\right. \\
&\quad \left. - s_W^2\frac{\Pi_{\gamma\gamma}(m_Z^2)}{m_Z^2} - 2sc\left(\frac{\Pi_{\gamma Z}(m_Z^2)}{m_Z^2} - \frac{\Pi_{\gamma Z}(0)}{m_Z^2}\right)\right]
\end{aligned} \tag{6.20}$$

The definitions of S, T, and U parameters could be chosen such that the Standard Model contributions at the tree level do not contribute to S, T, or U [62]. Also, any BSM correction which is indistinguishable from a redefinition of e, G_F and m_Z in the Standard Model proper at the tree level does not contribute Peskin-Takeuchi parameters.

Let us look at an example of an explicit BSM model contribution to *STU* parameters. The simplest possible extension of the Higgs sector of the SM is an addition of a read scalar S with hypercharge $Y = 0$. To further simplify the model, we can require the scalar field to be odd under an additional Z_2 symmetry, $S \to -S$, so we eliminate terms with the odd numbers of S fields in the scalar Higgs potential $V(\Phi, S)$,

$$\begin{aligned} \mathcal{L}_S &= (D_\mu \Phi)^\dagger + \frac{1}{2}(\partial_\mu S)^2 - V(\Phi, S), \\ V &= \mu^2 \Phi^\dagger \Phi + \frac{1}{2} m^2 S^2 + \lambda\left(\Phi^\dagger \Phi\right)^2 + y \Phi^\dagger \Phi S^2 + \frac{\lambda_S}{4} S^4, \end{aligned} \tag{6.21}$$

where Φ is the SM Higgs doublet. If both Φ and S develop vacuum expectation values (v and s, respectively) after spontaneous symmetry breaking, then the neutral component of the Higgs doublet ϕ_0 and S would mix into the physical states h and H with masses m_h and M_H,

$$\begin{pmatrix} h \\ H \end{pmatrix} = \begin{pmatrix} \cos\alpha & -\sin\alpha \\ \sin\alpha & \cos\alpha \end{pmatrix} \begin{pmatrix} \sqrt{2}\phi_0 - v \\ S - s \end{pmatrix} \tag{6.22}$$

If the singlet state does not directly couple to SM fermions or gauge bosons, it will only affect the observables via mixing with the Higgs. Such mixing affects the SM-like Higgs couplings to both fermions and gauge bosons in an identical fashion, so all SM couplings are suppressed by $\cos\alpha$. In the limit $M_W, M_Z \ll m_h, m_H$, the contribution to the STU parameters could be seen to be

$$\begin{aligned} S &= \frac{1}{12\pi} \sin^2\alpha \log\frac{M_H^2}{m_h^2}, \\ T &= -\frac{3}{16\pi c_W^2} \sin^2\alpha \log\frac{M_H^2}{m_h^2}, \\ U &= 0 \end{aligned} \tag{6.23}$$

If STU are experimentally measured, we can determine the BSM parameters for this model (M_H and α).

As the reader undoubtedly noticed, our discussion of the oblique corrections revolved around computing various vacuum polarization graphs. How would oblique corrections, and specifically, S, T, and U parameters be defined in SM EFT? While it is possible to write a set of operators that contribute to STU [97] one must be careful in interpreting the results of such analyses.

The main reason for such caution is the fact that Π_{AB} and Π'_{AB} computed in such a way are *not* invariant under field redefinitions! Since they are not physical observables, their values would be basis-dependent and constraints placed on them would be ambiguous [98, 166, 175]. This should not be surprising, as barring the ghost contributions, the self-energy graphs are not gauge-invariant and therefore cannot be physical observables*.

Experimental bounds on S, T, and U parameters are usually derived assuming that they capture the majority of NP effects, as we argued at the beginning of this chapter. This assumption does not work for the most general BSM models.

6.3 Z-boson decays

Decays of the Z-boson have been studied both theoretically and experimentally for decades [65]. Studies of various SM-dominated decays rates, such as $Z \to b\bar{b}$, as well as kinematic asymmetries, can be sensitive to possible NP contributions. Here we mention decays that, if observed, would immediately indicate the presence of BSM physics.

*An infinite sum of self-energy graphs, representing the position of the pole in the expression of the mass is physical, due to Ward identities. It can be explicitly seen in the calculations of self-energies in R_ξ gauge.

Lepton flavor violating Z decays

Lepton-flavor violating decays of Z receive a tiny contribution from the SM, being proportional to neutrino masses. This follows from the calculation of the penguin diagram with neutrino mass insertions to ensure flavor change. Thus, any NP contribution that is not proportional to neutrino masses would likely dominate the LFV Z-width,

$$\Gamma(Z^0 \to \ell_f^{\pm} \ell_i^{\mp}) = \frac{m_Z}{24\pi} \left(\frac{m_Z^2}{\Lambda^2} \left(\left|C_{fi}^{ZR}\right|^2 + \left|C_{fi}^{ZL}\right|^2 \right) + \left|\Gamma_{fi}^{ZR}\right|^2 + \left|\Gamma_{fi}^{ZL}\right|^2 \right), \tag{6.24}$$

where the coefficients C_{fi}^{ZL}, C_{fi}^{ZR} are defined as

$$\begin{aligned} \Gamma_{fi}^{ZL} &= \frac{e}{2s_W c_W} \left(\frac{v^2}{\Lambda^2} \left(C_{\phi\ell}^{(1)fi} + C_{\phi\ell}^{(3)fi} \right) + \left(1 - 2s_W^2\right) \delta_{fi} \right), \\ \Gamma_{fi}^{ZL} &= \frac{e}{2s_W c_W} \left(\frac{v^2}{\Lambda^2} C_{\phi\ell}^{fi} - 2s_W^2 \delta_{fi} \right), \\ C_{fi}^{ZR} &= C_{fi}^{ZL*} = -\frac{v}{\sqrt{2}\Lambda^2} \left(s_W C_{eB}^{fi} + c_W C_{eW}^{fi} \right) \equiv -\frac{v}{\sqrt{2}\Lambda^2} C_Z^{fi}. \end{aligned} \tag{6.25}$$

Experimental bounds on LFV decays of Z come from the high-energy colliders, such as LHC. Most "cleanest" data could come from high-luminosity Z-factory – an e^+e^- machine that is tuned to run at the "Z-pole", in other words, such that the $\sqrt{s} = m_Z$. One should note that it is often the case that the theoretical predictions, such as the ones in Eq. (6.24), are given for the decays $Z^0 \to \ell_i^+ \ell_j^-$ or $Z^0 \to \ell_i^- \ell_j^+$, while the experimental data is given for the averaged result, $Z^0 \to \left(\ell_i^+ \ell_j^- + \ell_i^- \ell_j^+\right)$. This means that the theoretical result must be multiplied by a factor of 2 in order to compare it to the experimental results.

6.4 Higgs boson decays

Similar to Sect. 6.3, flavor-diagonal decays of the Higgs bosons to fermion states can probe New Physics contributions [107]. There are, however, interesting decays channels that have enhanced sensitivity to NP. Those include invisible Higgs decays and flavor-violating decays. We will consider them below.

6.4.1 Invisible decays of the Higgs boson

Since, besides neutrinos, there are no "invisible" particles in the Standard Model, any unaccounted contribution to the invisible Higgs boson decays could be from the light BSM states.

As a simple example, let's consider the simplest model, given by the Lagrangian of Eq. (6.21). It is actually quite impressive that such a model could answer so many questions about Nature. Since S is odd under a Z_2 discrete symmetry, it is stable.

Other constraints on this model could come from the assumption that, as a stable neutral particle, S gives a contribution to the relic abundance of Dark Matter. Relic abundance can be calculated as a function of the various parameters of the theory including most importantly m, m_h, and y. There are accessible regions of parameter space where relic abundance is the value needed to be the cold Dark Matter of the universe. Since S does not get a vacuum expectation value, the Higgs boson of the Standard Model cannot mix with S. However, the SM Higgs state can decay into a pair of S particles leading to an invisible decay width of the Higgs boson,

$$\Gamma(h \to SS) = \frac{y^2 v^2}{8\pi m_h} \sqrt{1 - \frac{4m_S^2}{m_h^2}}, \tag{6.26}$$

where m_h and m_S are the SM Higgs and scalar S masses, respectively. Similar expressions can be obtained for light invisible particles of other spins.

6.4.2 Flavor-violating decays of the Higgs boson

One of the most striking signatures of BSM physics that one would hope to see at a high-energy collider is a flavor off-diagonal decay of a Higgs boson, $h \to \bar{f}_1 f_2$. Such signature might manifest BSM physics, as the SM Higgs only has flavor-diagonal couplings. There could be several sources for non-diagonal terms in Y_{ik} [110].

First, there could be only one Higgs, but non-diagonal terms could arise from higher-dimensional operators. Indeed, after electroweak symmetry breaking the fermionic mass terms and the couplings of the Higgs boson to fermion pairs in the mass basis can be written as

$$\mathcal{L}_Y = -m_i \bar{F}_L^i f_R^i - Y_{ij}(\bar{F}_L^i f_R^j)h + h.c., \tag{6.27}$$

where $F_L = q_L, \ell_L$ are the $SU(2)_L$ doublets, and $f_R = u_R, d_R, \nu_R, \ell_R$ the right-handed weak singlets. In the Standard Model Higgs couplings to fermions are diagonal, $Y_{ik} = (m_i/v)\delta_{ik}$, but in other models those matrices Y_{ik} could be non-diagonal.

Here, in the spirit of the EFT approach to the Standard Model, those non-diagonal terms could be induced from higher-dimensional operators due to the heavy degrees of freedom,

$$\Delta\mathcal{L}_Y = -\frac{\lambda'_{ij}}{\Lambda^2}(\bar{F}_L^i f_R^j)H(H^\dagger H) + h.c. \tag{6.28}$$

It is possible to show that the addition of dimension-6 terms above is sufficient to decouple the first (masses) term from the second (Yukawa couplings) one, so no higher-order terms are needed.

After SSB, one can see that diagonalization of the mass matrices leave the Yukawa of Eq. (6.27) matrices non-diagonal,

$$\sqrt{2}m = V_L\left[\lambda + \frac{v^2}{2\Lambda^2}\lambda'\right]V_R^\dagger v, \quad \sqrt{2}Y = V_L\left[\lambda + 3\frac{v^2}{2\Lambda^2}\lambda'\right]V_R^\dagger. \tag{6.29}$$

Indeed, if the matrices V_L and V_R diagonalize the mass term (left equation in Eq. 6.29), they do not necessarily do so for the Yukawa term (right equation in Eq. 6.29), as those matrices are, in general, different.

Then, in the mass basis we can write

$$Y_{ij} = \frac{m_i}{v}\delta_{ij} + \frac{v^2}{\sqrt{2}\Lambda^2}\hat{\lambda}_{ij}, \tag{6.30}$$

where $\hat{\lambda} = V_L \lambda' V_R$. In the decoupling limit $\Lambda \to \infty$ one obtains the SM, where the Yukawa matrix Y is diagonal, $Yv = m$ [110].

Another possibility involves a type-III two Higgs doublet model, i.e. a model with no natural flavor conservation. In this case, the Lagrangian take the following form,

$$\mathcal{L}_Y = -m_i \bar{F}_L^i f_R^i - Y_{ij}^a(\bar{F}_L^i f_R^j)h^a + h.c., \tag{6.31}$$

This is the case where masses receive contributions from different VEVs. A decay rate for the flavor-violating decay of the Higgs can be easily computed from Eq. (6.27),

$$\Gamma(h \to \ell_i \ell_k) = \frac{m_h}{8\pi}\left(|Y_{ik}|^2 + |Y_{ki}|^2\right). \tag{6.32}$$

With $\Gamma(h \to \ell_i \ell_k)$ computed, a branching ratio for this transition can be also defined,

$$\mathcal{B}(h \to \ell_i \ell_k) = \frac{\Gamma(h \to \ell_i \ell_k)}{\Gamma(h \to \ell_i \ell_k) + \Gamma_{SM}}, \tag{6.33}$$

where $\Gamma_{SM} \approx 4$ MeV is the SM Higgs decay width.

TABLE 6.1 Operators coupling WIMPs to SM particles. The D-operators: Dirac fermions as WIMPs

Name	Operator	Coefficient
D1	$\bar{\chi}\chi\bar{q}q$	m_q/M_*^3
D2	$\bar{\chi}\gamma^5\chi\bar{q}q$	im_q/M_*^3
D3	$\bar{\chi}\chi\bar{q}\gamma^5 q$	im_q/M_*^3
D4	$\bar{\chi}\gamma^5\chi\bar{q}\gamma^5 q$	m_q/M_*^3
D5	$\bar{\chi}\gamma^\mu\chi\bar{q}\gamma_\mu q$	$1/M_*^2$
D6	$\bar{\chi}\gamma^\mu\gamma^5\chi\bar{q}\gamma_\mu q$	$1/M_*^2$
D7	$\bar{\chi}\gamma^\mu\chi\bar{q}\gamma_\mu\gamma^5 q$	$1/M_*^2$
D8	$\bar{\chi}\gamma^\mu\gamma^5\chi\bar{q}\gamma_\mu\gamma^5 q$	$1/M_*^2$
D9	$\bar{\chi}\sigma^{\mu\nu}\chi\bar{q}\sigma_{\mu\nu} q$	$1/M_*^2$
D10	$\bar{\chi}\sigma_{\mu\nu}\gamma^5\chi\bar{q}\sigma_{\alpha\beta} q$	i/M_*^2
D11	$\bar{\chi}\chi G_{\mu\nu}G^{\mu\nu}$	$\alpha_s/4M_*^3$
D12	$\bar{\chi}\gamma^5\chi G_{\mu\nu}G^{\mu\nu}$	$i\alpha_s/4M_*^3$
D13	$\bar{\chi}\chi G_{\mu\nu}\tilde{G}^{\mu\nu}$	$i\alpha_s/4M_*^3$
D14	$\bar{\chi}\gamma^5\chi G_{\mu\nu}\tilde{G}^{\mu\nu}$	$\alpha_s/4M_*^3$

6.5 Collider searches for Dark Matter

If Dark Matter particles can be produced at particle accelerators, they will most likely leave no trace in the detector. In that sense, its signature is the same as in "invisible" decays that we considered in previous chapters. In that respect, DM particle does not have to be a part of some sophisticated model, we can write a *simplified model* with such particle, writing out the most general interactions allowed by present symmetries, but otherwise completely model-independently.

For such particle χ to be a DM candidate, it needs to be odd under some Z_2 symmetry [94,158] to be produced pairwise, and it must be weakly-interacting – so it can be classified as Weakly Interacting Massive Particle (WIMP). In the simplest case, the WIMP is a singlet under the SM gauge groups, and thus possesses no tree-level couplings to the electroweak gauge bosons. We also neglect couplings with Higgs bosons and postpone the discussion of Higgs signatures until the next section.

If we are two write a set of effective Lagrangians describing χ, we need to assume that the mediator of interactions between χ and SM fields is heavy so that their effects are manifested by higher-dimensional operators.

With such constraints, the only effective operators left to consider are the ones built out of WIMP pairs and either quark bilinears of the form $\bar{q}\Gamma q$, where Γ is one of the Dirac matrices of the complete set,

$$\Gamma = \left\{1, \gamma^5, \gamma^\mu, \gamma^\mu\gamma^5, \sigma^{\mu\nu}\right\} , \tag{6.34}$$

or QCD gluonic operators GG or $G\widetilde{G}$.

Altogether, these higher-dimensional operators define an effective field theory of the interactions of singlet WIMPs with hadrons at particle colliders. We list those operators in Tables 6.1 and 6.2. The scale M_* in those tables characterize both the strength of those interactions, but also shows a scale where the effective description breaks down.

We consider Higgs-related collider signature in the next section.

6.5.1 Mono-Higgs and Dark Matter

So far, we avoided a situation where Higgs boson is a part of the experimental signature for *something else*. It is time to do that as well, considering experimental signature "Higgs plus missing energy" [150]. Searches of that type, i.e. "*something* plus missing energy" are part of the standard toolbox of an experimentalist working on DM detection, in that respect, Higgs is no different from a jet or a high-energy photon.

TABLE 6.2 Operators coupling WIMPs to SM particles. The C- and R-operators: complex or real scalars as WIMPs.

Name	Operator	Coefficient
C1	$\chi^\dagger \chi \bar{q} q$	m_q/M_*^2
C2	$\chi^\dagger \chi \bar{q}\gamma^5 q$	im_q/M_*^2
C3	$\chi^\dagger \partial_\mu \chi \bar{q}\gamma^\mu q$	$1/M_*^2$
C4	$\chi^\dagger \partial_\mu \chi \bar{q}\gamma^\mu \gamma^5 q$	$1/M_*^2$
C5	$\chi^\dagger \chi G_{\mu\nu} G^{\mu\nu}$	$\alpha_s/4M_*^2$
C6	$\chi^\dagger \chi G_{\mu\nu} \tilde{G}^{\mu\nu}$	$i\alpha_s/4M_*^2$
R1	$\chi^2 \bar{q} q$	$m_q/2M_*^2$
R2	$\chi^2 \bar{q}\gamma^5 q$	$im_q/2M_*^2$
R3	$\chi^2 G_{\mu\nu} G^{\mu\nu}$	$\alpha_s/8M_*^2$
R4	$\chi^2 G_{\mu\nu} \tilde{G}^{\mu\nu}$	$i\alpha_s/8M_*^2$

Let us write a set of effective operators that generate such an experimental signature. We shall consider all possible operators, suppressed by no more than three powers of the new physics scale Λ, and study their implications for experimental signals at hadronic colliders. Similar to the previous section, we will label the DM field as χ, and for concreteness we have assumed that DM is a fermion and a singlet of the SM gauge group.

Because we aren't working with a complete model of DM and its associated physics, and we've posited only the existence of the DM field, the only interactions that DM can have with the SM fields are given by effective operators of dimension more than or equal to five.

The lowest dimension at which the Dark Matter field χ can interact with SM fields under these assumptions is five. These operators are

$$\mathcal{L}_5 = \frac{2C_1^{(5)}}{\Lambda} |H|^2 \overline{\chi}\chi + \frac{2C_2^{(5)}}{\Lambda} |H|^2 \overline{\chi}\gamma_5\chi. \tag{6.35}$$

Throughout, $C_i^{(n)}$ are the effective Wilson coefficients that characterize the strength of Higgs-DM interactions of dimension n in the effective theory and Λ characterizes the scale at which the EFT description breaks down. Below, we shall always expect that the scale Λ is always sufficiently high so that EFT expansion in terms of local operators is well defined. As we can see from Eq. (6.35), however, all dependence on short-distance physics in the effective Lagrangian is encoded in the Wilson coefficients $C_i^{(n)}$. Experimental constraints available for each operator are probing the "coupling constant", i.e. a *combination* $C_i^{(n)}/\Lambda^{n-4}$. This fact should always be kept in mind while interpreting the fit results, as $C_i^{(n)}$ are usually set to some value (usually $C_i^{(n)} = 1$) to constrain Λ.

In the Eq. (6.35) and below H represents the Higgs doublet field, which in the unitary gauge takes the usual form given in Eq. (2.17). In terms of the physical Higgs field h, and considering only the terms quadratic in the physical field, this operator can be written as

$$\mathcal{L}_5 = \frac{C_1^{(5)}}{\Lambda} h^2\overline{\chi}\chi + \frac{C_2^{(5)}}{\Lambda} h^2\overline{\chi}\gamma_5\chi. \tag{6.36}$$

One should not expect to have a strong bound on the scale Λ from those operators. First, Higgs production by itself is relatively rare at the LHC, with subsequent DM-Higgs interactions giving an additional suppression. Second, since the Higgs boson is in the s-channel, it has to be significantly off-shell in this process. We note that the differences between the scalar and pseudoscalar couplings to the DM pair, while very significant for dynamics in the low-velocity regime, are negligible at the LHC where all particles are produced relativistically.

Next, there are two operators of dimension six,

$$\mathcal{L}_6 = \frac{C_1^{(6)}}{\Lambda^2} H^\dagger \overleftrightarrow{D}_\mu H \; \overline{\chi}\gamma^\mu\chi + \frac{C_2^{(6)}}{\Lambda^2} H^\dagger \overleftrightarrow{D}_\mu H \; \overline{\chi}\gamma^\mu\gamma_5\chi, \tag{6.37}$$

where the covariant derivative D_μ is defined in Eq. (2.4). Once again, in the unitary gauge and in terms of the physical Higgs field h, we can rewrite Eq. (6.37) as

$$\mathcal{L}_6 = \frac{iC_1^{(6)} m_Z v}{2\Lambda^2} Z_\mu \, \overline{\chi}\gamma^\mu\chi + \frac{iC_2^{(6)} m_Z v}{2\Lambda^2} Z_\mu \, \overline{\chi}\gamma^\mu\gamma_5\chi + \frac{iC_1^{(6)} m_Z}{\Lambda^2} h Z_\mu \, \overline{\chi}\gamma^\mu\chi + \frac{iC_2^{(6)} m_Z}{\Lambda^2} h Z_\mu \, \overline{\chi}\gamma^\mu\gamma_5\chi + \frac{iC_1^{(6)} m_Z}{2v\Lambda^2} h^2 Z_\mu \, \overline{\chi}\gamma^\mu\chi + \frac{iC_2^{(6)} m_Z}{2v\Lambda^2} h^2 Z_\mu \, \overline{\chi}\gamma^\mu\gamma_5\chi. \tag{6.38}$$

Defining Z_μ in a standard way (see Eq. (2.10) and using Eq. (2.18) we generated new operators of Eq. (6.38), with each line giving rise to a different experimental signature.

The first line gives an effective coupling of DM to the Z boson, which could lead to a change in the Z boson invisible width for light enough DM. The second line gives the mono-Higgs signature we are interested in here, with an off-shell Z boson in the s-channel. Once again, the presence or absence of the γ_5 in the DM bilinear does not appreciably impact the collider phenomenology for this operator. The final line could lead to a final state with two Higgs bosons and missing energy, or a Z boson, Higgs boson, and missing energy.

There are similarly four operators of dimension seven that involve Higgs doublets and their derivatives,

$$\begin{aligned} \mathcal{L}_{7H} &= \frac{C_1'^{(7)}}{\Lambda^3}\left(H^\dagger H\right)^2 \overline{\chi}\chi + \frac{C_2'^{(7)}}{\Lambda^3}\left(H^\dagger H\right)^2 \overline{\chi}\gamma_5\chi, \\ &+ \frac{C_3'^{(7)}}{\Lambda^3}\left|D_\mu H\right|^2 \overline{\chi}\chi + \frac{C_4'^{(7)}}{\Lambda^3}\left|D_\mu H\right|^2 \overline{\chi}\gamma_5\chi. \end{aligned} \tag{6.39}$$

The part of $\mathcal{L}_{7H}$ that generates the mono-Higgs signature for hadron collider and can be written as

$$\begin{aligned} \mathcal{L}_{7H} &= \frac{3C_1'^{(7)}}{2}\frac{v^2}{\Lambda^3}h^2 \, \overline{\chi}\chi + \frac{3C_2'^{(7)}}{2}\frac{v^2}{\Lambda^3}h^2 \, \overline{\chi}\gamma_5\chi, \\ &+ \frac{C_3'^{(7)}}{2\Lambda^3}\left(\partial_\mu h\right)^2 \overline{\chi}\chi + \frac{C_4'^{(7)}}{2\Lambda^3}\left(\partial_\mu h\right)^2 \overline{\chi}\gamma_5\chi. \end{aligned} \tag{6.40}$$

We do not expect strong constraints on Λ from those operators, as they simply represent higher order $1/\Lambda$ corrections to the operators discussed above. We have listed them here for completeness but shall not consider them further.

There are also four operators of dimension seven that describe the coupling of Dark Matter to the SM fermions f,

$$\begin{aligned} \mathcal{L}_{7F} &= \frac{2\sqrt{2}C_1^{(7)}}{\Lambda^3} \, y_d \, \overline{Q}_L H d_R \, \overline{\chi}\chi + \frac{2\sqrt{2}C_1^{(7)}}{\Lambda^3} \, y_u \, \overline{Q}_L \widetilde{H} u_R \, \overline{\chi}\chi \\ &+ \frac{2\sqrt{2}C_2^{(7)}}{\Lambda^3} \, y_d \, \overline{Q}_L H d_R \, \overline{\chi}\gamma^5\chi + \frac{2\sqrt{2}C_2^{(7)}}{\Lambda^3} \, y_u \, \overline{Q}_L \widetilde{H} u_R \, \overline{\chi}\gamma^5\chi + h.c. \end{aligned} \tag{6.41}$$

Here $\widetilde{H} = i\sigma_2 H^*$ is the usual charge-conjugated Higgs field and we scaled the Wilson coefficients to introduce Yukawa couplings y_f for each fermion flavor $f = u, d$ of up (u) or down (d) type. Q_L is a standard electroweak doublet of left-handed fermions. This form of operators is invariant under the electroweak $SU(2)_L$ group and also naturally suppresses DM couplings to the light fermions, and is well motivated by the Minimal Flavor Violation paradigm discussed earlier. We assume the couplings $C_i^{(7)}$ are flavor-blind but permit them to be complex. In terms of the physical field h the Eq. (6.41) can be written as

$$\begin{aligned} \mathcal{L}_{7F} &= \frac{Re\left(C_1^{(7)}\right)}{\Lambda^3} y_f \left(\overline{f}f\right) h \left(\overline{\chi}\chi\right) + \frac{Im\left(C_1^{(7)}\right)}{\Lambda^3} i y_f \left(\overline{f}\gamma_5 f\right) h \left(\overline{\chi}\chi\right) \\ &+ \frac{Im\left(C_2^{(7)}\right)}{\Lambda^3} i y_f \left(\overline{f}f\right) h \left(\overline{\chi}\gamma_5\chi\right) + \frac{Re\left(C_2^{(7)}\right)}{\Lambda^3} y_f \left(\overline{f}\gamma_5 f\right) h \left(\overline{\chi}\gamma_5\chi\right) \end{aligned} \tag{6.42}$$

Note that these operators are identical to those which have traditionally been known as D1-D4 in Table 6.1 of the previous section [94] on effective theories of DM scattering and production, with the sole difference being that the implied Higgs vev has been replaced by the dynamical Higgs field in these operators. This is another case where the scalar versus pseudoscalar nature of the couplings is not important to the collider phenomenology. We expect the strongest constraints to come from this and the next set of operators, even though they are operators of relatively high dimensions.

There are also four operators that are formally of dimension 8 that describe DM couplings to the gluons and the physical Higgs [151],

$$\begin{aligned} \mathcal{L}_8 &= \frac{C_1^{(8)}}{\Lambda^3 M_{EW}} (\bar{\chi}\chi)\, h\, G^{a\mu\nu} G^a_{\mu\nu} + \frac{C_2^{(8)}}{\Lambda^3 M_{EW}} \left(\bar{\chi}\gamma^5\chi\right) h\, G^{a\mu\nu} G^a_{\mu\nu} \\ &+ \frac{C_3^{(8)}}{\Lambda^3 M_{EW}} (\bar{\chi}\chi)\, h\, G^{a\mu\nu} \widetilde{G}^a_{\mu\nu} + \frac{C_4^{(8)}}{\Lambda^3 M_{EW}} \left(\bar{\chi}\gamma^5\chi\right) h\, G^{a\mu\nu} \widetilde{G}^a_{\mu\nu} \end{aligned} \tag{6.43}$$

where we choose $M_{EW} = v$. Note that the presence of M_{EW} here makes these operators equivalent in power counting of the New Physics scale to the dimension-seven operators above. In fact, similarly to the operators in equation 6.42, these are equivalent to the well-known operators D11-D14 of Table 6.1 and [94] with a Higgs vev replaced by the dynamical field. Once again, the parity structure of the operator is largely irrelevant for collider experiments. It is important to note that these dimension seven operators mix with those in Eq. (6.42) due to diagrams analogous to those responsible for the gluon fusion production of the Higgs boson.

6.6 Notes for further reading

We have only touched upon the rich and beautiful physics of the Higgs and electroweak bosons. There are several reviews of electroweak oblique corrections [62, 109, 171, 180]. In our discussion, we mostly followed [180]. The S, T, and U parameters were introduced in [148]. A good review of the SM EFT can be found in [30]. An excellent review of the on-shell renormalization scheme used in NLO SM EFT computations in Higgs and EW boson decays is given in [63].

Problems for Chapter 6

1. **Problem 1. Tree-level relations for electroweak observables.** Using current experimental data from [1], check the consistency of tree-level relations given in Eq. (6.4). Use error propagation techniques to obtain statistical uncertainties of the predictions on the right-hand side of Eq. (6.4) to compare with the experimental uncertainties of the observables on the left-hand side. By how many sigmas do they differ? How large of a deviation did you expect to find?
2. **Problem 2. Invisible Higgs decays.** Using the Lagrangian in Eq. (6.21), derive the Higgs decay width in Eq. (6.26). You can use results from Section A.2.1 of the Appendix.

7

Conclusions

This book is an attempt to answer one basic question: can manifestations of New Physics be unambiguously seen in a low-energy experiment? The answer to this basic question is complicated in that we do not know *what* type of New Physics we should expect, and how to properly define "unambiguously". The first issue was addressed by the introduction of the Standard Model Effective Field Theory, while the second relied on a multitude of symmetry relations or considerations of some limiting cases where emerging symmetries allowed for the systematic perturbative expansions in the symmetry-breaking parameters. The examples of the former included flavor symmetries, while the latter involved heavy quark or chiral symmetries.

In this book, we discussed searches for indirect effects of New Physics in many systems, ranging from atomic interactions to collider physics. Applications of effective field theory techniques allowed us to approach problems in a *model-independent way*, properly identifying the relevant degrees of freedom, and cordoning off any additional assumptions into coefficients that can be measured, calculated, or in some way quantified. It is probably the most sensible way to approach a problem in physics, and, logically, it is the most robust.

There are many topics that we had not covered in this book. The list below is far from complete, and we cannot do these topics justice. One could only hope that the reader would get interested enough in the topics that we did touch upon in this book, so the list below would serve as an inspiration to keep studying.

Astrophysical searches for Dark Matter. As we argued in Chapter 2, there are light particles, such as axions or dark photons, that could play a role of DM, provided that they are stable or at least have "cosmological" lifetimes, i.e. those comparable to the lifetime of the Universe. If this is the case, such particles can be searched for in astrophysical experiments. This is a fast-developing sector of physics.

A modern approach to DM detection can be colloquially called "make it, shake it, break it", and is illustrated in Fig. 7.1. Light Dark Matter particles could be created in the interactions or decays of the SM particles ("make it"), they can be detected in direct scattering events with SM particles ("shake it"), or seen as astrophysical remnants of their annihilation into lighter SM particles ("break it"). While the first method of studies of light NP particles was discussed throughout the book, the last two have not been touched upon at all. These processes could provide an enormous amount of interesting information, as they all can be seen as different manifestations of the same process seen in the s, t, or u channel. Useful reviews of DM phenomenology can be found in [132].

Multi-messenger astronomy. We have not discussed *multi-messenger astronomy* (MMA) and its role in constraining BSM models. MMA can be defined as a coordinated approach to the

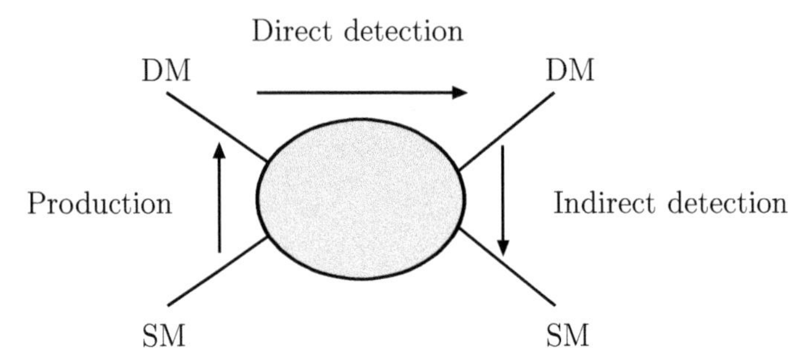

FIGURE 7.1 A cartoon illustrating the "Make it, shake it, break it" approach to Dark Matter detection. Arrows indicate the way one should look at the direction of a scattering/annihilation process.

observation of different classes of signals originating from the same astrophysical event. It allows studies of high-energy neutrinos and gravitational waves that could be originating from exotic sources, applied in concert with wide electromagnetic coverage. A simple introduction can be found in [141].

New Physics searches in atomic and molecular systems. We only touched upon the uses of atoms as means for detecting BSM physics, mainly in connection with EDM measurements. Compared to atoms, the larger polarizability and internal fields of molecules can result in increased sensitivity to electron's EDM by up to three orders of magnitude. Other atomic techniques, including those involving quantum sensors and high-accuracy atomic clocks, pave the way to other measurements sensitive to New Physics. Such measurements can address the most fundamental questions, such as to see if the speed of light is constant throughout the lifetime of the Universe [164].

Machine learning. Machine learning (ML) is a concept that involves both algorithm building and the development of modeling tools to deal with data processing. Various ML techniques, including artificial neural networks (ANNs), are now widely employed in analyses of experimental data and modeling of theoretical inputs. Their use in experimental particle physics for jet-finding algorithms and other applications are well-known [40]. Additionally, ANN ability to provide unbiased universal approximants to incomplete data allowed the application of ML to the modeling of non-perturbative QCD functions, such as form factors and parton distribution amplitudes. We did not have an opportunity to describe this powerful tool in this book.

Amplitude methods and EFTs. The standard way of doing QFT calculations, employed in this book, involves several steps. First, a Lagrangian is constructed based on a set of chosen fields and symmetries. Next, Feynman amplitudes are constructed from the Lagrangian to describe the scattering process. As we saw in this book, precision calculations often involve thousands of diagrams, making the computations extremely difficult. One of the newer methods to deal with the complexity of standard calculations is to introduce methods that directly involve on-shell scattering amplitudes. Similar to effective Lagrangian construction, the method of *on-shell amplitudes* [44, 71] is built on constructions that respect Lorentz and global internal symmetries, as well as locality and unitarity. The advantages of this method include the absence of gauge or operator redundancies, which greatly simplifies the calculations. Maybe the power of the on-shell technique can be used in an effective field theory context?

The answer to this question is indeed positive [66]; the techniques of on-shell amplitudes have been applied to the electroweak sector of the Standard Model and SM EFT. We encourage readers to follow these exciting developments.

Explicit New Physics models. Last, but not least important item that we barely touched upon in this book includes explicit NP model building. We avoided talking about explicit models of New Physics, choosing instead SM EFT techniques that encoded NP models in the matching expressions of Wilson coefficients. Such an approach has a huge advantage of covering a multitude of explicit NP models in terms of only a few effective operators.

Yet, explicit model building is important because both direct and indirect searches for new degrees of freedom can be done. It is the correlation of different observables that makes such analyses powerful. Even concentrating on the indirect searches allows for the studies of correlations between processes that cannot be meaningfully correlated if only effective operators are considered.

With new data constantly becoming available from the current and new experiments, particle phenomenology will continue to be an exciting field for years to come. We hope that the methods and techniques described in this book will motivate the readers to participate in this exercise.

Useful mathematics

A.1 Dimensional regularization: useful formulas

A.1.1 Frequently used formulas

Dimensional regularization (DR) is a wonderful regularization scheme that achieves control over divergent integrals by computing them in $d = 4 - 2\epsilon$ dimensions. Here ϵ can be thought of as an arbitrary parameter that is often taken as small. The integrals are analytically continued to d-dimensions, which requires the introduction of an arbitrary mass scale μ,

$$\int d^4 k \to \mu^{2\epsilon} \int d^{4-2\epsilon} k = \mu^{2\epsilon} \int d^d k. \tag{A.1}$$

An important feature of dimensional regularization is that this mass scale does not appear as a cut-off that can spoil momentum-dependent power counting of the effective theory. Both ultraviolet (UV) and infrared (IR) infinities appear as poles in ϵ and can be subtracted (redefined into relevant physical parameters) using a scheme of choice. One such convenient scheme is $\overline{MS}$, where one subtracts the poles in ϵ and an additional constant $\log(4\pi) - \gamma$ appearing from the expansion of Γ-functions. This scheme is the one that is most often used by effective field theorists (except for the cases when it is not!), but it is by no means unique.

Here we will not attempt to introduce the techniques of dimensional regularization, which is usually covered in any course in quantum field theory. Instead, we will provide some reference formulas needed to perform calculations described in this book and elaborate on some points of this technique that is relevant for calculations in effective field theories.

While there is no *unique* regularization scheme for dealing with divergent integrals appearing in calculations of quantum corrections in field theory (Nature does not care which method we use to calculate the integrals), dimensional regularization is certainly one of the most convenient schemes we know. It helps to regularize both UV and IR divergencies without introducing a hard scale or breaking important symmetries of the theory (such as gauge or chiral). Dimensional regularization is often used in conjunction with Feynman parameterization of integrals for combining denominators.

$$\frac{1}{A_1 A_2 ... A_n} = \int_0^1 dx_1 dx_2 ... dx_n \frac{(n-1)!\ \delta(1 - x_1 - ... - x_n)}{(x_1 A_1 + x_2 A_2 + ... + x_n A_n)^n}, \tag{A.2}$$

where $\delta(1 - x_1 - ... - x_n)$ is a Dirac delta function. Some particular cases are used most often, so we provide those particular examples here. Two denominators can be combined as

$$\frac{1}{A_1 A_2} = \int_0^1 \frac{dx}{\left(A_2 - x(A_2 - A_1)\right)^2}. \tag{A.3}$$

Taking a derivative with respect to A_1 or A_2 several times, we obtain

$$\frac{1}{A_1^n A_2^m} = \frac{\Gamma(n+m)}{\Gamma(n)\Gamma(m)} \int_0^1 dx \frac{x^{n-1}(1-x)^{m-1}}{\left(A_2 - x(A_2 - A_1)\right)^{n+m}}, \tag{A.4}$$

which is also often used. Three denominators can be combined as

$$\frac{1}{A_1 A_2 A_3} = \int_0^1 dx_1 \int_0^{1-x_1} \frac{2\, dx_2}{\left(A_3 - x_1(A_3 - A_1) - x_2(A_3 - A_2)\right)^3}. \tag{A.5}$$

Similar formulas can be obtained for different powers of A_i. We note that the parameterization above is not unique and sometimes not the most convenient one. For example, in HQET another formula is often used,

$$\frac{1}{A_1^n A_2^m} = \frac{2^m \Gamma(n+m)}{\Gamma(n)\Gamma(m)} \int_0^\infty dx \frac{x^{m-1}}{\left(A_1 + 2x A_2\right)^{n+m}}. \tag{A.6}$$

This formula is useful when the denominators have different dimensions, such as a heavy-quark propagator and a gluon propagator. In that case, the dummy integration variable caries dimensions and is integrated to infinity rather than over a finite range.

Let us also present several useful formulas needed for isolation of $1/\epsilon$ poles in calculations of one-loop integrals. The following tensor integral often appears in calculations of radiative corrections in effective theories,

$$\begin{aligned} \int \frac{d^d s}{(2\pi)^d} \left(\frac{\omega}{v \cdot s + \omega}\right)^\beta \frac{s^{\mu_1} ... s^{\mu_n}}{(-s^2)^\alpha} &= i \, (4\pi)^{d/2} \, I_n(\alpha, \beta) \\ &\times \; (-2\omega)^{d-2\alpha+n} K^{\mu_1 ... \mu_n}(v; \alpha), \end{aligned} \tag{A.7}$$

where we denote

$$I_n(\alpha, \beta) = \frac{\Gamma(d/2 - \alpha + n)\Gamma(2\alpha + \beta - d - n)}{\Gamma(\alpha)\Gamma(\beta)}. \tag{A.8}$$

The tensor structure can be written as follows,

$$\begin{aligned} K^{\mu_1 ... \mu_n}(v; \alpha) &= \sum_{j=0}^{[n/2]} (-1)^{n-j} C(n, j; \alpha) \\ &\times \sum_{\nu_i = \sigma(\mu_i)}{}' g^{\nu_1 \nu_2} ... g^{\nu_{2j-1} \nu_{2j}} v^{\nu_{2j+1}} ... v^{\nu_n}, \end{aligned} \tag{A.9}$$

where $[n/2]$ denotes the largest integer that is less than or equal to $n/2$, and the primed sums denote a summation over all permutations that lead to a different assignment of indices. Finally,

$$C(n, j; \alpha) = \prod_{k=1}^{j} \frac{1}{d + 2(n - k - \alpha)}. \tag{A.10}$$

Standard d-dimensional integrals also often appear in EFT calculations,

$$\begin{aligned} \int \frac{d^d k}{(2\pi)^d} \frac{1}{[k^2 - \Lambda^2]^n} &= \frac{i(-1)^n}{(4\pi)^{d/2}} \frac{\Gamma(n - d/2)}{\Gamma(n)} \left[\frac{1}{\Lambda^2}\right]^{n-d/2}, \\ \int \frac{d^d k}{(2\pi)^d} \frac{k^\mu k^\nu}{[k^2 - \Lambda^2]^n} &= \frac{i(-1)^{n-1}}{(4\pi)^{d/2}} \frac{g^{\mu\nu}}{2} \frac{\Gamma(n - 1 - d/2)}{\Gamma(n)} \left[\frac{1}{\Lambda^2}\right]^{n-1-d/2}, \end{aligned} \tag{A.11}$$

where $\Lambda = \Lambda(x_i, m_j^2, p_k \cdot p_n, ...)$ is a function of Feynman parameters x_i, masses m_j^2, and/or external momenta p_k. More identities can be obtained by tracing the indices with $g_{\mu\nu}$, one just has to remember that in d-dimensions $g^{\mu\nu}g_{\mu\nu} = d$. Also, rotational symmetry implies that one can always substitute $k^\mu k^\nu \to g^{\mu\nu}k^2/d$ under the integral. Integrals with more momenta in the numerator can be found in appendices of standard quantum field theory textbooks, such as [147] or [65].

Ultraviolet poles in ϵ appear from the expansion of Gamma functions of the type appearing in Eq. (A.11) around their poles at negative integers $n \in N$,

$$\Gamma(-n+\epsilon) = \frac{(-1)^n}{n!}\left[\frac{1}{\epsilon} + \sum_{k=1}^{n}\frac{1}{k} - \gamma_E\right] + \mathcal{O}(\epsilon), \tag{A.12}$$

where $\gamma_E \approx 0.5772$ is the Euler-Mascheroni constant. More identities can be obtained from the ones above by using one of the properties of the Gamma function, $\Gamma(z+1) = z\Gamma(z)$. It is also useful to remembers that

$$\begin{aligned} \Gamma(n) &= (n-1)! \\ \Gamma(1) &= 1 \\ \Gamma(2) &= 1 \\ \Gamma(1/2) &= \sqrt{\pi}. \end{aligned} \tag{A.13}$$

Finally, most expansions of Feynman integrals around poles in ϵ are obtained using the following trick,

$$a^b = \exp\log a^b = \exp(b\log(a)) = 1 + b\log(a) + \dots. \tag{A.14}$$

For example, in the expansion of the Feynman integral

$$\mu^{4-d}\left(\frac{1}{\Lambda^2}\right)^{2-d/2} \approx 1 - \epsilon\log\frac{\Lambda^2}{\mu^2} \tag{A.15}$$

for $4 - d = 2\epsilon$, where μ is some scale.

A.1.2 Passarino-Veltman functions

It is often convenient to rewrite one-loop integrals in terms of the Passarino-Veltman functions. Some tools for automated computations of Feynman amplitudes use these functions in computing loop integrals. These functions are often used in the output of applications doing automated high-energy physics calculations. We will list them here along with their expansions in dimensional regularization. Note that we will distinguish "ϵ"-parameter of dimensional regularization introduce above and "ε" – a small shift of the poles in integral loops [180].

$$A_0(m^2) = 16\pi^2\mu^{4-d}\int\frac{d^dk}{i(2\pi)^d}\frac{1}{k^2 - m^2 + i\varepsilon}, \tag{A.16}$$

$$B_0(p^2, m_1^2, m_2^2) = 16\pi^2\mu^{4-d}\int\frac{d^dk}{i(2\pi)^d}\frac{1}{[k^2 - m_1^2 + i\varepsilon][(k-p)^2 - m_2^2 + i\varepsilon]}, \tag{A.17}$$

$$p_\mu B_1(p^2, m_1^2, m_2^2) = 16\pi^2\mu^{4-d}\int\frac{d^dk}{i(2\pi)^d}\frac{k_\mu}{[k^2 - m_1^2 + i\varepsilon][(k-p)^2 - m_2^2 + i\varepsilon]}, \tag{A.18}$$

$$p_\mu p_\nu B_{21}(p^2, m_1^2, m_2^2) + g_{\mu\nu}B_{22}(p^2, m_1^2, m_2^2) = 16\pi^2\mu^{4-d}\int\frac{d^dk}{i(2\pi)^d}\frac{k_\mu k_\nu}{[k^2 - m_1^2 + i\varepsilon][(k-p)^2 - m_2^2 + i\varepsilon]} \tag{A.19}$$

Some of these functions are infinite with isolated poles at $d = 4$. As it is often done in the Modified Minimal Subtraction scheme ($\overline{MS}$), we define a coefficient

$$\frac{1}{\bar{\epsilon}} \equiv \frac{1}{4-d} - \gamma_E + \ln 4\pi. \tag{A.20}$$

The one-point and two-point functions have analytic solutions

$$\begin{aligned} A_0(m^2) &= m^2\left(\frac{1}{\bar{\epsilon}} + 1 - \ln m^2/\mu^2\right) = m^2\left(1 + B_0(0, m^2, m^2)\right), \\ B_0(p^2, m_1^2, m_2^2) &= \frac{1}{\bar{\epsilon}} - \int_0^1 \log\frac{(1-x)m_1^2 + xm_2^2 - x(1-x)p^2 - i\varepsilon}{\mu^2} \\ &= \frac{1}{\bar{\epsilon}} - \log(p^2/\mu^2) - I(x_+) - I(x_-) \end{aligned} \tag{A.21}$$

where we defined

$$\begin{aligned} x_\pm &= \frac{(p^2 - m_2^2 + m_1^2) \pm \sqrt{(p^2 - m_2^2 + m_1^2)^2 - 4p^2(m_1^2 - i\varepsilon)}}{2p^2}, \quad \text{and} \\ I(x) &= \log(1-x) - x\log\frac{x-1}{x} - 1. \end{aligned} \tag{A.22}$$

A.2 Phase space calculations

A.2.1 Two body decay

Two body decay rate is discussed at length in most QFT textbooks. It might be useful to present it here for reference before we consider decays with more particles in the final state. Let us consider a decay of a particle P with the mass M into a set of particles $a, b, c, ...$ with the masses $M_1, m_2, m_3, ...$ respectively. The decay rate $P(P) \to a(p_1)b(p_2)c(p_3)...$ can be written as

$$\Gamma(P \to abc...) = \frac{1}{2M}\int \frac{d^3\mathbf{p}_1}{(2\pi)^3 2E_1}\frac{d^3\mathbf{p}_2}{(2\pi)^3 2E_2}...(2\pi)^4\delta^{(4)}(P - p_1 - p_2 - ...)\left|A\right|^2, \tag{A.23}$$

where $|A|^2$ is the amplitude squared that governs the dynamics of the decay. We will denote everything on the right-hand side of this equation, with the exception of $|A|^2$, as $d\Phi(p_1, p_2, ...)$. If the spin of the decaying particle is not detected, we average over its spin states,

$$|A|^2 \to \left|\overline{A}\right|^2 = \frac{1}{2S+1}|A|^2. \tag{A.24}$$

For a two body decay $P(P) \to a(p_1)b(p_2)$,

$$\int d\Phi(p_1, p_2) = \int \frac{d^3\mathbf{p}_1}{(2\pi)^3 2E_1}\frac{d^3\mathbf{p}_2}{(2\pi)^3 2E_2}(2\pi)^4\delta^{(4)}(P - p_1 - p_2) \tag{A.25}$$

As we can see from the Eq. (A.25), there are six integrals and four delta functions. It is most convenient to consider the decay in the rest frame of the decaying particle P, or the center-of-mass (CM) frame, so $P_\mu = (M, 0, 0, 0)$. In this frame $\mathbf{p}_1 + \mathbf{p}_2 = 0$, so the integration over $\mathbf{p}_2$ is trivial,

$$\begin{aligned} \int d\Phi(p_1, p_2) &= \int \frac{d^3\mathbf{p}_1}{(2\pi)^3 2E_1}\frac{d^3\mathbf{p}_2}{(2\pi)^3 2E_2}(2\pi)^4\delta\left(M - E_1 - E_2\right)\delta^{(3)}(\mathbf{p}_1 + \mathbf{p}_2) \\ &= \frac{1}{4}\frac{1}{(2\pi)^2}\int \frac{p^2 dp}{\sqrt{p^2 + m_1^2}}\frac{d\cos\theta d\phi}{\sqrt{p^2 + m_2^2}}\delta\left(M - \sqrt{p^2 + m_1^2} - \sqrt{p^2 + m_2^2}\right), \end{aligned} \tag{A.26}$$

where we introduced $p \equiv |\mathbf{p}_1|$, the length of the vector $\mathbf{p}_1$. Note that $E_i = \sqrt{p^2 + m_i^2}$. We now need to resolve a delta function, which we can do with the help of the following formula,

$$\delta\left(f(p)\right) = \sum_i \frac{1}{|f'(p^*{}_i)|} \delta\left(p - p^*{}_i\right), \tag{A.27}$$

where f is some function, and the sum is over all possible roots p_i^* of $f(p)$. Here f' is a derivative of the function f. In our case, there is only one positive root for p, which comes as a result of the solution of

$$M - \sqrt{p^2 + m_1^2} - \sqrt{p^2 + m_2^2} = 0. \tag{A.28}$$

This equation is trivially solved and a solution can be written as

$$p^* = \frac{M}{2}\sqrt{1 - \frac{2(m_1^2 + m_2^2)}{M^2} + \frac{(m_1^2 - m_2^2)^2}{M^4}} = \frac{M}{2}\overline{\beta}. \tag{A.29}$$

The coefficient $\overline{\beta}$ can be called *pseudo threshold velocity*, as it reduces to threshold velocity for $m_1 = m_2$,

$$\overline{\beta} \xrightarrow{m_1 = m_2 = m} \beta = \sqrt{1 - \frac{4m^2}{M^2}}, \tag{A.30}$$

but in reality there is no special name for it. A derivative can also be taken, so

$$\frac{d}{dp}\left(M - \sqrt{p^2 + m_1^2} - \sqrt{p^2 + m_2^2}\right) = -p\frac{E_1 + E_2}{E_1 E_2}. \tag{A.31}$$

The expression in Eq. (A.31) needs to be evaluated at $p = p^*$. Resisting this temptation for just one step we are rewarded with simplifying cancellations. Putting everything together,

$$\begin{aligned} \int d\Phi(p_1, p_2) &= \frac{1}{4}\frac{1}{(2\pi)^2}\int \frac{p^2 dp}{E_1 E_2}\frac{E_1 E_2}{p(E_1 + E_2)} d(\cos\theta) d\phi\, \delta\left(p - \frac{M}{2}\overline{\beta}\right) \\ &= \frac{1}{4}\frac{1}{2\pi}\overline{\beta}\int \frac{d(\cos\theta)}{2}\frac{d\phi}{2\pi}, \end{aligned} \tag{A.32}$$

where we used the fact that $E_1 + E_2 = M$. There is another way to write this expression that is often provided in literature,

$$\int d\Phi(p_1, p_2) = \frac{1}{4\pi}\frac{p}{M}\int \frac{d(\cos\theta)}{2}\frac{d\phi}{2\pi}, \tag{A.33}$$

where p is written in terms of a so-called "triangle function" $\lambda(x, y, z) = x^2 + y^2 + z^2 - 2xy - 2xz - 2yz$* as

$$p = \frac{\lambda^{1/2}(s, m_1^2, m_2^2)}{2M}. \tag{A.34}$$

If there is no angular dependence in the matrix element, which might arise if we are looking for the decay of polarized particle, one can integrate over the angular variables to arrive at the final answer,

$$\Gamma(P \to ab) = \frac{1}{2M}\int d\Phi_2 \left|A\right|^2 = \frac{1}{16\pi M}\overline{\beta}\left|A\right|^2. \tag{A.35}$$

We shall come back to this expression in the next section.

---

*It is called so because $\sqrt{-\lambda(x, y, z)}/4$ is the area of the triangle with the sides $\sqrt{x}$, $\sqrt{y}$, and $\sqrt{z}$, which can be proven with the help of the Heron's formula. As can be seen from its definition, it is symmetric with respect to the interchange of any two of its variables.

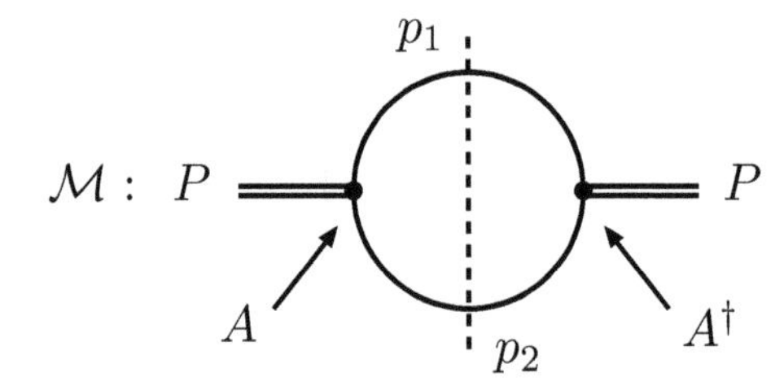

FIGURE A.1 Cut diagrams for a scalar particle decay. Note that the quantity to the right of the dashed line are Hermitian-conjugated of those to the left of the line.

A.2.2 Three body decay

Evaluation of the three body decay is rather similar to what we have already discussed above. For a two body decay $P(P) \to a(p_1)b(p_2)c(p_3)$,

$$\int d\Phi(p_1, p_2, p_3) = \int \frac{d^3\mathbf{p}_1}{(2\pi)^3 2E_1} \frac{d^3\mathbf{p}_2}{(2\pi)^3 2E_2} \frac{d^3\mathbf{p}_3}{(2\pi)^3 2E_3} (2\pi)^4 \delta^{(4)}(P - p_1 - p_2 - p_3). \tag{A.36}$$

As we can see from the Eq. (A.36), there are now nine integrals and four delta functions, which means that in general there are five variables left after the delta-function integrations. In the CM frame now $\mathbf{p}_1 + \mathbf{p}_2 + \mathbf{p}_3 = 0$. Integrating over $\mathbf{p}_3$ and p_2^0 we can write

$$\begin{aligned} \int d\Phi(p_1, p_2, p_3) &= \frac{1}{(4\pi)^2} \frac{|\mathbf{p}_1|^2 d|\mathbf{p}_1|}{(2\pi)^3 2E_1} d\Omega_1 \frac{|\mathbf{p}_2|}{m_{23}} d\Omega_2 \\ &= \frac{2}{(4\pi)^3} \lambda^{1/2}\left(1, \frac{m_2^2}{m_{23}^2}, \frac{m_3^2}{m_{23}^2}\right) |\mathbf{p}_1|\, dE_1 dx_2 dx_3 dx_4 dx_5, \end{aligned} \tag{A.37}$$

where we introduced the normalized angle variables $d\cos\theta_{1,2} = 2dx_{2,4}$, and $d\phi_{1,2} = 2\pi dx_{3,5}$. All angular variables x_i change in the range $0 \le x_i \le 1$. Note that the energy changes as to

$$E_1^{max} = \frac{s + m_1^2 - (m_2 + m_3)^2}{2M}. \tag{A.38}$$

The standard form of the Dalitz plot for the three-body decays can be found in [1].

A.3 Cut diagrams. Cutkosky rules

In 1960, Cutkosky published a paper [57] where he formulated the rules on how to determine the discontinuity of a Feynman amplitude across a branch cut. These discontinuities are connected to the imaginary part of the Feynman integral represented the amplitude. Just like the optical theorem, Cutkosky rules follow from the unitarity of the S-matrix, $SS^\dagger = 1$. They provide a useful tool in computing imaginary parts of Feynman diagrams and are quite instrumental in computing lifetimes and lifetime differences of mesons and baryons. Cutkosky rules are derived in general QFT courses (see, for example, [147]), so here we only provide a prescription for computing imaginary parts of Feynman graphs. As $S = 1 + iT$, where T is the nontrivial scattering matrix, the unitarity condition leads to

$$T - T^\dagger \equiv 2i\text{Im}T = iTT^\dagger, \quad \text{or} \quad 2\text{Im}T = TT^\dagger. \tag{A.39}$$

It is convenient to compute the diagrams for the amplitudes squared by "cutting" the diagram, i.e., by drawing a dashed line vertically through the diagram, completely separating the parts to the left and to the right of the cut and then summing over all possible cuts.

It is convenient to substitute "cut" propagators with their imaginary parts from the start. For example, for a scalar particle,

$$G(p) = \frac{i}{p^2 - m^2 + i\epsilon} \to (2\pi)\, \theta(p^0)\, \delta(p^2 - m^2), \tag{A.40}$$

which corresponds to $2\text{Im}G(p)$. Similar substitutions can be done for other on-shell particles in the "cut" amplitudes. For the fermions,

$$G_F(p) = \frac{i(\not p + m)}{p^2 - m^2 + i\epsilon} \to (2\pi)\, \theta(p^0)\, \delta(p^2 - m^2) \left(\not p + m\right), \tag{A.41}$$

while for the gauge particles, such as photon or gluon,

$$G_{\mu\nu}(p) = \frac{-id_{\mu\nu}}{p^2 + i\epsilon} \to (2\pi)\, \theta(p^0)\, \delta(p^2) d_{\mu\nu}. \tag{A.42}$$

Because of the delta-functions enforcing the on-shell condition $p^2 = m^2$ in the equations above, the cut lines are often referred to as being "on their mass surface."

An example might illustrate the concept. Consider, as an example, a "cut" diagram for the decay of a spin-0 particle into two spin-0 particles, as shown in Fig. A.1. The decay width can be written as

$$\Gamma = \frac{1}{M} \text{ Disc } \mathcal{M} \tag{A.43}$$

where the discontinuity is determined by a "cut diagram" in Fig. A.1 with the vertices on the right-hand side of the diagram, as determined by a dashed line, are Hermitian conjugated of those on the left side of the diagram. This is,

$$2\text{Disc } \mathcal{M} = \int \frac{d^4 p_1}{(2\pi)^4} \left|A\right|^2 (2\pi)^2 \delta(p_1^2 - m_1^2) \delta(p_2^2 - m_2^2) \theta(p_1^0) \theta(p_2^0), \tag{A.44}$$

where the theta functions ensure that only positive energy solutions are selected. This will affect how we solve the delta-function using the formula of Eq. (A.27). Selecting $p_1^0 \equiv E_1 = +\sqrt{\mathbf{p}_1^2 + m_1^2}$, we obtain

$$2\text{Disc } \mathcal{M} = \frac{1}{(2\pi)^2} \int \frac{d^4 p_1}{2E_1} \left|A\right|^2 (2\pi)^2 \delta\left((P - p_1)^2 - m_2^2 \right). \tag{A.45}$$

Solving a delta function and, again, introducing $p \equiv |\mathbf{p}_1|$ as the length of the vector $\mathbf{p}_1$, we obtain,

$$\delta\left((P - p_1)^2 - m_2^2 \right) = \frac{E_1}{2Mp} \delta\left(p - \frac{M}{2} \overline{\beta} \right), \tag{A.46}$$

with $\overline{\beta}$ defined in Eq. (A.29). This results in

$$\Gamma = \frac{1}{8M} \frac{1}{2\pi} \overline{\beta} \int \frac{d(\cos\theta)}{2} \frac{d\phi}{2\pi} \left|A\right|^2 = \frac{1}{16\pi M} \left|A\right|^2 \overline{\beta}, \tag{A.47}$$

which is consistent with Eq. (A.35), where the phase space function was computed using more direct methods.

A.4 Fierz relations

Fierz relations represent a useful tool for computing matrix elements of four-fermion operators. They are a consequence of the fact that Dirac matrices represent a complete set of matrices,

TABLE A.1 Fierz matrix with anti-commuting spinors.

	S	P	V	A	T
S	−1/4	−1/4	−1/4	1/4	1/4
P	−1/4	−1/4	1/4	−1/4	1/4
V	−1	1	1/2	1/2	0
A	1	−1	1/2	1/2	0
T	3/2	3/2	0	0	1/2

which could be used as a basis upon which one can expand any 4×4 matrix. Fiertz relations are usually written in terms of five independent constructions, S (scalar), P (pseudoscalar), V (vector), A (axial), and T (tensor) built out of four fermions fields a, b, c, and d. We follow [144] and define them as

$$\begin{aligned} S &= (\bar{a}b)(\bar{c}d), \\ P &= (\bar{a}\gamma_5 b)(\bar{c}\gamma_5 d), \\ V &= (\bar{a}\gamma_\mu b)(\bar{c}\gamma^\mu d), \\ A &= (\bar{a}\gamma_\mu\gamma_5 b)(\bar{c}\gamma^\mu\gamma_5 d), \\ T &= \frac{1}{2}(\bar{a}\sigma_{\mu\nu} b)(\bar{c}\sigma^{\mu\nu} d), \end{aligned} \tag{A.48}$$

What would happen if we interchange b and d spinors? It can be shown that constructions of that type can be expanded in terms of the basis constructions S, P, V, A, and T with the coefficients of expansion given in Table A.1. As an example, let us look at the expansion of $(\bar{a}\gamma_\mu b)(\bar{c}\gamma^\mu d)$. According to the third line of Table A.1,

$$\begin{aligned} (\bar{a}\gamma_\mu b)(\bar{c}\gamma^\mu d) &= -(\bar{a}d)(\bar{c}b) + (\bar{a}\gamma_5 d)(\bar{c}\gamma_5 b) \\ &+ \frac{1}{2}(\bar{a}\gamma_\mu d)(\bar{c}\gamma^\mu b) + \frac{1}{2}(\bar{a}\gamma_\mu\gamma_5 d)(\bar{c}\gamma^\mu\gamma_5 b). \end{aligned} \tag{A.49}$$

Now, recalling that the Fierz relations are relations derived from expanding in the basis of the Dirac matrices, we can use any kind of spinors. In particular, instead of $\bar{a}$, b, $\bar{c}$ and d we can write the same relations for the left or right-handed projections of those spinors. For example, substituting $\bar{a} \to \bar{a}_L$, $b \to b_L$, $\bar{c} \to \bar{c}_L$ and $d \to d_L$, we get a famous relation,

$$(\bar{a}_L\gamma_\mu b_L)(\bar{c}_L\gamma^\mu d_L) = (\bar{a}_L\gamma_\mu d_L)(\bar{c}_L\gamma^\mu b_L). \tag{A.50}$$

Other relations can be obtained in a similar way. We shall list them here for reference, employing left and right-handed operators,

$$\begin{aligned} (\bar{a}_L\gamma_\mu b_L)(\bar{c}_R\gamma^\mu d_R) &= -2(\bar{a}_L d_R)(\bar{c}_R b_L), \\ (\bar{a}_L b_R)(\bar{c}_L d_R) &= -\frac{1}{2}(\bar{a}_L d_R)(\bar{c}_L b_R) + \frac{1}{8}(\bar{a}_L\sigma_{\mu\nu} d_R)(\bar{c}_L\sigma^{\mu\nu} b_R). \end{aligned} \tag{A.51}$$

All other relations for the left and right-handed spinors can be obtained from the combinations of the relations in Eq. (A.50) and (A.51).

There is one extra point that one needs to keep in mind when dealing with four-quark operators, which is related to the fact that quarks are also charged under the strong interaction gauge group, i.e. carry a color index. This means that we need to track the color indices as well. For example, a quark version of the relation of Eq. (A.50) reads

$$(\overline{q_1}_L\gamma_\mu q_{2L})(\overline{q_3}_L\gamma^\mu q_{4L}) \equiv (\overline{q_1}^i_L\gamma_\mu q^i_{2L})(\overline{q_3}^k_L\gamma^\mu q^k_{4L}) = (\overline{q_1}^i_L\gamma_\mu q^k_{4L})(\overline{q_3}^k_L\gamma^\mu q^i_{2L}), \tag{A.52}$$

where Latin indices i and k represent color. If we now try to compute matrix elements of the last operator in Eq. (A.52) using vacuum saturation, we would realize that the matrix element of an operator with two untraced color indices between two color-neutral states is not so obviously defined, a non-color-neutral part needs to be projected out. This can be done with the help of the $SU(3)$ relation

$$\delta_{i\ell}\delta_{kj} = \frac{1}{3}\delta_{ij}\delta_{k\ell} + \frac{1}{2}\lambda^a_{ij}\lambda^a_{k\ell}, \tag{A.53}$$

where λ^a_{ij} is the $SU(3)$ Gell-Mann matrix. We will be using above relations extensively when computing matrix elements of operators generating muonium-antimuonium or meson-antimeson mixing.

References

1. Particle Data Group, The Review of Particle Physics. `https://pdg.lbl.gov`.
2. *Neutrino non-standard interactions: a status report*, volume 2, 2019.
3. *Proceedings, Les Houches summer school: EFT in Particle Physics and Cosmology: Les Houches (Chamonix Valley), France*, volume 108, 2020.
4. L.F. Abbott and M.B. Wise. The Effective Hamiltonian for nucleon decay. *Phys. Rev. D*, 22:2208, 1980.
5. R. Alonso, E.E. Jenkins, A.V. Manohar, and M. Trott. Renormalization group evolution of the Standard Model dimension six operators III: gauge coupling dependence and phenomenology. *JHEP*, 04:159, 2014.
6. A.A. Anselm and A.A. Johansen. Can electroweak theta term be observable? *Nucl. Phys. B*, 412:553–573, 1994.
7. T. Aoyama, M. Hayakawa, T. Kinoshita, and M. Nio. Complete tenth-order QED contribution to the muon $g-2$. *Phys. Rev. Lett.*, 109:111808, 2012.
8. T. Appelquist and J. Carazzone. Infrared singularities and massive fields. *Phys. Rev.*, D11:2856, 1975.
9. J.M. Arnold, B. Fornal, and M.B. Wise. Simplified models with baryon number violation but no proton decay. *Phys. Rev. D*, 87:075004, 2013.
10. D. Atwood, I. Dunietz, and A. Soni. Improved methods for observing CP violation in $B^+ \to KD$ and measuring the CKM phase gamma. *Phys. Rev. D*, 63:036005, 2001.
11. A. Badin and A.A. Petrov. Searching for light Dark Matter in heavy meson decays. *Phys. Rev. D*, 82:034005, 2010.
12. P. Ball and R. Zwicky. New results on $B \to \pi, K, \eta$ decay formfactors from light-cone sum rules. *Phys. Rev. D*, 71:014015, 2005.
13. R. Barbieri and E. Remiddi. Electron and muon $1/2(g-2)$ from vacuum polarization insertions. *Nucl. Phys.*, B90:233–266, 1975.
14. C.W. Bauer, S. Fleming, D. Pirjol, and I.W. Stewart. An Effective field theory for collinear and soft gluons: Heavy to light decays. *Phys. Rev. D*, 63:114020, 2001.
15. D. Becirevic and A.B. Kaidalov. Comment on the heavy $\to$ light form-factors. *Phys. Lett. B*, 478:417–423, 2000.
16. M. Beneke. Renormalons. *Phys. Rept.*, 317:1–142, 1999.
17. M. Beneke, G. Buchalla, and I. Dunietz. Width difference in the $B_s - \bar{B}_s$ system. *Phys. Rev. D*, 54:4419–4431, 1996. [Erratum: *Phys. Rev. D* 83, 119902 (2011)].
18. M. Beneke, G. Buchalla, M. Neubert, and C.T. Sachrajda. QCD factorization for $B \to \pi\pi$ decays: Strong phases and CP violation in the heavy quark limit. *Phys. Rev. Lett.*, 83:1914–1917, 1999.
19. W. Bernreuther and M. Suzuki. The electric dipole moment of the electron. *Rev. Mod. Phys.*, 63:313–340, 1991. [Erratum: *Rev. Mod. Phys.* 64, 633 (1992)].
20. R.H. Bernstein and P.S. Cooper. Charged lepton flavor violation: an experimenter's guide. *Phys. Rept.*, 532:27–64, 2013.
21. B. Bhattacharya, C.M. Grant, and A.A. Petrov. Invisible widths of heavy mesons. *Phys. Rev. D*, 99(9):093010, 2019.
22. B. Bhattacharya and A.A. Petrov. Hadronic decays of B_c mesons with flavor $SU(3)_F$ symmetry. *Phys. Lett. B*, 774:430–434, 2017.
23. I.I.Y. Bigi and A.I. Sanda. Notes on the observability of CP violations in B decays. *Nucl. Phys. B*, 193:85–108, 1981.
24. I.I.Y. Bigi and A.I. Sanda. On the other five KM triangles. *hep-ph/9909479*, 9 1999.
25. S.M. Bilenky and C. Giunti. Neutrinoless double-beta decay: a probe of physics beyond the Standard Model. *Int. J. Mod. Phys. A*, 30(04n05):1530001, 2015.

26. S.M. Bilenky, S.T. Petcov, and B. Pontecorvo. Lepton mixing, $\mu \to e\gamma$ decay and neutrino oscillations. *Phys. Lett. B*, 67:309, 1977.
27. C. Bouchiat and L. Michel. Theory of μ-meson decay with the hypothesis of nonconservation of parity. *Phys. Rev.*, 106:170–172, 1957.
28. G.C. Branco, P.M. Ferreira, L. Lavoura, M.N. Rebelo, M. Sher, and J.P. Silva. Theory and phenomenology of two-Higgs-doublet models. *Phys. Rept.*, 516:1–102, 2012.
29. G.C. Branco, L. Lavoura, and J.P. Silva. CP violation. *Int. Ser. Monogr. Phys.*, 103:1–536, 1999.
30. I. Brivio and M. Trott. The Standard Model as an effective field theory. *Phys. Rept.*, 793:1–98, 2019.
31. L.S. Brown and G. Gabrielse. Geonium theory: physics of a single electron or ion in a Penning trap. *Rev. Mod. Phys.*, 58:233, 1986.
32. G. Buchalla. Kaon and charm physics: Theory. In *Theoretical Advanced Study Institute in Elementary Particle Physics (TASI 2000): Flavor Physics for the Millennium*, pages 143–205, 3 2001.
33. G. Buchalla, A.J. Buras, and M.E. Lautenbacher. Weak decays beyond leading logarithms. *Rev. Mod. Phys.*, 68:1125–1144, 1996.
34. G. Buchalla et al. *B*, *D* and *K* decays. *Eur. Phys. J. C*, 57:309–492, 2008.
35. W. Buchmuller, R.D. Peccei, and T. Yanagida. Leptogenesis as the origin of matter. *Ann. Rev. Nucl. Part. Sci.*, 55:311–355, 2005.
36. W. Buchmuller and D. Wyler. Effective Lagrangian analysis of new interactions and flavor conservation. *Nucl. Phys.*, B268:621–653, 1986.
37. A.J. Buras. Weak Hamiltonian, CP violation and rare decays. In *Les Houches Summer School in Theoretical Physics, Session 68: Probing the Standard Model of Particle Interactions*, pages 281–539, 6 1998.
38. G. Burdman, E. Golowich, J.L. Hewett, and S. Pakvasa. Rare charm decays in the standard model and beyond. *Phys. Rev. D*, 66:014009, 2002.
39. W.B. Cairncross and J. Ye. Atoms and molecules in the search for time-reversal symmetry violation. *Nature Rev. Phys.*, 1(8):510–521, 2019.
40. G. Carleo, I. Cirac, K. Cranmer, L. Daudet, M. Schuld, N. Tishby, et al. Machine learning and the physical sciences. *Rev. Mod. Phys.*, 91(4):045002, 2019.
41. A.B. Carter and A.I. Sanda. CP violation in B meson decays. *Phys. Rev. D*, 23:1567, 1981.
42. G. Cavoto, A. Papa, F. Renga, E. Ripiccini, and C. Voena. The quest for $\mu \to e\gamma$ and its experimental limiting factors at future high intensity muon beams. *Eur. Phys. J. C*, 78(1):37, 2018.
43. L.N. Chang, O. Lebedev, and J.N. Ng. On the invisible decays of the Upsilon and J/Ψ resonances. *Phys. Lett. B*, 441:419–424, 1998.
44. C. Cheung. *TASI lectures on scattering amplitudes*, pages 571–623. 2018.
45. C.-W. Chiang, Z. Luo, and J.L. Rosner. Two-body Cabibbo suppressed charmed meson decays. *Phys. Rev. D*, 67:014001, 2003.
46. V. Cirigliano, W. Dekens, J. De Vries, et al. Renormalized approach to neutrinoless double-β decay. *Phys. Rev. C*, 100(5):055504, 2019.
47. V. Cirigliano, R. Kitano, Y. Okada, and P. Tuzon. On the model discriminating power of $\mu \to e$ conversion in nuclei. *Phys. Rev. D*, 80:013002, 2009.
48. A.G. Cohen, D.B. Kaplan, and A.E. Nelson. Progress in electroweak baryogenesis. *Ann. Rev. Nucl. Part. Sci.*, 43:27–70, 1993.
49. T. Cohen. As scales become separated: lectures on effective field theory. *PoS*, TASI2018:011, 2019.

50. P. Colangelo and A. Khodjamirian. QCD sum rules, a modern perspective. pages 1495–1576, 10 2000.
51. D. Colladay and V.A. Kostelecky. CPT violation and the Standard Model. *Phys. Rev. D*, 55:6760–6774, 1997.
52. D. Colladay and V.A. Kostelecky. Lorentz violating extension of the Standard Model. *Phys. Rev. D*, 58:116002, 1998.
53. R. Conlin and A.A. Petrov. Muonium-antimuonium oscillations in effective field theory. *Phys. Rev. D*, 102(9):095001, 2020.
54. R. Contino, M. Ghezzi, C. Grojean, M. Muhlleitner, and M. Spira. Effective Lagrangian for a light Higgs-like scalar. *JHEP*, 07:035, 2013.
55. R.J. Crewther, P. Di Vecchia, G. Veneziano, and E. Witten. Chiral estimate of the electric dipole moment of the neutron in Quantum Chromodynamics. *Phys. Lett. B*, 88:123, 1979. [Erratum: *Phys. Lett. B* 91, 487 (1980)].
56. A. Crivellin, S. Najjari, and J. Rosiek. Lepton flavor violation in the Standard Model with general dimension-six operators. *JHEP*, 04:167, 2014.
57. R.E. Cutkosky. Singularities and discontinuities of Feynman amplitudes. *J. Math. Phys.*, 1:429–433, 1960.
58. A. Czarnecki, G.P. Lepage, and W.J. Marciano. Muonium decay. *Phys. Rev. D*, 61:073001, 2000.
59. A. Czarnecki, W.J. Marciano, and K. Melnikov. Coherent muon electron conversion in muonic atoms. *AIP Conf. Proc.*, 435(1):409–418, 1998.
60. G. D'Ambrosio, G.F. Giudice, G. Isidori, and A. Strumia. Minimal flavor violation: an effective field theory approach. *Nucl. Phys. B*, 645:155–187, 2002.
61. S. Dar. The neutron EDM in the SM: a review. 8 2000.
62. S. Dawson. Electroweak symmetry breaking and effective field theory. In *Theoretical Advanced Study Institute in Elementary Particle Physics: Anticipating the Next Discoveries in Particle Physics*, pages 1–63, 12 2017.
63. A. Denner. Techniques for calculation of electroweak radiative corrections at the one loop level and results for W physics at LEP-200. *Fortsch. Phys.*, 41:307–420, 1993.
64. M.J. Dolinski, A.W.P. Poon, and W. Rodejohann. Neutrinoless double-beta decay: status and prospects. *Ann. Rev. Nucl. Part. Sci.*, 69:219–251, 2019.
65. J.F. Donoghue, E. Golowich, and B.R. Holstein. Dynamics of the standard model. *Camb. Monogr. Part. Phys. Nucl. Phys. Cosmol.*, 2:1–540, 1992. [Camb. Monogr. Part. Phys. Nucl. Phys. Cosmol.35(2014)].
66. G. Durieux, T. Kitahara, Y. Shadmi, and Y. Weiss. The electroweak effective field theory from on-shell amplitudes. *JHEP*, 01:119, 2020.
67. P. Ekstrom and D. Wineland. The isolated electron. *Sci. Am.*, 243:90–101, 1980.
68. J. Elias-Miro, J.R. Espinosa, E. Masso, and A. Pomarol. Higgs windows to new physics through $d = 6$ operators: constraints and one-loop anomalous dimensions. *JHEP*, 11:066, 2013.
69. J. Elias-Mir, J.R. Espinosa, E. Masso, and A. Pomarol. Renormalization of dimension-six operators relevant for the Higgs decays $h \to \gamma\gamma, \gamma Z$. *JHEP*, 08:033, 2013.
70. J. Elias-Mir, C. Grojean, R.S. Gupta, and D. Marzocca. Scaling and tuning of EW and Higgs observables. *JHEP*, 05:019, 2014.
71. H. Elvang and Y.-T. Huang. *Scattering amplitudes in gauge theory and gravity.* Cambridge University Press, 4 2015.
72. J. Engel, M.J. Ramsey-Musolf, and U. van Kolck. Electric dipole moments of nucleons, nuclei, and atoms: the Standard Model and beyond. *Prog. Part. Nucl. Phys.*, 71:21–74, 2013.
73. Z. Epstein, G. Paz, and J. Roy. Model independent extraction of the proton magnetic radius from electron scattering. *Phys. Rev. D*, 90(7):074027, 2014.

74. A.F. Falk, Y. Grossman, Z. Ligeti, and A.A. Petrov. SU(3) breaking and D^0 - anti-D^0 mixing. *Phys. Rev. D*, 65:054034, 2002.
75. A.F. Falk and A.A. Petrov. Measuring gamma cleanly with CP tagged B_s and B_d decays. *Phys. Rev. Lett.*, 85:252–255, 2000.
76. A. Falkowski, M. Gonzlez-Alonso, and K. Mimouni. Compilation of low-energy constraints on 4-fermion operators in the SMEFT. *JHEP*, 08:123, 2017.
77. S. Faller, T. Mannel, and S. Turczyk. Limits on New Physics from exclusive $B \to D^{(*)}\ell\bar{\nu}$ decays. *Phys. Rev. D*, 84:014022, 2011.
78. G. Feinberg and S. Weinberg. Conversion of muonium into antimuonium. *Phys. Rev.*, 123:1439–1443, 1961.
79. T. Feldmann, M. Jung, and T. Mannel. Sequential flavour symmetry breaking. *Phys. Rev. D*, 80:033003, 2009.
80. N. Fernandez, I. Seong, and P. Stengel. Constraints on light Dark Matter from single-photon decays of heavy quarkonium. *Phys. Rev. D*, 93(5):054023, 2016.
81. M. Fertl. Next generation muon $g-2$ experiment at FNAL. *Hyperfine Interact.*, 237(1):94, 2016.
82. P. Fileviez Perez. New paradigm for baryon and lepton number violation. *Phys. Rept.*, 597:1–30, 2015.
83. V.V. Flambaum and A. Kozlov. Extension of the Schiff theorem to ions and molecules. *Phys. Rev. A*, 85:022505, 2012.
84. G. Gabadadze and M. Shifman. QCD vacuum and axions: What's happening? *Int. J. Mod. Phys. A*, 17:3689–3728, 2002.
85. S. Gardner and J. Shi. Patterns of CP violation from mirror symmetry breaking in the $\eta \to \pi^+\pi^-\pi^0$ Dalitz plot. *Phys. Rev. D*, 101(11):115038, 2020.
86. A. Giri, Y. Grossman, A. Soffer, and J. Zupan. Determining gamma using $B^\pm \to DK^\pm$ with multibody D decays. *Phys. Rev. D*, 68:054018, 2003.
87. G.F. Giudice, C. Grojean, A. Pomarol, and R. Rattazzi. The strongly-interacting light Higgs. *JHEP*, 06:045, 2007.
88. C. Giunti and C.W. Kim. *Fundamentals of Neutrino Physics and Astrophysics.* Oxford University Press, 4 2007.
89. C. Giunti and A. Studenikin. Neutrino electromagnetic interactions: a window to new physics. *Rev. Mod. Phys.*, 87:531, 2015.
90. E. Golowich, J.L. Hewett, S. Pakvasa, and A.A. Petrov. Relating D^0-anti-D^0 mixing and $D^0 \to l^+l^-$ with New Physics. *Phys. Rev. D*, 79:114030, 2009.
91. E. Golowich, J.L. Hewett, S. Pakvasa, A.A. Petrov, and G.K. Yeghiyan. Relating B_s mixing and $B_s \to \mu^+\mu^-$ with New Physics. *Phys. Rev. D*, 83:114017, 2011.
92. E. Golowich and A.A. Petrov. Short distance analysis of D^0 - anti-D^0 mixing. *Phys. Lett. B*, 625:53–62, 2005.
93. R. Golub and K. Lamoreaux. Neutron electric dipole moment, ultracold neutrons and polarized He-3. *Phys. Rept.*, 237:1–62, 1994.
94. J. Goodman, M. Ibe, A. Rajaraman, W. Shepherd, T.M.P. Tait, and H.-B. Yu. Constraints on Dark Matter from colliders. *Phys. Rev. D*, 82:116010, 2010.
95. C.M. Grant, A. Gunawardana, and A.A. Petrov. Semileptonic decays of heavy mesons with artificial neural networks. *Phys. Rev. D*, 102(3):034003, 2020.
96. B. Grinstein and R.F. Lebed. SU(3) decomposition of two-body B decay amplitudes. *Phys. Rev. D*, 53:6344–6360, 1996.
97. B. Grinstein and M.B. Wise. Operator analysis for precision electroweak physics. *Phys. Lett. B*, 265:326–334, 1991.
98. C. Grojean, W. Skiba, and J. Terning. Disguising the oblique parameters. *Phys. Rev. D*, 73:075008, 2006.

99. M. Gronau, O.F. Hernandez, D. London, and J.L. Rosner. Decays of B mesons to two light pseudoscalars. *Phys. Rev. D*, 50:4529–4543, 1994.
100. M. Gronau and D. London. Isospin analysis of CP asymmetries in B decays. *Phys. Rev. Lett.*, 65:3381–3384, 1990.
101. M. Gronau and D. London. How to determine all the angles of the unitarity triangle from $B_d^0 \to DK_s$ and $B_s^0 \to D^0\phi$. *Phys. Lett. B*, 253:483–488, 1991.
102. M. Gronau and D. Wyler. On determining a weak phase from CP asymmetries in charged B decays. *Phys. Lett. B*, 265:172–176, 1991.
103. M. Gronau and J. Zupan. On measuring alpha in $B(t) \to \rho^{\pm}\pi^{\mp}$. *Phys. Rev. D*, 70:074031, 2004.
104. Y. Grossman. Nonstandard neutrino interactions and neutrino oscillation experiments. *Phys. Lett. B*, 359:141–147, 1995.
105. Y. Grossman and Y. Nir. $K_L \to \pi^0\nu\bar{\nu}$ beyond the Standard Model. *Phys. Lett. B*, 398:163–168, 1997.
106. B. Grzadkowski, M. Iskrzynski, M. Misiak, and J. Rosiek. Dimension-six terms in the Standard Model Lagrangian. *JHEP*, 10:085, 2010.
107. J.F. Gunion, H.E. Haber, G.L. Kane, and S. Dawson. *The Higgs hunter's guide*, volume 80. CRC Press, 2000.
108. K. Hagiwara, A. Keshavarzi, A. D. Martin, D. Nomura, and T. Teubner. $g-2$ of the muon: status report. *Nucl. Part. Phys. Proc.*, 287-288:33–38, 2017.
109. Z. Han. Effective theories and electroweak precision constraints. *Int. J. Mod. Phys. A*, 23:2653–2685, 2008.
110. R. Harnik, J. Kopp, and J. Zupan. Flavor violating Higgs decays. *JHEP*, 03:026, 2013.
111. D.E. Hazard and A.A. Petrov. Lepton flavor violating quarkonium decays. *Phys. Rev. D*, 94(7):074023, 2016.
112. J. Heeck and V. Takhistov. Inclusive nucleon decay searches as a frontier of baryon number violation. *Phys. Rev. D*, 101(1):015005, 2020.
113. R.J. Hill. The Modern description of semileptonic meson form factors. *eConf*, C060409:027, 2006.
114. R.J. Hill and G. Paz. Model independent extraction of the proton charge radius from electron scattering. *Phys. Rev. D*, 82:113005, 2010.
115. T. Inami and C.S. Lim. Effects of superheavy quarks and leptons in low-energy weak processes $K_L \to \mu anti - \mu$, $K^+ \to \pi^+\nu\bar{\nu}$ and K^0 - anti-K^0. *Prog. Theor. Phys.*, 65:297, 1981. [Erratum: *Prog. Theor. Phys.* 65, 1772 (1981)].
116. N. Isgur and M.B. Wise. Weak transition form-factors between heavy mesons. *Phys. Lett. B*, 237:527–530, 1990.
117. F. Jegerlehner and A. Nyffeler. The muon $g-2$. *Phys. Rept.*, 477:1–110, 2009.
118. E.E. Jenkins, A.V. Manohar, and M. Trott. Renormalization group evolution of the Standard Model dimension six operators I: formalism and lambda dependence. *JHEP*, 10:087, 2013.
119. E.E. Jenkins, A.V. Manohar, and M. Trott. Renormalization group evolution of the Standard Model dimension six operators II: Yukawa dependence. *JHEP*, 01:035, 2014.
120. A. Khodjamirian, T. Mannel, and A.A. Petrov. Direct probes of flavor-changing neutral currents in e^+e^--collisions. *JHEP*, 11:142, 2015.
121. T. Kinoshita and A. Sirlin. Radiative corrections to Fermi interactions. *Phys. Rev.*, 113:1652–1660, 1959.
122. R. Kitano, M. Koike, and Y. Okada. Detailed calculation of lepton flavor violating muon electron conversion rate for various nuclei. *Phys. Rev. D*, 66:096002, 2002. [Erratum: *Phys. Rev. D* 76, 059902 (2007)].
123. T.S. Kosmas and J.D. Vergados. (μ^-, e^-) conversion: a symbiosis of particle and nuclear physics. *Phys. Rept.*, 264:251–266, 1996.

124. V.A. Kostelecky. Gravity, Lorentz violation, and the Standard Model. *Phys. Rev. D*, 69:105009, 2004.
125. Y. Kuno and Y. Okada. Muon decay and physics beyond the Standard Model. *Rev. Mod. Phys.*, 73:151–202, 2001.
126. L. Lehman. Extending the Standard Model Effective Field Theory with the complete set of dimension-7 operators. *Phys. Rev.*, D90(12):125023, 2014.
127. R. Lehnert. CPT symmetry and its violation. *Symmetry*, 8(11):114, 2016.
128. A. Lenz and U. Nierste. Theoretical update of $B_s - \bar{B}_s$ mixing. *JHEP*, 06:072, 2007.
129. A. Lenz and U. Nierste. Numerical updates of lifetimes and mixing parameters of B mesons. In *6th International Workshop on the CKM Unitarity Triangle*, 2 2011.
130. P. Lipari. Introduction to neutrino physics. In *1st CERN-CLAF School of High-Energy Physics*, pages 115–199, 5 2001.
131. H.J. Lipkin, Y. Nir, H.R. Quinn, and A. Snyder. Penguin trapping with isospin analysis and CP asymmetries in B decays. *Phys. Rev. D*, 44:1454–1460, 1991.
132. M. Lisanti. Lectures on Dark Matter physics. In *Theoretical Advanced Study Institute in Elementary Particle Physics: New Frontiers in Fields and Strings*, pages 399–446, 2017.
133. M.A. Luty. Baryogenesis via leptogenesis. *Phys. Rev. D*, 45:455–465, 1992.
134. A.V. Manohar. Introduction to effective field theories. In *Les Houches summer school: EFT in Particle Physics and Cosmology Les Houches, Chamonix Valley, France, July 3-28, 2017*, 2018.
135. A.V. Manohar and M.B. Wise. *Heavy quark physics*, volume 10. Cambridge University Press, 2000.
136. W.J. Marciano and A. Sirlin. Electroweak radiative corrections to tau decay. *Phys. Rev. Lett.*, 61:1815–1818, 1988.
137. D. Melikhov and B. Stech. Weak form-factors for heavy meson decays: an update. *Phys. Rev. D*, 62:014006, 2000.
138. E. Mereghetti and U. van Kolck. Effective field theory and time-reversal violation in light nuclei. *Ann. Rev. Nucl. Part. Sci.*, 65:215–243, 2015.
139. S. Mller, U. Nierste, and S. Schacht. Topological amplitudes in D decays to two pseudoscalars: A global analysis with linear $SU(3)_F$ breaking. *Phys. Rev. D*, 92(1):014004, 2015.
140. P. Nath and P. Fileviez Perez. Proton stability in grand unified theories, in strings and in branes. *Phys. Rept.*, 441:191–317, 2007.
141. A. Neronov. Introduction to multi-messenger astronomy. *J. Phys. Conf. Ser.*, 1263(1):012001, 2019.
142. Y. Nir and H.R. Quinn. Theory of CP violation in B decays. pages 362–392, 8 1991.
143. Y. Nomura and J. Thaler. Dark Matter through the axion portal. *Phys. Rev. D*, 79:075008, 2009.
144. L.B. Okun. *Leptons and quarks.* North Holland, USA, 1985.
145. G. Panico, A. Pomarol, and M. Riembau. EFT approach to the electron Electric Dipole Moment at the two-loop level. *JHEP*, 04:090, 2019.
146. R.D. Peccei and H.R. Quinn. CP conservation in the presence of instantons. *Phys. Rev. Lett.*, 38:1440–1443, 1977.
147. M.E. Peskin and D.V. Schroeder. *An Introduction to quantum field theory.* Addison-Wesley, Reading, USA, 1995.
148. M.E. Peskin and T. Takeuchi. Estimation of oblique electroweak corrections. *Phys. Rev. D*, 46:381–409, 1992.
149. A.A. Petrov and A.E. Blechman. *Effective field theories.* World Scientific Publishing, 2016.

150. A.A. Petrov and W. Shepherd. Searching for Dark Matter at LHC with mono-Higgs production. *Phys. Lett. B*, 730:178–183, 2014.
151. A.A. Petrov and D.V. Zhuridov. Lepton flavor-violating transitions in effective field theory and gluonic operators. *Phys. Rev. D*, 89(3):033005, 2014.
152. II Phillips, D.G. et al. Neutron-antineutron oscillations: theoretical status and experimental prospects. *Phys. Rept.*, 612:1–45, 2016.
153. B. Pontecorvo. Mesonium and anti-mesonium. *Sov. Phys. JETP*, 6:429, 1957.
154. M. Pospelov and A. Ritz. Theta vacua, QCD sum rules, and the neutron electric dipole moment. *Nucl. Phys. B*, 573:177–200, 2000.
155. M. Pospelov and A. Ritz. Electric dipole moments as probes of new physics. *Annals Phys.*, 318:119–169, 2005.
156. G.M. Pruna and A. Signer. The $\mu \to e\gamma$ decay in a systematic effective field theory approach with dimension 6 operators. *JHEP*, 10:014, 2014.
157. M. Raidal et al. Flavour physics of leptons and dipole moments. *Eur. Phys. J. C*, 57:13–182, 2008.
158. A. Rajaraman, W. Shepherd, T.M.P. Tait, and A.M. Wijangco. LHC bounds on interactions of Dark Matter. *Phys. Rev. D*, 84:095013, 2011.
159. S. Rao and R.E. Shrock. Six fermion $(B-L)$ violating operators of arbitrary generational structure. *Nucl. Phys.*, B232:143–179, 1984.
160. M.E. Rose. *Relativistic electron theory.* John Wiley, New York, USA, 1961.
161. J.L. Rosner. Final state phases in charmed meson two-body nonleptonic decays. *Phys. Rev. D*, 60:114026, 1999.
162. G.G. Ross. *Grand unified theories.* Reading, USA: Benjamin/Cummings (1984) 497 P. (Frontiers In Physics, 60), 1985.
163. L.H. Ryder. *Quantum field theory.* Cambridge University Press, 6 1996.
164. M.S. Safronova, D. Budker, D. DeMille, et al. Search for New Physics with Atoms and Molecules. *Rev. Mod. Phys.*, 90(2):025008, 2018.
165. A.D. Sakharov. Violation of CP invariance, C asymmetry, and baryon asymmetry of the universe. *Sov. Phys. Usp.*, 34(5):392–393, 1991.
166. G. Sanchez-Colon and J. Wudka. Effective operator contributions to the oblique parameters. *Phys. Lett. B*, 432:383–389, 1998.
167. P.G.H. Sandars. The electric dipole moment of an atom. *Phys. Lett.*, 14:194–196, 1965.
168. M.J. Savage. SU(3) violations in the nonleptonic decay of charmed hadrons. *Phys. Lett. B*, 257:414–418, 1991.
169. M.J. Savage and M.B. Wise. SU(3) predictions for nonleptonic B meson decays. *Phys. Rev. D*, 39:3346, 1989. [Erratum: *Phys. Rev. D* 40, 3127 (1989)].
170. L.I. Schiff. Measurability of nuclear electric dipole moments. *Phys. Rev.*, 132:2194–2200, 1963.
171. W. Skiba. Effective field theory and precision electroweak measurements. In *Theoretical Advanced Study Institute in Elementary Particle Physics: Physics of the Large and the Small*, pages 5–70, 2011.
172. R. Slansky. Group theory for unified model building. *Phys. Rept.*, 79:1–128, 1981.
173. A.E. Snyder and H.R. Quinn. Measuring CP asymmetry in $B \to \rho\pi$ decays without ambiguities. *Phys. Rev. D*, 48:2139–2144, 1993.
174. V. Takhistov et al. Search for nucleon and dinucleon decays with an invisible particle and a charged lepton in the final state at the Super-Kamiokande experiment. *Phys. Rev. Lett.*, 115(12):121803, 2015.
175. M. Trott. On the consistent use of Constructed Observables. *JHEP*, 02:046, 2015.
176. T. van Ritbergen and R.G. Stuart. Complete two loop quantum electrodynamic contributions to the muon lifetime in the Fermi model. *Phys. Rev. Lett.*, 82:488–491, 1999.

177. M.B. Voloshin and M.A. Shifman. On the annihilation constants of mesons consisting of a heavy and a light quark, and $B^0 \leftrightarrow \overline{B}^0$ oscillations. *Sov. J. Nucl. Phys.*, 45:292, 1987.
178. S. Weinberg. Baryon and lepton nonconserving processes. *Phys. Rev. Lett.*, 43:1566–1570, 1979.
179. H.A. Weldon and A. Zee. Operator analysis of New Physics. *Nucl. Phys. B*, 173:269–290, 1980.
180. J.D. Wells. TASI lecture notes: Introduction to precision electroweak analysis. In *Theoretical Advanced Study Institute in Elementary Particle Physics: Physics in $D \geqq 4$*, pages 41–64, 12 2005.
181. F. Wilczek and A. Zee. Operator analysis of nucleon decay. *Phys. Rev. Lett.*, 43:1571–1573, 1979.
182. S. Willenbrock. Symmetries of the Standard Model. In *Physics in $D \geq 4$. Proceedings, Theoretical Advanced Study Institute in elementary particle physics, TASI 2004, Boulder, USA, June 6-July 2, 2004*, pages 3–38, 2004.
183. L. Willmann et al. New bounds from searching for muonium to anti-muonium conversion. *Phys. Rev. Lett.*, 82:49–52, 1999.
184. L. Wolfenstein. Parametrization of the Kobayashi-Maskawa matrix. *Phys. Rev. Lett.*, 51:1945, 1983.
185. G.K. Yeghiyan. Upsilon decays into light scalar Dark Matter. *Phys. Rev. D*, 80:115019, 2009.
186. A. Zee. *Quantum field theory in a nutshell.* Princeton, UK: Princeton Univ. Pr. (2010) 576 p, 2003.
187. A. Zee. *Group theory in a nutshell for physicists.* Princeton University Press, USA, 2016.
188. D. Zeppenfeld. SU(3) relations for B meson decays. *Z. Phys. C*, 8:77, 1981.
189. A.S. Zhevlakov, M. Gorchtein, A.N. Hiller Blin, T. Gutsche, and V.E. Lyubovitskij. Bounds on rare decays of η and η' mesons from the neutron EDM. *Phys. Rev. D*, 99(3):031703, 2019.

Index

For Product Safety Concerns and Information please contact our EU
representative GPSR@taylorandfrancis.com
Taylor & Francis Verlag GmbH, Kaufingerstraße 24, 80331 München, Germany

www.ingramcontent.com/pod-product-compliance
Ingram Content Group UK Ltd.
Pitfield, Milton Keynes, MK11 3LW, UK
UKHW051833180425
457613UK00022B/1242